VHDL

VHDL

Douglas L. Perry

Third Edition

McGraw-Hill
New York • San Francisco • Washington, D.C. • Auckland
Bogotá • Caracas • Lisbon • London • Madrid • Mexico City
Milan • Montreal • New Delhi • San Juan • Singapore
Sydney • Tokyo • Toronto

McGraw-Hill

A Division of The McGraw-Hill Companies

VHDL
International Editions 1999

Exclusive rights by McGraw-Hill Book Co – Singapore for manufacture and export. This book cannot be re-exported from the country to which it is consigned by McGraw-Hill.

2 3 4 5 6 7 8 9 0 KKP UPE 2 0

P/N 049452-5
PART OF ISBN 0-07-049436-3

The sponsoring editor of this book was Steve Chapman and the production supervisor was Pamela Pelton. It was set in Century Schoolbook by Douglas & Gayle Limited.

Library of Congress Cataloging-in-Publication Data

Perry, Douglas L.
    VHDL / Douglas Perry. – 3rd ed.
        p.   cm.
    Includes index.
    ISBN 0-07-04936-3
    1. VHDL (Computer hardware description language)   I. Title.
    TK7885.7.P47   1998
    621.39'2–dc21   98-16663
                                                CIP

When ordering this title, use ISBN 0-07-116673-4

Printed in Singapore

ACKNOWLEDGMENTS

This book would not have been possible without the help of a number of people, and I would like to express my gratitude to all of them.

Rod Farrow, Cary Ussery, Alec Stanculescu, and Ken Scott answered a multitude of questions about some of the vagaries of VHDL.

Mark Beardslee and Derek Palmer reviewed of parts of the third edition. Their comments were both helpful and insightful.

Pual Krol developed the chart in Chapter 7 that describes generics. Keith Irwin helped define the style of some of the chapters. Hoa Dinh and David Emrich answered a lot of questions about FPGA synthesis.

Thanks to John Ott for making the ModelSim software available during the writing of this book and for the software on the CD.

Thanks to Derek Palmer and Robert Blake of Altera for making the MaxPlus II software available and for answering questions.

CONTENTS

Contents

Contents

PREFACE

This is the third version of the book. This version now not only provides VHDL language coverage but design methodology information as well. This version guides the reader through the process of creating a VHDL design, simulating the design, synthesizing the design, placing and routing the design, and using VITAL simulation to verify the final result. The design example in this version has been updated to reflect the new focus on the design methodology.

This book was written to help hardware design engineers learn how to write good VHDL design descriptions. The goal is to provide enough VHDL and design methodology information to enable a designer to quickly write good VHDL designs and be able to verify the results. This book also attempts to bring the designer with little or no knowledge of VHDL to the level of writing complex VHDL descriptions. It is not intended to show every possible construct of VHDL in every possible use but rather to show the designer how to write concise, efficient, and correct VHDL descriptions of hardware designs.

The book is organized into three logical sections. The first section of the book introduces the VHDL language; the second section walks through a VHDL-based design process including simulation, syntheses, place and route, and VITAL simulation; and the third section walks through a design example of a small CPU design from VHDL capture to final gate-level implementation. A number of appendices containing useful information about the language and examples used throughout the book are included at the back.

VHDL features are introduced one or more at a time in the first section. As each feature is introduced, one or more real examples are given to show how the feature would be used. The first section consists of Chapters 1 through 8, and each chapter introduces a basic description capability of VHDL. Chapter 1 discusses how VHDL design relates to schematic-based design and introduces the basic terms of the language. Chapter 2 describes some of the basic concepts of VHDL including the different delay mechanisms available, how to use instance-specific data, and defines VHDL drivers. Chapter 2 discusses concurrent statements while Chapter 3 introduces VHDL sequential statements. Chapter 4 talks about the wide range of types available for use in a real example. In Chapter 5 the concepts of subprograms and packages are introduced. The different uses for functions are given, as well as the features available in VHDL packages.

Chapter 6 introduces the five kinds of VHDL attributes. Each attribute type has examples describing how to use the specific attribute to the designer's best advantage. Examples are given describing the purpose of each of the attributes.

Chapters 7 and 8 introduce some of the more advanced VHDL features. Chapter 7 discusses how VHDL configurations can be used to construct and manage complex VHDL designs. Each of the different configuration styles are discussed along with examples showing usage. Chapter 8 introduces more of the VHDL advanced topics with discussions of overloading, user defined attributes, generate statements, and TextIO.

The second section of the book consists of Chapters 9 through 11. Chapters 9 and 10 discuss the synthesis process and how to write synthesizeable designs. These two chapters describe the basics of the synthesis process including how to write synthesizeable VHDL; what is a technology library; what does the synthesis process look like; what are constraints and attributes; and what does the optimization process look like. Chapter 11 discusses the complete high level design flow from VHDL capture through VITAL simulation.

The third section of the book walks through a description of a small CPU design from the VHDL capture through simulation, synthesis, place and route, and VITAL simulation. Chapter 12 describes the top level of the CPU design from a functional point of view. In Chapter 13 the RTL description of the CPU is presented and discussed from a synthesis point of view. Chapter 14 begins with a discussion of VHDL testbenches and how they are used to verify functionality. Chapter 14 finishes the discussion by describing the simulation of the CPU design. In Chapter 15 the verified design is synthesized to a target technology. Chapter 16 takes the synthesized design and places and routes the design to a target device. Chapter 17 begins with a discussion of VITAL and ends with the VITAL simulation of the placed and routed CPU design.

Finally there are four appendices at the end of the book to provide reference information. Appendix A provides the gate level netlists from the vending machine synthesis process in Chapter 12. Appendix B is a listing of the IEEE 1164 STD_LOGIC package used throughout the book. Appendix C is a set of useful tables that condense some of the information in the rest of the book into quick reference tables. Finally, Appendix D describes how to read the Bachus-Naur format (BNF) descriptions found in the VHDL Language Reference Manual. I can only hope that you will have as much fun reading this book and working with VHDL as I did in writing it.

1

Introduction to VHDL

The VHSIC Hardware Description Language is an industry standard language used to describe hardware from the abstract to the concrete level. VHDL is rapidly being embraced as the universal communication medium of design. Computer-aided engineering workstation vendors, FPGA vendors, and ASIC vendors throughout the industry are standardizing on VHDL as input and output from their tools. These include simulation tools, synthesis tools, place and route tools, and so on.

In this chapter, we examine the basics of VHDL. The history of VHDL is presented and some basic terms are defined. Finally, VHDL design is contrasted with traditional design methods.

VHSIC Program

VHDL is an offshoot of the *very high speed integrated circuit* (VHSIC) program that was funded by the Department of Defense in the late 1970s and early 1980s. The goal of the VHSIC program was to produce the next generation of integrated circuits. Program participants were urged to push technology limits in every phase of the design and manufacture of integrated circuits.

The goals were accomplished admirably but, in the process of developing these extremely complex integrated circuits, the designers found out that the tools used to create these large designs were inadequate for the task. The tools that were available to the designers were mostly based at the gate level. Creating designs of hundreds of thousands of gates using gate level tools was an extremely challenging task, and therefore a new method of description was in order. Initially, VHDL was designed to be a documentation and simulation language. Later, VHDL was embraced by synthesis tools to describe hardware design at the Register Transfer Level.

VHDL as a Standard

A new hardware description language was proposed in 1981 called the *VHSIC Hardware Description Language*, or as we know it now, VHDL. The goals of this new language were twofold. First, the designers wanted a language that could handle the complex circuits they were trying to describe. Second, they wanted a language that was a standard so that all of the players in the VHSIC program could distribute designs to other players in a standard format. Also, any subcontractors needed to be able to talk to their main contractors with a single standard format.

In 1986, VHDL was proposed as an IEEE standard. It went through a number of revisions and changes until it was adopted as the IEEE 1076 standard in December 1987. The IEEE 1076−1987 standard VHDL is the VHDL used in this book. (Appendix D contains a brief description of VHDL 1076-1993). All of the examples have been described in IEEE 1076 VHDL and compiled and simulated with the VHDL simulation environment from Model Technology Inc. The synthesis examples were synthesized with the Exemplar Logic Inc. synthesis tools.

Learning VHDL

VHDL can be a difficult language to learn by reading the VHDL *Language Reference Manual* (also called the LRM) from cover to cover. The LRM describes VHDL for the VHDL implementor and was never intended to be a VHDL user's guide. VHDL itself is a large language, however, and learning all of it can be quite a task. The entire language does not need to be learned to write useful models. A subset of the language can be learned initially, and, as more complex models are required, the more complex features can be learned and used.

VHDL contains levels of representations that can be used to represent all levels of description from the bidirectional switch level to the system level, and any level in between. The best way to approach VHDL is to learn enough of the language to try some small designs. When you become familiar enough with this subset of VHDL that you feel comfortable writing it, move on to some of the other features of the language and try new things.

VHDL Terms

Before we go any further, let's define some of the terms that we will be using throughout the book. These are the basic VHDL building blocks that are used in almost every description, along with some terms that are redefined in VHDL to mean something different to the average designer.

- *Entity*—All designs are expressed in terms of entities. An entity is the most basic building block in a design. The uppermost level of the design is the top-level entity. If the design is hierarchical, then the top-level description will have lower-level descriptions contained in it. These lower-level descriptions will be lower-level entities contained in the top-level entity description.

- *Architecture*—All entities that can be simulated have an architecture description. The architecture describes the behavior of the entity. A single entity can have multiple architectures. One architecture might be behavioral while another might be a structural description of the design.

- *Configuration*—A configuration statement is used to bind a component instance to an entity-architecture pair. A configuration can be considered like a parts list for a design. It describes which behavior to use for each entity—much like a parts list describes which part to use for each part in the design.

- *Package*—A package is a collection of commonly used data types and subprograms used in a design. Think of a package as a toolbox that contains tools used to build designs.

- *Bus*—The term bus usually brings to mind a group of signals or a particular method of communication used in the design of hardware. In VHDL, a bus is a special kind of signal that may have its drivers turned off.

- *Driver*—This is a source on a signal. If a signal is driven by two tristate inverters, when both inverters are active, the signal will have two drivers.

- *Attribute*—An attribute is data that is attached to VHDL objects or predefined data about VHDL objects. Examples are the current drive capability of a buffer or the maximum operating temperature of the device.

- *Generic*—A generic is VHDL's term for a parameter that passes information to an entity. For instance, if an entity is a gate level model with a rise and a fall delay, values for the rise and fall delays could be passed into the entity with generics.

- *Process*—A process is the basic unit of execution in VHDL. All operations that are performed in a simulation of a VHDL description are broken into single or multiple processes.

Traditional Design Methods

Today, when a design engineer develops a new piece of hardware, it is probably designed with *Computer Aided Engineering* (CAE) software. To create a design using typical CAE tools, the designer creates a schematic for the design.

A typical schematic consists of symbols representing the basic units of the design connected together with signals. The symbols come from a library of parts that the designer uses to build the schematic. The type of symbols available depends on the type of design that the designer is creating. If the designer is creating the schematic for a board design that uses standard logic parts, then the symbols used in the schematic represent the standard parts the designer has available. If the designer is creating a schematic for an *application-specific integrated circuit* (ASIC), then the symbols available are the library macros available for use on this specific type of ASIC.

The symbols are wired together using signals (or nets, short for networks). The interconnection of the symbols by signals creates the connections needed to specify the design such that a netlist can be derived from the connections. A netlist can be used to create a simulation model of the design to verify the design before it is built. After the design has been verified, the netlist can be used to provide the information needed by a routing software package to complete the actual design. The routing software creates the physical connection data to either create the trace information needed for a PC board or the layer information needed for an ASIC.

Figure 1-1 illustrates an example of a symbol used in a schematic. It is the symbol for a *reset-set flip-flop* (RSFF). The symbol describes the following pieces of information to the designer:

- The number of input pins for the device. In the example shown in Figure 1-1, the number is 2, SET and RESET.
- The number of output pins for the device. In Figure 1-1, the number of output pins is 2, Q and QB.
- The function of the device. In this example, the function of the device is described by the name of the symbol. In the case of simple gates, the function of the symbol is described by the shape of the symbol.

Figure 1-1
RSFF Symbol.

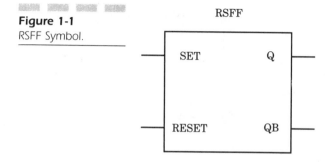

Symbols specify the interface and the function to the designer. When the symbols are placed on a schematic sheet and wired together with signals, a schematic for the design is formed. An example of a simple schematic for an RSFF is shown in Figure 1-2.

Traditional Schematics

The schematic contains two NAND gate symbol instances and four port instances. Four nets connect the symbols and ports together to form the RS flip-flop. Each port has a unique name that also specifies the name of the signal (net) connected to it. Each symbol instance has a unique instance identifier (U1, U2) . The instance identifier is used to uniquely identify an instance for reference.

When this schematic is compiled into a typical gate level simulator, it functions as an RSFF. A '0' level on the reset port causes the reset signal to have the value '0'. This causes NAND gate U2 to output a '1' value on signal QB independent of the value on the other input of the NAND gate. The '1' value on QB feeds back to one input of the NAND gate U1. If the set input is at an inactive value ('1'), then NAND gate U1 will have two '1' values as input, causing it to output a value of '0' on the output Q. The RSFF will have been reset.

How would this same design look in VHDL? First of all, how do you represent a symbol in VHDL? What does a schematic look like in VHDL?

Figure 1-2
RSFF Schematic.

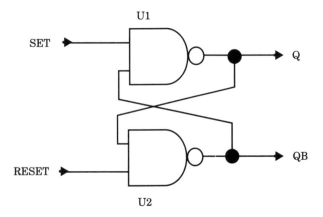

Symbols Versus Entities

All designs are created from entities. An entity in VHDL corresponds directly to a symbol in the traditional CAE design methodology. Let's look at the top-level entity for the RSFF symbol described earlier. An entity for the RSFF would look like this:

```
ENTITY rsff    IS
   PORT ( set, reset : IN BIT;
          q, qb : INOUT BIT);
END rsff;
```

The keyword **ENTITY** signifies that this is the start of an entity statement. In the descriptions shown throughout the book, keywords of the language and types provided with the STANDARD package are shown in ALL CAPITAL letters. For instance, in the preceding example, the keywords are **ENTITY, IS, PORT, IN, INOUT**, and so on. The standard type provided is **BIT**.

The name of the entity is **rsff**, as was the name of the symbol described earlier. The entity has four ports in the **PORT** clause. Two ports are of mode **IN** and two ports are of mode **INOUT**. The reason for port mode **INOUT** instead of just **OUT** will become clear later in the book. The two input ports correspond directly to the two input pins on the symbol from the CAE software. The two inout ports correspond directly to the two output ports for the symbol. All of the ports have a type of **BIT**.

The entity describes the interface to the outside world. It specifies the number of ports, the direction of the ports, and the type of the ports. A lot more information can be put into the entity than is shown here, but this gives us a foundation upon which we can later build.

Schematics Versus Architectures

The schematic for the **rsff** component also has a counterpart in VHDL. It is called an architecture. An architecture is always related to an entity and describes the behavior of that entity. An architecture for the **rsff** device described earlier would look like this:

```
ARCHITECTURE netlist OF rsff IS
   COMPONENT nand2
  PORT ( a, b : IN BIT;
         c    : OUT BIT);
```

```
END COMPONENT;
BEGIN
U1: nand2
 PORT MAP (set, qb, q);

U2: nand2
 PORT MAP (reset, q, qb);
END netlist;
```

The keyword **ARCHITECTURE** signifies that this statement describes an architecture for an entity. The architecture name is **netlist**. The entity the architecture is describing is called **rsff**.

The reason for the connection between the architecture and the entity is that an entity can have multiple architectures describing the behavior of the entity. For instance, one architecture could be a behavioral description, and another, like the one shown in the preceding example, could be a structural description.

The textual area between the keyword **ARCHITECTURE** and the keyword **BEGIN** is where local signals and components are declared for later use. In this example, there are two instances of a **nand2** gate placed in the architecture. The compiler needs to know what the interface to the components placed in the architecture looks like. The component declaration statement describes that information to the compiler.

The statement area of the architecture starts with the keyword **BEGIN**. All statements between the **BEGIN** and the **END** netlist statement are called concurrent statements, because all of the statements execute concurrently.

Component Instantiation

In the statement area are two component instantiation statements. Each statement creates an instance of a component in the model. In this example, each statement creates an instance of a **nand2** gate. The first instance, U1, corresponds directly with the **nand2** gate U1 in the schematic in Figure 1-2. The way to read the component instantiation statement is as follows: The first statement creates an instance called U1 of a **nand2** component, with the first port connected to signal **set**, the second port connected to signal **qb**, and the last port connected to signal **q**.

If we look again at the **nand2** component declaration, we see that the first port is an **IN** port called **a**, the second port is an **IN** port called **b**, and the last port is an **OUT** port called **c**. Therefore, the component instantiation would connect the **a** port of the **nand2** component to signal **set**, the **b** port to signal **qb**, and the **c** port to signal **q**. We have thus matched the actual parameters **set**, **qb**, and **q** of the instantiation with the corresponding formal parameters **a**, **b**, and **c** of the declaration.

There is another way to map the ports, if the ports are not in a particular order. Component instantiation statements, as demonstrated in the following, can be used:

```
U1: nand2 port map (a => set, b => qb, c => q );
```

This form is called named association and maps the ports directly without concern for the order. In fact, the following statement would work perfectly well also:

```
U1: nand2 port map ( b => qb, c => q, a => set);
```

The second component instantiation statement of architecture netlist creates another instance of the **nand2** component with an instance identifier, U2. These instances correspond directly with the instance identifiers of the schematic in Figure 1-2. This type of VHDL representation is called a structural model, or structural representation. What makes this description structural is the fact that it has components instantiated in it. In the next few chapters, we discuss structural, behavioral, and mixed structural-behavioral descriptions.

The structural architecture netlist is very similar to a netlist in a typical CAE software simulator. Some of the examples in this book use structural parts to describe the functionality, but behavioral modeling and synthesizable RTL descriptions are the main focus.

Behavioral Descriptions

Another way to describe this same circuit is by using a behavioral architecture. An example of a behavioral architecture is shown in the following architecture example. This architecture uses concurrent signal assignment statements. As the name implies, the statements contained in the model assign values to signals. What makes these statements different from assignment statements in typical programming languages is the fact that these statements execute in parallel (concurrently), not serially:

```
ARCHITECTURE behave OF rsff IS
BEGIN
q <= NOT( qb AND set ) AFTER 2 ns;
```

```
qb <= NOT( q AND reset ) AFTER 2 ns;
END behave;
```

Concurrent Signal Assignment

In a typical programming language such as C or Pascal, each assignment statement executes one after the other and in a specified order. The order of execution is determined by the order of the statements in the source file. Inside a VHDL architecture, there is no specified ordering of the assignment statements. The order of execution is solely specified by events occurring on signals that the assignment statements are sensitive to.

Examine the first assignment statement from architecture **behave**, as shown here:

```
q <= NOT( qb AND set ) AFTER 2 ns;
```

A signal assignment is identified by the symbol **<=**. The logical **AND** of **qb** and **set** is complemented and assigned to signal **q**. This statement is executed whenever either **qb** or **set** has an event occur on it. An event on a signal is a change in the value of that signal. A signal assignment statement is said to be sensitive to changes on any signals that are to the right of the **<=** symbol. This signal assignment statement is sensitive to **qb** and **set**. The other signal assignment statement in architecture **behave** is sensitive to signals **q** and **reset**.

The **AFTER** clause in the signal assignment is used to emulate propagation delay in the circuit. Any event (change in value) on **qb** or **set** may cause a change on signal **q** 2 nanoseconds later.

Let's take a look at how these statements actually work. Suppose that we have a steady state condition where both **set** and **reset** are at a '1' value, and signal **q** is currently at a '0' value. Signal **qb** is at a '1' value because it is opposite of **q**, except when both **set** and **reset** are at a '0' value. Now assume that we place an event on the **set** signal that causes its value to change to a '0'. When this happens, the first signal assignment statement wakes up and executes. This happens because **set** is on the right side of the **<=** and is therefore in the implied sensitivity list for the first signal assignment statement.

When the first statement executes, it computes the new value to be assigned to **q** from the current value of the signal expression on the right side of the **<=** symbol. The expression value calculation uses the current values for all signals contained in it.

What will the signal expression calculate? Signal **set** is now equal to '0' because its value just changed. Signal **qb** is equal to '1' because it did not change. The new value for signal **q** is the complement of the two values **AND**ed together. This results in a '1' value to be assigned to signal **q**.

Event Scheduling

The assignment to signal **q** does not happen instantly. The **AFTER** clause discussed earlier delays the assignment of the new value to **q** by 2 nanoseconds. The mechanism for delaying the new value is called scheduling an event. By assigning **q** a new value, an event was scheduled 2 nanoseconds in the future that contains the new value for signal **q**. When the event matures (2 nanoseconds in the future), signal **q** receives the new value.

Statement Concurrency

The first assignment is the only statement to execute at the current time when the event on **set** happens. The second signal assignment statement does not execute until the event on signal **q** happens, or an event happens on signal **reset**. If no event happens on signal **reset**, then the second assignment statement does not execute for 2 nanoseconds. This is when the event on signal **q** that was scheduled by the first signal assignment occurs.

When the event on signal **q** occurs, the second signal assignment statement executes. It calculates a new value from **q** and **reset**. Assuming **reset** stays at a '1', then the new value for **qb** is a '0'. This value is scheduled to occur on signal **qb** 2 nanoseconds in the future.

When the event occurs on signal **qb**, the first signal assignment wakes up, executes again, and calculates the new value for **q** based on the values of **qb** and **set**. If **set** is still a '0', then the '0' value of **qb** does not change the value of **q**. Signal **q** is already at a '1' value from the **set** signal at a '0'. No new event is scheduled on the **q** output signal, because signal **q** does not change its value.

The two signal assignment statements in architecture **behave** form a behavioral model, or architecture, for the **rsff** entity. The **behave** architecture contains no structure. There are no components instantiated in the architecture. There is no further hierarchy, and this architecture can be considered a leaf node in the hierarchy of the design.

Sequential Behavior

There is yet another way to describe the functionality of an **rsff** device in VHDL. The fact that VHDL has so many possible representations for similar functionality is what makes learning the entire language a big task. The third way to describe the functionality of the **rsff** is to use a process statement to describe the functionality in an algorithmic representation. This is shown in architecture sequential, as shown in the following:

```
ARCHITECTURE sequential OF rsff IS
BEGIN
PROCESS (set, reset )
BEGIN
 IF set = '1' AND reset = '0' THEN
   q <= '0' AFTER 2 ns;
   qb <= '1' AFTER 4 ns;
 ELSIF set = '0' AND reset = '1' THEN
   q <= '1' AFTER 4 ns;
   qb <='0' AFTER 2 ns;
 ELSIF set = '0' AND reset = '0' THEN
   q <= '1' AFTER 2 ns;
   qb <= '1' AFTER 2 ns;
 END IF;
END PROCESS;
END sequential;
```

The architecture contains only one statement, called a process statement. It starts at the line beginning with the keyword **PROCESS** and ends with the line that contains **END PROCESS**. All of the statements between these two lines are considered part of the process statement.

Process Statements

The process statement consists of a number of parts. The first part is called the sensitivity list; the second part is called the process declarative part; and the third is the statement part. In the preceding example, the list of signals in parentheses after the keyword **PROCESS** is called the sensitivity list. This list enumerates exactly which signals cause the process statement to be executed. In this example, the list consists of **set** and **reset**. Only events on these signals cause the process statement to be executed.

Process Declarative Region

The process declarative part consists of the area between the end of the sensitivity list and the keyword **BEGIN**. In this example, the declarative part is empty. This area is used to declare local variables or constants that can be used only inside the process.

Process Statement Part

The statement part of the process starts at the keyword **BEGIN** and ends at the **END PROCESS** line. All of the statements enclosed by the process are sequential statements. This means that any statements enclosed by the process are executed one after the other in a sequential order just like a typical programming language. Remember that the order of the statements in the architecture did not make any difference; however, inside the process, this is not true. The order of execution is the order of the statements in the process statement.

Process Execution

Let's see how this works by walking through the execution of the example in architecture **sequential**, line by line. To be consistent, let's assume that **set** changes to a '0' and **reset** remains at a '1'. Because **set** is in the sensitivity list for the process statement, the process is invoked. Each statement in the process is then executed sequentially. In this example, however, there is only one **IF** statement. Each check that the **IF** statement performs is done sequentially starting with the first in the model.

The first check is to see if **set** is equal to a '1' and **reset** is equal to a '0'. This statement fails because **set** is equal to a '0' and **reset** is equal to a '1'. The two signal assignment statements that follow the first check are not executed. Instead, the next check is performed. This check succeeds and the signal assignment statements following the check for **set = '0'** and **reset = '1'** are executed. These statements are as follows:

```
q  <= '1' AFTER 2 ns;
qb <= '0' AFTER 2 ns;
```

Sequential Statements

These two statements execute sequentially. They may look exactly the same as previous signal assignment statements we have examined, but because of the context (they are inside the process statement), they are different. These two assignment statements are called sequential signal assignment statements, and they execute one after the other inside the process statement. The first statement may schedule an event on signal **q**, and then the second statement may schedule an event on signal **qb**. However, if signal **q** is already at a '1' value, no change in value occurs and therefore no event is scheduled.

After these two statements are executed, the next check of the **IF** statement is not performed. Whenever a check succeeds, no other checks are done. Because the **IF** statement was the only statement inside the process, the process terminates.

Architecture Selection

So far, three architectures have been described for one entity. Which architecture should be used to model the **rsff** device? It depends on the accuracy wanted and if structural information is required. If the model is going to be used to drive a layout tool, then the structural architecture netlist is probably most appropriate. If a structural model is not wanted for some other reason, then a more efficient model can be used. Either of the other two methods (architectures **behave** and **sequential**) are probably more efficient in memory space required and speed of execution. How to choose between these two methods may come down to a question of programming style. Would the modeler rather write concurrent or sequential VHDL code? If the modeler wants to write concurrent VHDL code, then the style of architecture **behave** is the way to go; otherwise, architecture **sequential** should be chosen. Typically, modelers are more familiar with sequential coding styles, but concurrent statements are very powerful tools to write small efficient models.

We will also look at yet another architecture that can be written for an entity. This is the architecture that can be used to drive a synthesis tool. Synthesis tools convert a *Register Transfer Level* (RTL) VHDL description into an optimized gate level description. Synthesis tools can offer greatly enhanced productivity compared to manual methods. The synthesis process is discussed in Chapters 9, "Synthesis" and 10, "VHDL Synthesis."

Configuration Statements

An entity can have more than one architecture, but how does the modeler choose which architecture to use in a given simulation? The configuration statement maps component instantiations to entities. With this powerful statement, the modeler can pick and choose which architectures are used to model an entity at every level in the design.

Let's look at a configuration statement using the netlist architecture of the **rsff** entity. The following is an example configuration:

```
CONFIGURATION rsffcon1 OF rsff IS
  FOR netlist
    FOR U1,U2 : nand2 USE ENTITY WORK.mynand(version1);
    END FOR;
  END FOR;
END rsffcon1;
```

The function of the configuration statement is to spell out exactly which architecture to use for every component instance in the model. This occurs in a hierarchical fashion. The highest-level entity in the design needs to have the architecture to use specified, as well as any components instantiated in the design.

The preceding configuration statement reads as follows: This is a configuration named **rsffcon1** for entity **rsff**. Use architecture **netlist** as the architecture for the topmost entity, which is **rsff**. For the two component instances **U1** and **U2** of type **nand2** instantiated in the **netlist** architecture, use entity **mynand**, architecture **version1** from the library called **WORK**. All of the entities now have architectures specified for them. Entity **rsff** has architecture **netlist**, and component **nand2** has entity **mynand** and architecture **version1**.

Power of Configurations

By compiling the entities, architectures, and the configuration specified earlier, you can create a simulatable model. But what if you did not want to simulate at the gate level? What if you really wanted to use architecture **BEHAVE** instead? The power of the configuration is that you do not need to recompile your complete design; you only need to recompile the new configuration. Following is an example configuration:

```
CONFIGURATION rsffcon2 OF rsff IS
  FOR behave
```

```
END FOR;
END rsffcon2;
```

This is a configuration named **rsffcon2** for entity **rsff**. Use architecture **behave** for the topmost entity, which is **rsff**. By compiling this configuration, the architecture **behave** is selected for entity **rsff** in this simulation.

This configuration is not necessary in standard VHDL, but gives the designer the freedom to specify exactly which architecture will be used for the entity. The default architecture used for the entity is the last one compiled into the working library.

In this chapter, we have had a basic introduction to VHDL and how it can be used to model the behavior of devices and designs. The first example showed how a larger design can be made of smaller designs—in this case an RSFF was modeled using NAND gates. The first example provided a structural view of VHDL.

The second example showed a more abstract view that did not include the gate level view of the RSFF, but had more of a dataflow flavor to it. In the third example, an algorithmic or behavioral view of the RSFF was presented. All of these views of the RSFF successfully model the functionality of an RSFF and all can be simulated with a VHDL simulator. Ultimately, however, a designer will want to use the model to facilitate building a piece of hardware. The most common use of VHDL in actually building hardware today is through synthesis tools. Therefore, the focus of the rest of the book is not only on the simulation of VHDL, but also the synthesis of VHDL.

Behavioral Modeling

In Chapter 1, we discussed structural modeling with traditional CAE systems and touched briefly on behavioral modeling. In this chapter, we discuss behavioral modeling more thoroughly, as well as some of the issues relating to the simulation and synthesis of VHDL models.

Introduction to Behavioral Modeling

The signal assignment statement is the most basic form of behavioral modeling in VHDL. Following is an example:

```
a <= b;
```

This statement is read as follows: **a** gets the value of **b**. The effect of this statement is that the current value of signal **b** is assigned to signal **a**. This statement is executed whenever signal **b** changes value. Signal **b** is in the sensitivity list of this statement. Whenever a signal in the sensitivity list of a signal assignment statement changes value, the signal assignment statement is executed. If the result of the execution is a new value that is different from the current value of the signal, then an event is scheduled for the target signal. If the result of the execution is the same value, then no event is scheduled but a transaction is still generated (transactions are discussed in Chapter 3, "Sequential Processing"). A transaction is always generated when a model is evaluated, but only signal value changes cause events to be scheduled.

The next example shows how to introduce a nonzero delay value for the assignment:

```
a <= b after 10 ns;
```

This statement is read as follows: **a** gets the value of **b** when 10 nanoseconds of time have elapsed.

Both of the preceding statements are concurrent signal assignment statements. Both statements are sensitive to changes in the value of signal **b**. Whenever **b** changes value, these statements execute and new values are assigned to signal **a**.

Using a concurrent signal assignment statement, a simple AND gate can be modeled, as follows:

```
ENTITY and2 IS
   PORT ( a, b : IN BIT;
          c : OUT BIT );
END and2;

ARCHITECTURE and2_behav OF and2 IS
BEGIN
   c <= a AND b AFTER 5 ns;
END and2_behav;
```

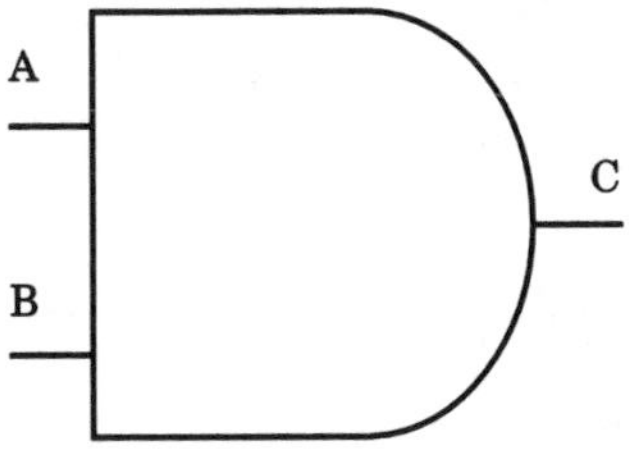

Figure 2-1
AND Gate Symbol.

The AND gate has two inputs **a, b** and one output **c**, as shown in Figure 2-1. The value of signal **c** may be assigned a new value whenever either **a** or **b** changes value. With an AND gate, if **a** is a 0 and **b** changes from a 1 to a 0, output **c** does not change. If the output does change value, then a transaction occurs which causes an event to be scheduled on signal **c**; otherwise, a transaction occurs on signal **c**.

The entity design unit describes the ports of the **and2** gate. There are two inputs **a** and **b**, as well as one output **c**. The architecture **and2_behav** for entity **and2** contains one concurrent signal assignment statement. This statement is sensitive to both signal **a** and signal **b** by the fact that the expression to calculate the value of **c** includes both **a** and **b** signal values.

The value of the expression **a** and **b** is calculated first, and the resulting value from the calculation is scheduled on output **c**, 5 nanoseconds from the time the calculation is completed.

The next example shows more complicated signal assignment statements and demonstrates the concept of concurrency in greater detail. In Figure 2-2, the symbol for a four-input multiplexer is shown.

This is the behavioral model for the mux:

```
LIBRARY IEEE;
USE IEEE.std_logic_1164.ALL;

ENTITY mux4 IS
PORT ( i0, i1, i2, i3, a, b : IN std_logic;
                        q : OUT std_logic);
END mux4;

ARCHITECTURE mux4 OF mux4 IS
SIGNAL sel: INTEGER;
BEGIN
WITH sel SELECT
  q <= i0 AFTER 10 ns WHEN 0,
       i1 AFTER 10 ns WHEN 1,
       i2 AFTER 10 ns WHEN 2,
       i3 AFTER 10 ns WHEN 3,
       'X' AFTER 10 ns WHEN OTHERS;
```

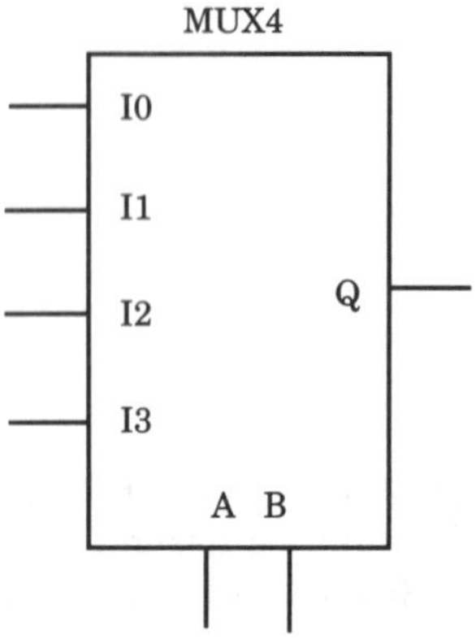

Figure 2-2
Mux4 Symbol.

```
sel <= 0 WHEN a = '0' AND b = '0' ELSE
        1 WHEN a = '1' AND b = '0' ELSE
        2 WHEN a = '0' AND b = '1' ELSE
        3 WHEN a = '1' AND b = '1' ELSE
        4 ;
END mux4;
```

The entity for this model has six input ports and one output port. Four of the input ports (**i0, i1, i2, i3**) represent signals that will be assigned to the output signal **q**. Only one of the signals will be assigned to the output signal **q** based on the value of the other two input signals **a** and **b**. The truth table for the multiplexer is shown in Figure 2-3.

To implement the functionality described in the preceding, we use a conditional signal assignment statement and a selected signal assignment.

The second statement type in this example is called a conditional signal assignment statement. This statement assigns a value to the target signal based on conditions that are evaluated for each statement. The statement **WHEN** conditions are executed one at a time in sequential order until the conditions of a statement are met. The first statement that matches the conditions required assigns the value to the target signal. The target signal for this example is the local signal **sel**. Depending on the values of signals **a** and **b**, the values 0 through 4 are assigned to **sel**.

If more than one statement's conditions match, the first statement that matches does the assign, and the other matching statements values are ignored.

Figure 2-3
Mux Functional
Table.

A	B	Q
0	0	I0
1	0	I1
0	1	I2
1	1	I3

The first statement is called a selected signal assignment and selects among a number of options to assign the correct value to the target signal. The target signal in this example is the signal **q**.

The expression (the value of signal **sel** in this example) is evaluated, and the statement that matches the value of the expression assigns the value to the target signal. All of the possible values of the expression must have a matching choice in the selected signal assignment (or an **OTHERS** clause must exist).

Each of the input signals can be assigned to output **q**, depending on the values of the two select inputs, **a** and **b**. If the values of **a** or **b** are unknown values, then the last value, 'X' (unknown), is assigned to output **q**. In this example, when one of the select inputs is at an unknown value, the output is set to unknown.

Looking at the model for the multiplexer, it looks like the model will not work as written. It seems that the value of signal **sel** is used before it is computed. This impression is received from the fact that the second statement in the architecture is the statement that actually computes the value for **sel**. The model does work as written, however, because of the concept of concurrency.

The second statement is sensitive to signals **a** and **b**. Whenever either **a** or **b** changes value, the second statement is executed, and signal **sel** is updated. The first statement is sensitive to signal **sel**. Whenever signal **sel** changes value, the first signal assignment is executed.

If this example is processed by a synthesis tool, the resulting gate structure created resembles a 4 to 1 multiplexer. If the synthesis library contains a 4 to 1 multiplexer primitive, that primitive may be generated based on the sophistication of the synthesis tool and the constraints put on the design.

Transport Versus Inertial Delay

In VHDL, there are two types of delay that can be used for modeling behaviors. Inertial delay is the most commonly used, while transport delay is used where a wire delay model is required.

Inertial Delay

Inertial delay is the default in VHDL. If no delay type is specified, inertial delay is used. Inertial delay is the default because, in most cases, it behaves similarly to the actual device.

In an inertial delay model, the output signal of the device has inertia, which must be overcome for the signal to change value. The inertia value is equal to the delay through the device. If there are any spikes, pulses, and so on that have periods where a signal value is maintained for less than the delay through the device, the output signal value does not change. If a signal value is maintained at a particular value for longer than the delay through the device, the inertia is overcome and the device changes to the new state.

Figure 2-4 is an example of a very simple buffer symbol. The buffer has a single input A and a single output B. The waveforms are shown for input A and the output B. Signal A changes from a '0' to a '1' at 10 nanoseconds and from a '1' to a '0' at 20 nanoseconds. This creates a pulse or spike that is 10 nanoseconds in duration. The buffer has a 20 nanosecond delay through the device.

The '0' to '1' transition on signal A causes the buffer model to be executed and schedules an event with the value '1' to occur on output B at time 30 nanoseconds. At time 20 nanoseconds, the next event on signal A occurs. This executes the buffer model again. The buffer model predicts a new event on output B of a 0 value at time 40 nanoseconds. The event scheduled on output B for time 30 nanoseconds still has not occurred. The new event predicted by the buffer model clashes with the currently scheduled event, and the simulator preempts the event at 30 nanoseconds.

The effect of the preemption is that the spike is swallowed. The reason for the cancellation is that, according to the inertial delay model, the first event at 30 nanoseconds did not have enough time to overcome the inertia of the output signal.

The inertial delay model is by far the most commonly used in all currently available simulators. This is partly because, in most cases, the inertial delay model is accurate enough for the designer's needs. One more

Figure 2-4
Inertial Delay Buffer
Waveforms.

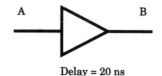

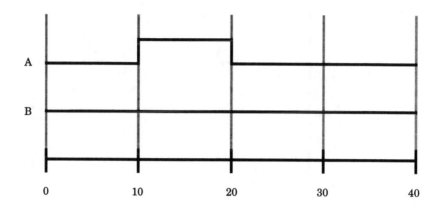

reason for the widespread use of inertial delay is that it prevents prolific propagation of spikes throughout the circuit. In most cases, this is the behavior wanted by the designer.

Transport Delay

Transport delay is not the default in VHDL and must be specified. It represents a wire delay in which any pulse, no matter how small, is propagated to the output signal delayed by the delay value specified. Transport delay is especially useful for modeling delay line devices, wire delays on a PC board, and path delays on an ASIC.

If we look at the same buffer circuit that was shown in Figure 2-4, but replace the inertial delay waveforms with the transport delay waveforms, we get the result shown in Figure 2-5. The same waveform is input to signal A, but the output from signal B is quite different. With transport delay, the spikes are not swallowed, but the events are ordered before propagation.

At time 10 nanoseconds, the buffer model is executed and schedules an event for the output to go to a 1 value at 30 nanoseconds. At time 20 nanoseconds, the buffer model is reinvoked and predicts a new value for

Figure 2-5
Transport Delay
Buffer Waveforms.

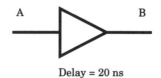

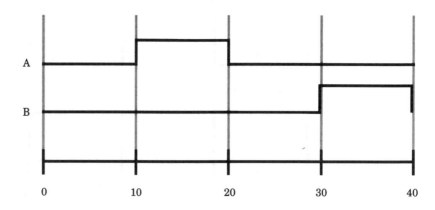

the output at time 40 nanoseconds. With the transport delay algorithm, the events are put in order. The event for time 40 nanoseconds is put in the list of events after the event for time 30 nanoseconds. The spike is not swallowed but is propagated intact after the delay time of the device.

Inertial Delay Model

The following model shows how to write an inertial delay model. It is the same as any other model we have been looking at. The default delay type is inertial; therefore, it is not necessary to specify the delay type to be inertial:

```
LIBRARY IEEE;
USE IEEE.std_logic_1164.ALL;
ENTITY buf IS
PORT ( a : IN std_logic;
       b : OUT std_logic);
END buf;

ARCHITECTURE buf OF buf IS
BEGIN
  b <= a AFTER 20 ns;
END buf;
```

Transport Delay Model

Following is an example of a transport delay model. It is similar in every respect to the inertial delay model except for the keyword **TRANSPORT** in the signal assignment statement to signal **b**. When this keyword exists, the delay type used in the statement is the transport delay mechanism:

```
LIBRARY IEEE;
USE IEEE.std_logic_1164.ALL;
ENTITY delay_line IS
PORT ( a : IN std_logic;
       b : OUT std_logic);
END delay_line;

ARCHITECTURE delay_line OF delay_line IS
BEGIN
   b <= TRANSPORT a AFTER 20 ns;
END delay_line;
```

Simulation Deltas

Simulation deltas are used to order some types of events during a simulation. Specifically, zero delay events must be ordered to produce consistent results. If zero delay events are not properly ordered, results can be disparate between different simulation runs. An example of this is shown using the circuit shown in Figure 2-6. This circuit could be part of a clocking scheme in a complex device being modeled. It probably would not be the entire circuit, but only a part of the circuit used to generate the clock to the D flip-flop.

The circuit consists of an inverter, a NAND gate, and an AND gate driving the clock input of a flip-flop component. The NAND gate and AND gate are used to gate the clock input to the flip-flop.

Let's examine the circuit operation, using a delta delay mechanism and another mechanism. By examining the two delay mechanisms, we will better understand how a delta delay orders events.

To use delta delay, all of the circuit components must have zero delay specified. The delay for all three gates is specified as zero. (Real circuits do not exhibit such characteristics, but sometimes modeling is easier if all of the delay is concentrated at the outputs.) Let's examine the non-delta delay mechanism first.

Figure 2-6
Simulation Delta
Circuit.

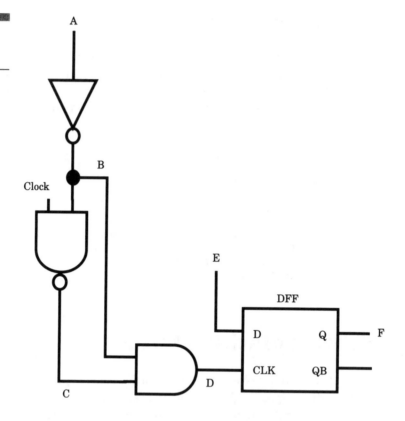

When a falling edge occurs on signal A, the output of the inverter changes in 0 time. Let's assume that such an event occurs at time 10 nanoseconds. The output of the inverter, signal B, changes to reflect the new input value. When signal B changes, both the AND gate and the NAND gate are reevaluated. For this example, the clock input is assumed to be a constant value `1`. If the NAND gate is evaluated first, its new value is `0`.

When the AND gate evaluates, signal B is a `0`, and signal C is a `1`; therefore, the AND gate predicts a new value of `0`. But what happens if the AND gate evaluates first? The AND gate sees a `1` value on signal B, and a `1` value on signal C before the NAND gate has a chance to reevaluate. The AND gate predicts a new value of `1`.

The NAND gate reevaluates and calculates its new value as `0`. The change on the output of the NAND gate causes the AND gate to reevaluate again. The AND gate now sees the value of B, a `1` value, and the new value of signal C, a `0` value. The AND gate now predicts a `0` on its output. This process is summarized in Figure 2-7.

Figure 2-7
Comparison of Two
Evaluation Mecha-
nisms.

AND First	NAND First
evaluate inverter	evaluate inverter
B <= 1	B <= 1
evaluate AND (C = 1)	evaluate NAND
D <= 1	C <= 0
evaluate NAND	evaluate AND
C <= 0	D <= 0
evaluate AND	
D <= 0	

Both circuits arrive at the same value for signal D. However, when the AND gate is evaluated first, a rising edge, one delta delay wide, occurs on signal D. This rising edge can clock the flip-flop, depending on how the flip-flop is modeled.

The point of this discussion is that without a delta synchronization mechanism, the results of the simulation can depend on how the simulator data structures are built. For instance, compiling the circuit the first time might make the AND gate evaluate first, while compiling again might make the NAND gate evaluate first—clearly not desirable results; simulation deltas prevent this behavior from occurring.

The same circuit evaluated using the VHDL delta delay mechanism would evaluate as shown in Figure 2-8.

The evaluation of the circuit does not depend on the order of evaluation of the NAND gate or AND gate. The sequence in Figure 2-8 occurs irrespective of the evaluation order of the AND or NAND gate.

During the first delta time point of time 10 nanoseconds, signal A receives the value `0`. This causes the inverter to reevaluate with the new value. The inverter calculates the new value for signal B, which is the value `1`. This value is not propagated immediately, but is scheduled for the next delta time point (delta 2).

The simulator then begins execution of delta time point 2. Signal B is updated to a `1` value, and the AND gate and NAND gate are reevaluated.

Figure 2-8
Delta Delay Evalua-
tion Mechanism.

Time	Delta	Activity
10 ns	(1)	A <= 0 evaluate inverter
	(2)	B <= 1 evaluate AND evaluate NAND
	(3)	D <= 1 C <= 0 evaluate AND
	(4)	D <= 0
11 ns		

Both the AND gate and NAND gate now schedule their new values for the next delta time point (delta 3).

When delta 3 occurs, signal D receives a `1` value, and signal C receives a `0` value. Because signal C also drives the AND gate, the AND gate is reevaluated and schedules its new output for delta time point 4.

To summarize, simulation deltas are an infinitesimal amount of time used as a synchronization mechanism when 0 delay events are present. Delta delay is used whenever 0 delay is specified, as shown in the following:

```
a <= b AFTER 0 ns;
```

Another case for using delta delay is when no delay is specified. For example:

```
a <= b;
```

In both cases, whenever signal **b** changes value from an event, signal **a** has a delta-delayed signal assignment to it.

An equivalent VHDL model of the circuit shown in Figure 2-6, except for the flip-flop, is shown in the following:

```
ENTITY reg IS
  PORT( a, clock : in bit
        d : out bit);
END reg;
```

```
ARCHITECTURE test OF reg IS
SIGNAL b, c : bit;
BEGIN
b <= NOT(a);   -- notice no delay
c <= NOT( clock AND b);
d <= c AND b;
END test;
```

Drivers

VHDL has a unique way of handling *multiply driven signals*. Multiply driven signals are very useful for modeling a data bus, a bidirectional bus, and so on. Correctly modeling these kinds of circuits in VHDL requires the concept of signal drivers. A VHDL driver is one contributor to the overall value of a signal.

A multiply driven signal has many drivers. The values of all of the drivers are resolved together to create a single value for the signal. The method of resolving all of the contributors into a single value is through a *resolution function* (resolution functions are discussed in Chapter 5, "Subprograms and Packages"). A resolution function is a designer-written function that is called whenever a driver of a signal changes value.

Driver Creation

Drivers are created by signal assignment statements. A concurrent signal assignment inside of an architecture produces one driver for each signal assignment. Therefore, multiple signal assignments produce multiple drivers for a signal. Consider the following architecture:

```
ARCHITECTURE test OF test IS
BEGIN
  a <= b AFTER 10 ns;
  a <= c AFTER 10 ns;
END test;
```

Signal **a** is being driven from two sources, **b** and **c**. Each concurrent signal assignment statement creates a driver for signal **a**. The first statement creates a driver that contains the value of signal **b** delayed by 10 nanoseconds. The second statement creates a driver that contains the value of signal **c** delayed by 10 nanoseconds. How these two drivers are resolved is left to the designer. The designers of VHDL did not want to arbitrarily add language constraints to signal behavior. Synthesizing the preceding example would short **c** and **b** together.

Bad Multiple Driver Model

Let's look at a model that looks correct at first glance, but does not function as the user intended. The model is for the 4 to 1 multiplexer discussed earlier:

```
USE WORK.std_logic_1164.ALL;
ENTITY mux IS
PORT (i0, i1, i2, i3, a, b: IN std_logic;
       q : OUT std_logic);
END mux;

ARCHITECTURE bad OF mux IS
BEGIN
  q <= i0 WHEN a = '0' AND b = '0' ELSE '0';
  q <= i1 WHEN a = '1' AND b = '0' ELSE '0';
  q <= i2 WHEN a = '0' AND b = '1' ELSE '0';
  q <= i3 WHEN a = '1' AND b = '1' ELSE '0';
END BAD;
```

This model assigns **i0** to **q** when **a** is equal to a 0 and **b** is equal to a 0; **i1** when **a** is equal to a 1 and **b** is equal to a 0; and so on. At first glance, the model looks like it works. However, each assignment to signal **q** creates a new driver for signal **q**. Four drivers to signal **q** are created by this model.

Each driver drives either the value of one of the **i0**, **i1**, **i2**, **i3** inputs or '0'. The value driven is dependent on inputs **a** and **b**. If **a** is equal to '0' and **b** is equal to '0', the first assignment statement puts the value of **i0** into one of the drivers of **q**. The other three assignment statements do not have their conditions met and, therefore, are driving the value '0'. Three drivers are driving the value '0', and one driver is driving the value of **i0**. Typical resolution functions would have a difficult time predicting the desired output on **q**, which is the value of **i0**.

A better way to write this model is to create only one driver for signal **q**, as shown in the following:

```
ARCHITECTURE better OF mux IS
BEGIN
  q <= i0 WHEN a = '0' AND b = '0' ELSE
       i1 WHEN a = '1' AND b = '0' ELSE
       i2 WHEN a = '0' AND b = '1' ELSE
```

```
                    i3 WHEN a = '1' AND b = '1' ELSE
                    'X';        --- unknown
             END better;
```

■ ■ Generics

Generics are a general mechanism used to pass information to an instance of an entity. The information passed to the entity can be of most types allowed in VHDL. (Types are covered in detail later in Chapter 4, "Data Types.")

Why would a designer want to pass information to an entity? The most obvious, and probably most used, information passed to an entity is delay times for rising and falling delays of the device being modeled. Generics can also be used to pass any user-defined data types, including information such as load capacitance, resistance, and so on. For synthesis parameters such as datapath widths, signal widths, and so on, can be passed in as generics.

All of the data passed to an entity is instance-specific information. The data values pertain to the instance being passed the data. In this way, the designer can pass different values to different instances in the design.

The data passed to an instance is static data. After the model has been elaborated (linked into the simulator), the data does not change during simulation. Generics cannot be assigned information as part of a simulation run. The information contained in generics passed into a component instance or a block can be used to alter the simulation results, but results cannot modify the generics.

The following is an example of an entity for an AND gate that has three generics associated with it:

```
ENTITY and2 IS
   GENERIC(rise, fall : TIME; load : INTEGER);
PORT( a, b : IN BIT;
      c : OUT BIT);
END AND2;
```

This entity allows the designer to pass in a value for the rise and fall delays, as well as the loading that the device has on its output. With this information, the model can correctly model the AND gate in the design. Following is the architecture for the AND gate:

```
ARCHITECTURE load_dependent OF and2 IS
  SIGNAL internal : BIT;
BEGIN
internal <= a AND b;
c <= internal AFTER (rise + (load * 2 ns)) WHEN internal = `1'
  ELSE internal AFTER (fall + (load * 3 ns));

END load_dependent;
```

The architecture declares a local signal called **internal** to store the value of the expression **a** and **b**. Precomputing values used in multiple instances is a very efficient method for modeling.

The generics **rise**, **fall**, and **load** contain the values that were passed in by the component instantiation statement. Let's look at a piece of a model that instantiates the components of type **AND2** in another model:

```
LIBRARY IEEE;
USE IEEE.std_logic_1164.ALL;
ENTITY test IS
  GENERIC(rise, fall : TIME; load : INTEGER);
  PORT ( ina, inb, inc, ind : IN std_logic;
         out1, out2 : OUT std_logic);
END test;

ARCHITECTURE test_arch OF test IS
  COMPONENT AND2
    GENERIC(rise, fall : TIME; load : INTEGER);
    PORT ( a, b : IN std_logic;
           c : OUT std_logic);
  END COMPONENT;
BEGIN
  U1: AND2 GENERIC MAP(10 ns, 12 ns, 3 )
    PORT MAP (ina, inb, out1 );

  U2: AND2 GENERIC MAP(9 ns, 11 ns, 5 )
    PORT MAP (inc, ind, out2 );
END test_arch;
```

The architecture statement first declares any components that will be used in the model. In this example, component **AND2** is declared. Next, the body of the architecture statement contains two of component instantiation statements for components **U1** and **U2**. Port **a** of component **U1** is mapped to signal **ina**, port **b** is mapped to signal **inb**, and port **c** is mapped to **out1**. In the same way, component **U2** is mapped to signals **inc**, **ind**, and **out2**.

Generic **rise** of instance **U1** is mapped to 10 nanoseconds, generic **fall** is mapped to 12 nanoseconds, and generic **load** is mapped to 3. The generics for component **U2** are mapped to values 9 and 11 nanoseconds and value 5.

Generics can also have default values that are overridden if actual values are mapped to the generics. The next example shows two instances of component type **AND2**.

In instance **U1**, actual values are mapped to the generics, and these values are used in the simulation. In instance **U2**, no values are mapped to the instance, and the default values are used to control the behavior of the simulation if specified; otherwise an error occurs:

```
LIBRARY IEEE;
USE IEEE.std_logic_1164.ALL;
ENTITY test IS
  GENERIC(rise, fall : TIME;
          load : INTEGER);
  PORT ( ina, inb, inc, ind : IN std_logic;
         out1, out2 : OUT std_logic);
END test;

ARCHITECTURE test_arch OF test IS
  COMPONENT and2
    GENERIC(rise, fall : TIME := 10 NS;
            load : INTEGER := 0);
    PORT ( a, b : IN std_logic;
           c : OUT std_logic);
  END COMPONENT;
BEGIN

  U1: and2 GENERIC MAP(10 ns, 12 ns, 3 )
    PORT MAP (ina, inb, out1 );

  U2: and2 PORT MAP (inc, ind, out2 );

END test_arch;
```

As we have seen, generics have many uses. The uses of generics are limited only by the creativity of the model writer.

Block Statements

Blocks are a partitioning mechanism within VHDL that allow the designer to logically group areas of the model. The analogy with a typical CAE system is a schematic sheet. In a typical CAE system, a level or a portion of the design can be represented by a number of schematic sheets. The reason for partitioning the design may relate to design standards

about how many components are allowed on a sheet, or it may be a logical grouping that the designer finds more understandable.

The same analogy holds true for block statements. The statement area
in an architecture can be broken into a number of separate logical areas.
For instance, if you are designing a CPU, one block might be an ALU, another a register bank, and another a shifter.

Each block represents a self-contained area of the model. Each block
can declare local signals, types, constants, and so on. Any object that can
be declared in the architecture declaration section can be declared in the
block declaration section. Following is an example:

```
LIBRARY IEEE;
USE IEEE.std_logic_1164.ALL;
PACKAGE bit32 IS
  TYPE tw32 IS ARRAY(31 DOWNTO 0) OF std_logic;
END bit32;

LIBRARY IEEE;
USE IEEE.std_logic_1164.ALL;
USE WORK.bit32.ALL;
ENTITY cpu IS
  PORT( clk, interrupt : IN std_logic;
        addr : OUT tw32; data : INOUT tw32 );
END cpu;

ARCHITECTURE cpu_blk OF cpu IS
  SIGNAL ibus, dbus : tw32;
BEGIN
  ALU : BLOCK
   SIGNAL qbus : tw32;
  BEGIN
   -- alu behavior statements
  END BLOCK ALU;

  REG8 : BLOCK
   SIGNAL zbus : tw32;
  BEGIN
   REG1: BLOCK
     SIGNAL qbus : tw32;
   BEGIN
   -- reg1 behavioral statements
   END BLOCK REG1;

   -- more REG8 statements

  END BLOCK REG8;
END cpu_blk;
```

Entity **cpu** is the outermost entity declaration of this model. (This is
not a complete model, only a subset.) Entity **cpu** declares four ports that

are used as the model interface. Ports **clk** and **interrupt** are input ports, **addr** is an output port, and **data** is an inout port. All of these ports are visible to any block declared in an architecture for this entity. The input ports can be read from and the output ports can be assigned values.

Signals **ibus** and **dbus** are local signals declared in architecture **cpu_blk**. These signals are local to architecture **cpu_blk** and cannot be referenced outside of the architecture. However, any block inside of the architecture can reference these signals. Any lower-level block can reference signals from a level above, but upper-level blocks cannot reference lower-level local signals.

Signal **qbus** is declared in the block declaration section of block **ALU**. This signal is local to block **ALU** and cannot be referenced outside of the block. All of the statements inside of block **ALU** can reference **qbus**, but statements outside of block **ALU** cannot use **qbus**.

In exactly the same fashion, signal **zbus** is local to block **REG8**. Block **REG1** inside of block **REG8** has access to signal **zbus**, and all of the other statements in block **REG8** also have access to signal **zbus**.

In the declaration section for block **REG1**, another signal called **qbus** is declared. This signal has the same name as the signal **qbus** declared in block **ALU**. Doesn't this cause a problem? To the compiler, these two signals are separate, and this is a legal, although confusing, use of the language. The two signals are declared in two separate declarative regions and are valid only in those regions; therefore, they are considered to be two separate signals with the same name. Each **qbus** can be referenced only in the block that has the declaration of the signal, except as a fully qualified name, discussed later in this section.

Another interesting case is shown here:

```
BLK1 : BLOCK
   SIGNAL qbus : tw32;
BEGIN

BLK2 : BLOCK
  SIGNAL qbus : tw32;
BEGIN
-- blk2 statements
END BLOCK BLK2;

-- blk1 statements

END BLOCK BLK1;
```

In this example, signal **qbus** is declared in two blocks. The interesting feature of this model is that one of the blocks is contained in the other. It would seem that **BLK2** has access to two signals called **qbus**—the first from the local declaration of **qbus** in the declaration section of **BLK2** and the second from the declaration section of **BLK1**. **BLK1** is also the parent block of **BLK2**. However, **BLK2** sees only the **qbus** signal from the declaration in **BLK2**. The **qbus** signal from **BLK1** has been overridden by a declaration of the same name in **BLK2**.

The **qbus** signal from **BLK1** can be seen inside of **BLK2**, if the name of signal **qbus** is qualified with the block name. For instance, in this example, to reference signal **qbus** from **BLK1**, use **BLK1.qbus**.

In general, this can be a very confusing method of modeling. The problem stems from the fact that you are never quite sure which **qbus** is being referenced at a given time without fully analyzing all of the declarations carefully.

As mentioned earlier, blocks are self-contained regions of the model. But blocks are unique because a block can contain ports and generics. This allows the designer to remap signals and generics external to the block to signals and generics inside the block. But why, as designers, would we want to do that?

The capability of ports and generics on blocks allows the designer to reuse blocks written for another purpose in a new design. For instance, let's assume that you are upgrading a CPU design and need extra functionality in the ALU section. Let's also assume that another designer has a new ALU model that performs the functionality needed. The only trouble with the new ALU model is that the interface port names and generic names are different than the names that exist in the design being upgraded. With the port and generic mapping capability within blocks, this is no problem. Map the signal names and the generic parameters in the design being upgraded to ports and generics created for the new ALU block. Following is an example illustrating this:

```
PACKAGE math IS
  TYPE tw32 IS ARRAY(31 DOWNTO 0) OF std_logic;
  FUNCTION tw_add(a, b : tw32) RETURN tw32;
  FUNCTION tw_sub(a, b : tw32) RETURN tw32;
END math;

USE WORK.math.ALL;
LIBRARY IEEE;
USE IEEE.std_logic_1164.ALL;
ENTITY cpu IS
PORT( clk, interrupt : IN std_logic;
      addr : OUT tw32; cont : IN INTEGER;
      data : INOUT tw32 );
```

```
END cpu;

ARCHITECTURE cpu_blk OF cpu IS
  SIGNAL ibus, dbus : tw32;
BEGIN
  ALU : BLOCK
    PORT( abus, bbus : IN tw32;
          d_out : OUT tw32;
          ctbus : IN INTEGER);
    PORT MAP ( abus => ibus, bbus => dbus, d_out => data,
              ctbus => cont);
    SIGNAL qbus : tw32;
    BEGIN
      d_out <= tw_add(abus, bbus)   WHEN ctbus = 0 ELSE
               tw_sub(abus, bbus)   WHEN ctbus = 1 ELSE
               abus;
    END BLOCK ALU;
END cpu_blk;
```

Basically, this is the same model shown earlier except for the port and port map statements in the **ALU** block declaration section. The port statement declares the number of ports used for the block, the direction of the ports, and the type of the ports. The port map statement maps the new ports with signals or ports that exist outside of the block. Port **abus** is mapped to architecture **CPU_BLK** local signal **ibus**; port **bbus** is mapped to **dbus**. Ports **d_out** and **ctbus** are mapped to external ports of the entity.

Mapping implies a connection between the port and the external signal such that, whenever there is a change in value on the signal connected to a port, the port value changes to the new value. If a change occurs in the signal **ibus**, the new value of **ibus** is passed into the ALU block and port **abus** obtains the new value. The same is true for all ports.

Guarded Blocks

Block statements have another interesting behavior known as *guarded blocks*. A guarded block contains a guard expression that can enable and disable drivers inside the block. The guard expression is a boolean expression: when true, drivers contained in the block are enabled, and when false, the drivers are disabled. Let's look at the following example to show some more of the details:

```
LIBRARY IEEE;
USE IEEE.std_logic_1164.ALL;
ENTITY latch IS
PORT( d, clk : IN std_logic;
```

```
     q, qb : OUT std_logic);
END latch;

ARCHITECTURE latch_guard OF latch IS
  BEGIN
  G1 : BLOCK( clk = '1')
  BEGIN
   q <= GUARDED d AFTER 5 ns;
   qb <= GUARDED NOT(d) AFTER 7 ns;
  END BLOCK G1;
END latch_guard;
```

This model illustrates how a latch model could be written using a guarded block. This is a very simple-minded model; however, more complex and more accurate models will be shown later. The entity declares the four ports needed for the latch, and the architecture has only one statement in it. The statement is a guarded block statement. A guarded block statement looks like a typical block statement, except for the guard expression after the keyword **BLOCK**. The guard expression in this example is (clk = '1'). This is a boolean expression that returns **TRUE** when clk is equal to a '1' value and **FALSE** when clk is equal to any other value.

When the guard expression is true, all of the drivers of guarded signal assignment statements are enabled, or turned on. When the guard expression is false, all of the drivers of guarded signal assignment statements are disabled, or turned off. There are two guarded signal assignment statements in this model: One is the statement that assigns a value to q and the other is the statement that assigns a value to qb. A guarded signal assignment statement is recognized by the keyword **GUARDED** between the <= and the expression part of the statement.

When port clk of the entity has the value '1', the guard expression is true, and the value of input d is scheduled on the q output after 5 nanoseconds, and the NOT value of d is scheduled on the qb output after 7 nanoseconds. When port clk has the value '0' or any other legal value of the type, outputs q and qb turn off and the output value of the signal is determined by the default value assigned by the resolution function. When clk is not equal to '1', the drivers created by the signal assignments for q and qb in this architecture are effectively turned off. The drivers do not contribute to the overall value of the signal.

Signal assignments can be guarded by using the keyword **GUARDED**. A new signal is implicitly declared in the block whenever a block has a guard expression. This signal is called **GUARD**. Its value is the value of the guard expression. This signal can be used to trigger other processes to occur.

Blocks are useful for partitioning the design into smaller, more manageable units. They allow the designer the flexibility to create large designs from smaller building blocks and provide a convenient method of controlling the drivers on a signal.

SUMMARY

In the first chapter, concepts of structurally building models were discussed. This chapter is the first of many that discusses behavioral modeling. In this chapter, we discussed:

- How signal assignments are the most basic form of behavioral modeling
- Signal assignment statements can be selected or conditional
- Signal assignment statements can contain delays
- VHDL contains inertial delay and transport delay
- Simulation delta time points are used to order events in time
- Drivers on a signal are created by signal assignment statements
- Generics are used to pass data to entities
- Block statements allow grouping within an entity
- Guarded block statements allow the capability of turning off drivers within a block

Sequential Processing

In Chapter 2, we examined behavioral modeling using concurrent statements. We discussed concurrent signal assignment statements, as well as block statements and component instantiation. In this chapter, we focus on sequential statements. Sequential statements are statements that execute serially one after the other. Most programming languages, such as C and Pascal, support this type of behavior. In fact, VHDL has borrowed the syntax for its sequential statements from ADA.

Process Statement

In an architecture for an entity, all statements are concurrent. So where do sequential statements exist in VHDL? There is a statement called the *process statement* that contains only sequential statements. The process statement is itself a concurrent statement. A process statement can exist in an architecture and define regions in the architecture where all statements are sequential.

A process statement has a declaration section and a statement part. In the declaration section, types, variables, constants, subprograms, and so on can be declared (for a full list, see the LRM). The statement part contains only sequential statements. Sequential statements consist of **CASE** statements, **IF THEN ELSE** statements, **LOOP** statements, and so on. We examine these statements later in this chapter. First, let's look at how a process statement is structured.

Sensitivity List

The process statement can have an explicit sensitivity list. This list defines the signals that cause the statements inside the process statement to execute whenever one or more elements of the list change value. The sensitivity list is a list of the signals that the process is sensitive to changes on. The process has to have an explicit sensitivity list or, as we discuss later, a **WAIT** statement.

As of this writing, synthesis tools have a difficult time with sensitivity lists that are not fully specified. Synthesis tools think of process statements as either describing sequential logic or combinational logic. If a process contains a partial sensitivity list, one that does not contain every input signal used in the process, there is no way to map that functionality to either sequential or combinational logic.

Process Example

Let's look at an example of a process statement in an architecture to see how the process statement fits into the big picture, and discuss some more details of how it works. Following is a model of a two-input NAND gate:

```
LIBRARY IEEE;
USE IEEE.std_logic_1164.ALL;
```

```
ENTITY nand2 IS
  PORT( a, b : IN std_logic;
        c : OUT std_logic);
END nand2;

ARCHITECTURE nand2 OF nand2 IS
BEGIN
  PROCESS( a, b )
   VARIABLE temp : std_logic;
  BEGIN
   temp := NOT (a and b);

   IF (temp = '1') THEN
     c <= temp AFTER 6 ns;
   ELSIF (temp = '0') THEN
     c <= temp AFTER 5 ns;
   ELSE
     c <= temp AFTER 6 ns;
   END IF;

  END PROCESS;
END nand2;
```

This example shows how to write a model for a simple two-input NAND gate using a process statement. The **USE** statement declares a VHDL package that provides the necessary information to allow modeling this NAND gate with 9 state logic. (This package is described in Appendix A, "Standard Logic Package".) We discuss packages later in Chapter 5, "Subprograms and Packages." The **USE** statement was included so that the model could be simulated with a VHDL simulator without any modifications.

The entity declares three ports for the **nand2** gate. Ports **a** and **b** are the inputs to the **nand2** gate and port **c** is the output. The name of the architecture is the same name as the entity name. This is legal and can save some of the headaches of trying to generate unique names.

The architecture contains only one statement, a concurrent process statement. The process declarative part starts at the keyword **PROCESS** and ends at the keyword **BEGIN**. The process statement part starts at the keyword **BEGIN** and ends at the keywords **END PROCESS**. The process declaration section declares a local variable named **temp**. The process statement part has two sequential statements in it; a variable assignment statement:

```
temp := NOT (a AND b);
```

and an **IF THEN ELSE** statement:

```
IF (temp = '1') THEN
```

```
    c <= temp AFTER 6 ns;
ELSIF (temp = '0') THEN
    c <= temp AFTER 5 ns;
ELSE
    c <= temp AFTER 6 ns;
END IF;
```

The process contains an explicit sensitivity list with two signals contained in it:

```
PROCESS( a, b )
```

The process is sensitive to signals **a** and **b**. In this example, **a** and **b** are input ports to the model. Input ports create signals that can be used as inputs; output ports create signals that can be used as outputs; and inout ports create signals that can be used as both. Whenever port **a** or **b** has a change in value, the statements inside of the process are executed. Each statement is executed in serial order starting with the statement at the top of the process statement and working down to the bottom. After all of the statements have been executed once, the process waits for another change in a signal or port in its sensitivity list.

The process declarative part declares one variable called **temp**. Its type is **std_logic**. This type is explained in Appendix A, "Standard Logic Package," as it is used throughout the book. For now, assume that the type defines a signal that is a single bit and can assume the values 0, 1, and X. Variable **temp** is used as temporary storage in this model to save the precomputed value of the expression (**a** and **b**). The value of this expression is precomputed for efficiency.

Signal Assignment Versus Variable Assignment

The first statement inside of the process statement is a variable assignment that assigns a value to variable **temp**. In the previous chapter, we discussed how signals received values that were scheduled either after an amount of time or after a delta delay. A variable assignment happens immediately when the statement is executed. For instance, in this model, the first statement has to assign a value to variable **temp** for the second statement to use. Variable assignment has no delay; it happens immediately.

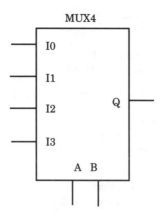

Figure 3-1
Four Input Mux Symbol and Function.

A	B	Q
0	0	I0
1	0	I1
0	1	I2
1	1	I3

Let's look at two examples that illustrate this point more clearly. Both examples are models of a 4 to 1 multiplexer device. The symbol and truth table for this device are shown in Figure 3-1. One of the four input signals is propagated to the output depending on the values of inputs A and B.

The first model for the multiplexer is an incorrect model, and the second is a corrected version of the model.

Incorrect Mux Example

The incorrect model of the multiplexer has a flaw in it that causes the model to produce incorrect results. This is shown by the following model:

```
LIBRARY IEEE;
USE IEEE.std_logic_1164.ALL;
ENTITY mux IS
PORT (i0, i1, i2, i3, a, b : IN std_logic;
```

```
         q : OUT std_logic);
END mux;

ARCHITECTURE wrong of mux IS
  SIGNAL muxval : INTEGER;
BEGIN
  PROCESS ( i0, i1, i2, i3, a, b )
  BEGIN
   muxval <= 0;
   IF (a = '1') THEN
     muxval <= muxval + 1;
   END IF;

   IF (b = '1') THEN
     muxval <= muxval + 2;
   END IF;

   CASE muxval IS
     WHEN 0 =>
       q <= I0 AFTER 10 ns;
     WHEN 1 =>
       q <= I1 AFTER 10 ns;
     WHEN 2 =>
       q <= I2 AFTER 10 ns;
     WHEN 3 =>
       q <= I3 AFTER 10 ns;
     WHEN OTHERS =>
       NULL;
   END CASE;
  END PROCESS;
END wrong;
```

Whenever one of the input signals in the process sensitivity list changes value, the sequential statements in the process are executed. The process statement in the first example contains four sequential statements. The first statement initializes the local signal **muxval** to a known value (0). The subsequent statements add values to the local signal depending on the value of the **a** and **b** input signals. Finally, the case statement chooses an input to propagate to the output based on the value of signal **muxval**. This model has a significant flaw, however. The first statement:

```
muxval <= 0;
```

causes the value 0 to be scheduled as an event for signal **muxval**. In fact, the value 0 is scheduled in an event for the next simulation delta because no delay was specified. When the second statement:

```
IF (a = '1') THEN
  muxval <= muxval + 1;
END IF;
```

is executed, the value of signal **muxval** is whatever was last propagated to it. The new value scheduled from the first statement has not propagated yet. In fact, when multiple assignments to a signal occur within the same process statement, the last assigned value is the value propagated.

The signal **muxval** has a garbage value when entering the process. Its value is not changed until the process has completed execution of all sequential statements contained in the process. In fact, if signal **b** is a '1' value, then whatever garbage value the signal had when entering the process will have the value 2 added to it.

A better way to implement this example is shown in the next example. The only difference between the next model and the previous one is the declaration of **muxval** and the assignments to **muxval**. In the previous model, **muxval** was a signal, and signal assignment statements were used to assign values to it. In the next example, **muxval** is a variable, and variable assignments are used to assign to it.

Correct Mux Example

In this example, the incorrect model is rewritten to reflect a solution to the problems with the last model:

```
LIBRARY IEEE;
USE IEEE.std_logic_1164ALL;
ENTITY mux IS
PORT (i0, i1, i2, i3, a, b : IN std_logic;
      q : OUT std_logic);
END mux;

ARCHITECTURE better OF mux IS
BEGIN
  PROCESS ( i0, i1, i2, i3, a, b )
   VARIABLE muxval : INTEGER;
  BEGIN
   muxval := 0;
   IF (a = '1') THEN
     muxval := muxval + 1;
   END IF;

   IF (b = '1') THEN
     muxval := muxval + 2;
   END IF;

   CASE muxval IS
     WHEN 0 =>
       q <= I0 AFTER 10 ns;
     WHEN 1 =>
```

```
          q <= I1 AFTER 10 ns;
        WHEN 2 =>
          q <= I2 AFTER 10 ns;
        WHEN 3 =>
          q <= I3 AFTER 10 ns;
        WHEN OTHERS =>
          NULL;
      END CASE;
    END PROCESS;
  END better;
```

This simple coding difference makes a tremendous operational difference. When the first statement:

```
muxval := 0;
```

is executed, the value 0 is placed in variable **muxval** immediately. The value is not scheduled because **muxval**, in this example, is a variable, not a signal. Variables represent local storage as opposed to signals, which represent circuit interconnect. The local storage is updated immediately, and the new value can be used later in the model for further computations.

Because **muxval** is initialized to 0 immediately, the next two statements in the process use 0 as the initial value and add appropriate numbers, depending on the values of signals **a** and **b**. These assignments are also immediate, and therefore when the **CASE** statement executes, variable **muxval** contains the correct value. From this value, the correct input signal can be propagated to the output.

Sequential Statements

Sequential statements exist inside the boundaries of a process statement as well as in subprograms. In this chapter, we are most concerned with sequential statements inside process statements. In Chapter 5, we discuss subprograms and the statements contained within them.

The sequential statements that we discuss are:

- IF
- CASE
- LOOP
- ASSERT
- WAIT

IF Statements

In Appendix A of the LRM, all VHDL constructs are described using a variant of the *Bachus-Naur format* (BNF) that is used to describe typical programming languages. If you are not familiar with BNF, Appendix C gives a cursory description. Becoming familiar with the BNF will help you better understand how to construct complex VHDL statements.

The BNF description of the **IF** statement is listed in of the LRM and looks like this:

```
if_statement ::=
  IF condition THEN
 sequence_of_statements
{ELSIF condition THEN
 sequence_of_statements}
[ELSE
 sequence_of_statements]
  END IF;
```

From the BNF description, we can conclude that the **IF** statement starts with the keyword **IF** and ends with the keywords **END IF** spelled out as two separate words. There are also two optional clauses: the **ELSIF** clause and the **ELSE** clause. The **ELSIF** clause is repeatable—more than one **ELSIF** clause is allowed; but the **ELSE** clause is optional, and only one is allowed. The condition construct in all cases is a boolean expression. This is an expression that evaluates to either true or false. Whenever a condition evaluates to a true value, the sequence of statements following is executed. If no condition is true, then the sequence of statements for the **ELSE** clause is executed, if one exists. Let's analyze a few examples to get a better understanding of how the BNF relates to the VHDL code.

The first example shows how to write a simple **IF** statement:

```
IF (x < 10) THEN
  a := b;
END IF;
```

The **IF** statement starts with the keyword **IF**. Next is the condition (**x < 10**), followed by the keyword **THEN**. The condition is true when the value of **x** is less than 10; otherwise it is false. When the condition is true, the statements between the **THEN** and **END IF** are executed. In this example, the assignment statement (**a := b**) is executed whenever **x** is less than 10. What happens if **x** is greater than or equal to 10? In this example, there is no **ELSE** clause, so no statements are executed in the **IF** statement. Instead, control is transferred to the statement after the **END IF**.

Let's look at another example where the **ELSE** clause is useful:

```
IF (day = sunday) THEN
   weekend := TRUE;
ELSIF (day = saturday) THEN
   weekend := TRUE;
ELSE
   weekday := TRUE;
END IF;
```

In this example, there are two variables—**weekend** and **weekday**—that are set depending on the value of a signal called **day**. Variable **weekend** is set to **TRUE** whenever **day** is equal to **saturday** or **sunday**. Otherwise, variable **weekday** is set to **TRUE**. The execution of the **IF** statement starts by checking to see if variable **day** is equal to **sunday**. If this is true, then the next statement is executed and control is transferred to the statement following **END IF**. Otherwise, control is transferred to the **ELSIF** statement part, and **day** is checked for **saturday**. If variable **day** is equal to **saturday**, then the next statement is executed and control is again transferred to the statement following the **END IF** statement. Finally, if **day** is not equal to **sunday** or **saturday**, then the **ELSE** statement part is executed.

The **IF** statement can have multiple **ELSIF** statement parts, but only one **ELSE** statement part. More than one sequential statement can exist between each statement part.

CASE Statements

The **CASE** statement is used whenever a single expression value can be used to select between a number of actions. Following is the BNF for the **CASE** statement:

```
case_statement ::=
  CASE expression IS
  case_statement_alternative
  {case_statement_alternative}
END CASE;

case_statement_alternative ::=
WHEN choices =>
```

```
sequence_of_statements

sequence_of_statements ::=
{sequential_statement}

choices ::=
choice{| choice}

choice ::=
SIMPLE_expression|
discrete_range|
ELEMENT_simple_name|
   OTHERS
```

A **CASE** statement consists of the keyword **CASE** followed by an expression and the keyword **IS**. The expression either returns a value that matches one of the **CHOICES** in a **WHEN** statement part, or matches an **OTHERS** clause. If the expression matches the **CHOICE** part of a **WHEN** **choices** => clause, the **sequence_of_statements** following is executed. After these statements are executed, control is transferred to the statement following the **END CASE** clause.

Either the **CHOICES** clause must enumerate every possible value of the type returned by the expression, or the last choice must contain an **OTHERS** clause.

Let's look at some examples to reinforce what the BNF states:

```
CASE instruction IS
  WHEN load_accum =>
    accum <= data;
  WHEN store_accum =>
    data_out <= accum;
  WHEN load|store =>
    process_IO(addr);
  WHEN OTHERS =>
    process_error(instruction);
END CASE;
```

The **CASE** statement executes the proper statement depending on the value of input instruction. If the value of instruction is one of the choices listed in the **WHEN** clauses, then the statement following the **WHEN** clause is executed. Otherwise, the statement following the **OTHERS** clause is executed. In this example, when the value of instruction is **load_accum**, the first assignment statement is executed. If the value of instruction is **load** or **store**, the **process_IO** procedure is called.

If the value of instruction is outside the range of the choices given, then the **OTHERS** clause matches the expression and the statement following the

OTHERS clause is executed. It is an error if an OTHERS clause does not exist, and the choices given do not cover every possible value of the expression type.

In the next example, a more complex type is returned by the expression. (Types are discussed in Chapter 4, "Data Types.") The CASE statement uses this type to select among the choices of the statement:

```
TYPE vectype IS ARRAY(0 TO 1) OF BIT;
VARIABLE bit_vec : vectype;
  .
  .

CASE bit_vec IS
  WHEN "00" =>
    RETURN 0;
  WHEN "01" =>
    RETURN 1;
  WHEN "10" =>
    RETURN 2;
  WHEN "11" =>
    RETURN 3;
END CASE;
```

This example shows one way to convert an array of bits into an integer. When both bits of variable bit_vec contain '0' values, the first choice "00" matches and the value 0 is returned. When both bits are '1' values, the value 3, or "11", is returned. This CASE statement does not need an OTHERS clause because all possible values of variable bit_vec are enumerated by the choices.

LOOP Statements

The LOOP statement is used whenever an operation needs to be repeated. LOOP statements are used when powerful iteration capability is needed to implement a model. Following is the BNF for the LOOP statement:

```
loop_statement ::=
  [LOOP_label : ] [iteration_scheme] LOOP
 sequence_of_statements
END LOOP[LOOP_label];

iteration_scheme ::=
WHILE condition | FOR LOOP_parameter_specification
```

```
LOOP_parameter_specification ::=
    identifier IN discrete_range
```

The **LOOP** statement has an optional label, which can be used to identify the **LOOP** statement. The **LOOP** statement has an optional **iteration_scheme** that determines which kind of **LOOP** statement is being used. The **iteration_scheme** includes two types of **LOOP** statements: a **WHILE**   condition **LOOP** statement and a "**FOR** identifier **IN** discrete_range" statement. The **FOR** loop loops as many times as specified in the **discrete_range**, unless the loop is exited. The **WHILE condition LOOP** statement loops as long as the condition expression is **TRUE**.

Let's look at a couple of examples to see how these statements work:

```
WHILE (day = weekday) LOOP
    day := get_next_day(day);
END LOOP;
```

This example uses the **WHILE condition** form of the **LOOP** statement. The condition is checked each time before the loop is executed. If the condition is **TRUE**, the **LOOP** statements are executed. Control is then transferred back to the beginning of the loop. The condition is checked again. If **TRUE**, the loop is executed again; if not, statement execution continues on the statement following the **END LOOP** clause.

The other version of the **LOOP** statement is the **FOR** loop:

```
FOR i IN 1 to 10 LOOP
    i_squared(i) := i * i;
END LOOP;
```

This loop executes 10 times whenever execution begins. Its function is to calculate the squares from 1 to 10 and insert them into the **i_squared** signal array. The index variable **i** starts at the leftmost value of the range and is incremented until the rightmost value of the range.

In some languages, the loop index (in this example, **i**) can be assigned a value inside the loop to change its value. VHDL does not allow any assignment to the loop index. This also precludes the loop index existing as the return value of a function, or as an out or inout parameter of a procedure.

Another interesting point about **FOR LOOP** statements is that the index value **i** is locally declared by the **FOR** statement. The variable **i** does not need to be declared explicitly in the process, function, or procedure. By virtue of the **FOR LOOP** statement, the loop index is declared locally. If another variable of the same name exists in the process, function, or procedure, then these two variables are treated as separate variables and are accessed by context. Let's look at an example to illustrate this point:

```
PROCESS(i)
BEGIN

x <= i + 1; -- x is a signal

FOR i IN 1 to a/2 LOOP
  q(i) := a; -- q is a variable
END LOOP;

END PROCESS;
```

Whenever the value of the signal i in the process sensitivity list changes value, the process will be invoked. The first statement schedules the value i + 1 on the signal x. Next, the **FOR** loop is executed. The index value i is not the same object as the signal i that was used to calculate the new value for signal x. These are separate objects that are each accessed by context. Inside the **FOR** loop, when a reference is made to i, the local index is retrieved. But outside the **FOR** loop, when a reference is made to i, the value of the signal i in the sensitivity list of the process is retrieved.

The values used to specify the range in the **FOR** loop need not be specific integer values, as has been shown in the examples. The range can be any discrete range. A **discrete_range** can be expressed as a **subtype_indication** or a range statement. Let's look at a few more examples of how **FOR** loops can be constructed with ranges:

```
PROCESS(clk)
   TYPE day_of_week IS (sun, mon, tue, wed, thur, fri,
      sat);
BEGIN
   FOR i IN day_of_week LOOP
   IF i = sat THEN
     son <= mow_lawn;
   ELSIF i = sun THEN
     church <= family;
   ELSE
     dad <= go_to_work;
   END IF;
   END LOOP;
END PROCESS;
```

In this example, the range is specified by the type. By specifying the type as the range, the compiler determines that the leftmost value is **sun**, and the rightmost value is **sat**. The range then is determined as from **sun** to **sat**.

If an ascending range is desired, use the **to** clause. The **downto** clause can be used to create a descending range. Here is an example:

```
PROCESS(x, y)
BEGIN
   FOR i IN x downto y LOOP
     q(i) := w(i);
   END LOOP;
END PROCESS;
```

When different values for **x** and **y** are passed in, different ranges of the array **w** are copied to the same place in array **q**.

NEXT Statement

There are cases when it is necessary to stop executing the statements in the loop for this iteration and go to the next iteration. VHDL includes a construct that accomplishes this. The **NEXT** statement allows the designer to stop processing this iteration and skip to the successor. When the **NEXT** statement is executed, processing of the model stops at the current point and is transferred to the beginning of the **LOOP** statement. Execution begins with the first statement in the loop, but the loop variable is incremented to the next iteration value. If the iteration limit has been reached, processing stops. If not, execution continues.

Following is an example showing this behavior:

```
PROCESS(A, B)
   CONSTANT max_limit : INTEGER := 255;
BEGIN
   FOR i IN 0 TO max_limit LOOP
    IF (done(i) = TRUE) THEN
      NEXT;
    ELSE
      done(i) := TRUE;
    END IF;

    q(i) <= a(i) AND b(i);

   END LOOP;
END PROCESS;
```

The process statement contains one **LOOP** statement. This **LOOP** statement logically "and"s the bits of arrays **a** and **b** and puts the results in array **q**. This behavior continues whenever the flag in array **done** is not true. If the **done** flag is already set for this value of index **i**, then the **NEXT** statement is executed. Execution continues with the first statement of the loop, and index **i** has the value **i + 1**. If the value of the **done** array is not true, then the **NEXT** statement is not executed, and execution continues with the statement contained in the **ELSE** clause for the **IF** statement.

The **NEXT** statement allows the designer the ability to stop execution of this iteration and go on to the next iteration. There are other cases when the need exists to stop execution of a loop completely. This capability is provided with the **EXIT** statement.

EXIT Statement

During the execution of a **LOOP** statement, it may be necessary to jump out of the loop. This can occur because a significant error has occurred during the execution of the model or all of the processing has finished early. The VHDL **EXIT** statement allows the designer to exit or jump out of a **LOOP** statement currently in execution. The **EXIT** statement causes execution to halt at the location of the **EXIT** statement. Execution continues at the statement following the **LOOP** statement.

Here is an example illustrating this point:

```
PROCESS(a)
  variable int_a : integer;
BEGIN
  int_a := a;

  FOR i IN 0 TO max_limit LOOP
   IF (int_a <= 0) THEN -- less than or
     EXIT;                -- equal to
   ELSE
     int_a := int_a -1;
     q(i) <= 3.1416 / REAL(int_a * i); -- signal
   END IF;                             -- assign
  END LOOP;

  y <= q;

END PROCESS;
```

Inside this process statement, the value of int_a is always assumed to be a positive value greater than 0. If the value of int_a is negative or zero, then an error condition results and the calculation should not be completed. If the value of int_a is less than or equal to 0, then the IF statement is true and the EXIT statement is executed. The loop is immediately terminated, and the next statement executed is the assignment statement to y after the LOOP statement.

If this were a complete example, the designer would also want to alert the user of the model that a significant error had occurred. A method to accomplish this function would be with an ASSERT statement, which is discussed later in this chapter.

The EXIT statement has three basic types of operations. The first involves an EXIT statement without a loop label, or a WHEN condition. If these conditions are true, then the EXIT statement behaves as follows.

The EXIT statement only exits from the most current LOOP statement encountered. If an EXIT statement is inside a LOOP statement that is nested inside another LOOP statement, the EXIT statement only exits the inner LOOP statement. Execution still remains in the outer LOOP statement. The exit statement only exits from the most recent LOOP statement. This case is shown in the previous example.

If the EXIT statement has an optional loop label, then the EXIT statement, when encountered, completes the execution of the loop specified by the loop label. Therefore, the next statement executed is the one following the END LOOP of the labeled loop. Here is an example:

```
PROCESS(a)
BEGIN
   first_loop: FOR i IN 0 TO 100 LOOP
    second_loop:FOR j IN 1 TO 10 LOOP
       . . .  . . .
       EXIT second_loop; -- exits the second loop only
       . . .  . . .
       EXIT first_loop; -- exits the first loop and second
                        -- loop
    END LOOP;
END LOOP;
END PROCESS;
```

The first EXIT statement only exits the innermost loop because it completes execution of the loop labeled second_loop. The last EXIT statement completes execution of the loop labeled first_loop, which exits from the first loop and the second loop.

If the **EXIT** statement has an optional **WHEN** condition, then the **EXIT** statement only exits the loop if the condition specified is true. The next statement executed depends on whether the **EXIT** statement has a loop label specified or not. If a loop label is specified, the next statement executed is contained in the **LOOP** statement specified by the loop label. If no loop label is present, the next statement executed is in the next outer loop. Here is an example of an **EXIT** statement with a **WHEN** condition:

```
EXIT first_loop WHEN (i < 10);
```

This statement completes the execution of the loop labeled **first_loop** when the expression **i < 10** is true.

The **EXIT** statement provides a quick and easy method of exiting a **LOOP** statement when all processing is finished or an error or warning condition occurs.

ASSERT Statement

The **ASSERT** statement is a very useful statement for reporting textual strings to the designer. The **ASSERT** statement checks the value of a boolean expression for true or false. If the value is true, the statement does nothing. If the value is false, the **ASSERT** statement outputs a user-specified text string to the standard output to the terminal.

The designer can also specify a severity level with which to output the text string. The four levels are, in increasing level of severity, note, warning, error, and failure. The severity level gives the designer the ability to classify messages into proper categories.

The note category is useful for relaying information to the user about what is currently happening in the model. For instance, if the model had a giant loop that took a long time to execute, an assertion of severity level note could be used to notify the designer when the loop was 10 percent complete, 20 percent complete, 30 percent complete, and so on.

Assertions of category warning can be used to alert the designer of conditions that, although not catastrophic, can cause erroneous behavior later. For instance, if a model expected a signal to be at a known value while some process was executing, but the signal was at a different value, it may not be an error as in the **EXIT** statement example, but a warning to the user that results may not be as expected.

Assertions of severity level error are used to alert the designer of conditions that will cause the model to work incorrectly, or not work at all. If the result of a calculation was supposed to return a positive value, but instead returned a negative value, depending on the operation, this could be considered an error.

Assertions of severity level failure are used to alert the designer of conditions within the model that can have disastrous effects. An example of such a condition was discussed in the **EXIT** statement section. Division by 0 is an example of an operation that could cause a failure in the model. Another is addressing beyond the end of an array. In both cases, the severity level failure can let the designer know that the model is behaving incorrectly.

The severity level is a good method for classifying assertions into informational messages to the designer that can describe conditions during execution of the model.

The **ASSERT** statement is currently ignored by synthesis tools. Because the **ASSERT** statement is used mainly for exception handling while writing a model, no hardware is built.

Assertion BNF

Following is the BNF description for the **ASSERT** statement:

```
assert_statement ::=
  ASSERT condition
[REPORT expression]
  [SEVERITY expression];
```

The keyword **ASSERT** is followed by a boolean-valued expression called a condition. The condition determines whether the text expression specified by the **REPORT** clause is output or not. If false, the text expression is output; if true, the text expression is not output.

There are two optional clauses in the **ASSERT** statement. The first is the **REPORT** clause. The **REPORT** clause gives the designer the ability to specify the value of a text expression to output. The second is the **SEVERITY** clause. The **SEVERITY** clause allows the designer to specify the severity level of the **ASSERT** statement. If the report clause is not specified, the default value for the **ASSERT** statement is assertion violation. If the severity clause is not specified, the default value is error.

Let's look at a practical example of an **ASSERT** statement to illustrate how it works. The example performs a data setup check between two signals that control a D flip-flop. Most flip-flops require the **din** (data) input to be at a stable value a certain amount of time before a clock edge appears. This time is called the setup time and guarantees that the **din** value will be clocked into the flip-flop if the setup time is met. This is shown in the following model. The assertion example issues an error message to the designer if the setup time is violated (assertion is false):

```
PROCESS(clk, din)
  VARIABLE last_d_change   : TIME := 0 ns;
  VARIABLE last_d_value    : std_logic := 'X';
  VARIABLE last_clk_value  : std_logic := 'X';
BEGIN
  IF (last_d_value /= din) THEN - /= is
   last_d_change := NOW;       - not equal
   last_d_value := din;
  END IF;

  IF (last_clk_value /= clk) THEN
   last_clk_value := clk;

   IF  (clk = '1') THEN
     ASSERT (NOW - last_d_change >= 20 ns)
       REPORT "setup violation"
       SEVERITY WARNING;
   END IF;
  END IF;
END PROCESS;
```

The process makes use of three local variables to record the time and last value of signal **din** as well as the value of the **clk** signal. By storing the last value of **clk** and **din**, we can determine if the signal has changed value or not. By recording the last time that **din** changed, we can measure from the current time to the last **din** transition to see if the setup time has been violated or not. (An easier method using attributes is shown in Chapter 5, "Subprograms and Packages.")

Whenever **din** or **clk** changes, the process is invoked. The first step in the process is to see if the **din** signal has changed. If it has, the time of the transition is recorded using the predefined function **NOW**. This function returns the current simulation time. Also, the latest value of **din** is stored for future checking.

The next step is to see if signal **clk** has made a transition. If the **last_clk_value** variable is not equal to the current value of **clk**, then we know that a transition has occurred. If signal **clk** is a '1' value, then we know that a rising edge has occurred. Whenever a rising edge occurs on signal **clk**, we need to check the setup time for a violation. If the last transition on signal **d** was less than 20 nanoseconds ago, then the expression:

```
(NOW - last_D_change)
```

returns a value that is less than 20 nanoseconds. The **ASSERT** statement triggers and reports the assertion message setup violation as a warning to the designer. If the last transition on signal **d** occurred more than 20 nanoseconds in the past, then the expression returns a value larger than 20 nanoseconds and the **ASSERT** statement does not write out the message. Remember, the **ASSERT** statement writes out the message when the assert condition is false.

The message reported to the user has, at a minimum, the user string and the error classification. Some simulators also include the time of the assertion report as well as the line number in the file of the assertion.

The **ASSERT** statement used in this example was a sequential **ASSERT** statement, because it was included inside a **PROCESS** statement. A concurrent version of the **ASSERT** statement also exists. It has exactly the same format as the sequential **ASSERT** statement and only exists outside a **PROCESS** statement or subprogram.

The concurrent **ASSERT** statement executes whenever any signals that exist inside of the condition expression have an event upon them. This is as opposed to the sequential **ASSERT** statement in which execution occurs when the sequential **ASSERT** statement is reached inside the **PROCESS** statement or subprogram.

WAIT Statements

The **WAIT** statement gives the designer the ability to suspend the sequential execution of a process or subprogram. The conditions for resuming execution of the suspended process or subprogram can be specified by the following three different means:

- **WAIT ON** signal changes
- **WAIT UNTIL** an expression is true
- **WAIT FOR** a specific amount of time

WAIT statements can be used for a number of different purposes. The most common use today is for specifying clock inputs to synthesis tools. The **WAIT** statement specifies the clock for a process statement that is read by synthesis tools to create sequential logic such as registers and flip-flops. Other uses are to delay process execution for an amount of time or to modify the sensitivity list of the process dynamically.

Let's take a look at a process statement with an embedded **WAIT** statement that is used to generate sequential logic:

```
PROCESS
BEGIN
  WAIT UNTIL clock = '1' AND clock'EVENT;
  q <= d;
END PROCESS;
```

This process is used to generate a flip-flop that clocks the value of **d** into **q** when the clock input has a rising edge. The attribute `'EVENT` attached to input clock is true whenever the clock input has had an event during the current delta timepoint. (`'EVENT` is discussed in great detail in Chapter 5). The combination of looking for a `'1'` value and a change on **clock** creates the necessary functionality to look for a rising edge on input **clock**. The effect is that the process is held at the **WAIT** statement until the clock has a rising edge. Then the current value of **d** is assigned to **q**.

Reading this description into a synthesis tool creates a D flip-flop without a **set** or **reset** input. A synchronous reset can be created by the following:

```
PROCESS
BEGIN
  WAIT UNTIL clock = '1' AND clock'EVENT;
  IF (reset = '1') THEN
      q <= '0';
  ELSE
      q <= d;
  END IF;
END PROCESS;
```

When the clock occurs, the **reset** signal is tested first. If it is active, then the **reset** value (`'0'`) is assigned to **q**; otherwise, the **d** input is assigned.

Finally, an asynchronous **reset** can be added as follows:

```
PROCESS
BEGIN
  IF (reset = '1') THEN
```

```
   q <= '0';
ELSIF clock'EVENT AND clock = '1' THEN
   q <= d;
END IF;

WAIT ON reset, clock;
END PROCESS;
```

This process statement contains a **WAIT ON** statement that causes the process to halt execution until an event occurs on either **reset** or **clock**. The **IF** statement is then executed and, if **reset** is active, the flip-flop is asynchronously **reset**; otherwise, the clock is checked for a rising edge with which to transfer the **d** input to the **q** output of the flip-flop.

A **WAIT** statement can also be used to control the signals a process or subprogram is sensitive to at any point in the execution. Here is an example:

```
PROCESS
BEGIN
WAIT ON a; -- 1.
  .
  .
WAIT ON b; -- 2.
  .
  .
END PROCESS;
```

Execution of the statements in the **PROCESS** statement proceeds until point 1 in the VHDL fragment shown in the preceding. The **WAIT** statement causes the process to halt execution at that point. The process does not continue execution until an event occurs on signal **a**. The process is therefore sensitive to changes in signal **a** at this point in the execution. When an event occurs on signal **a**, execution starts again at the statement directly after the **WAIT** statement at point 1. Execution proceeds until the **WAIT** statement at point 2 is encountered. Once again, execution is halted, and the process is now sensitive to events on signal **b**. Therefore, by adding in two **WAIT** statements, we can alter the process sensitivity list dynamically.

Next, let's discuss the three different options available to the **WAIT** statement:

- **WAIT ON signal [,signal]**
- **WAIT UNTIL boolean_expression**
- **WAIT FOR time_expression**

WAIT ON Signal

We have already seen an example of the first type in the previous process example. The **WAIT ON** signal clause specifies a list of one or more signals that the **WAIT** statement will wait for events upon. If any signal in the signal list has an event occur on it, execution continues with the statement following the **WAIT** statement. Here is an example:

```
WAIT ON a, b;
```

When an event occurs on either **a** or **b**, the process resumes with the statement following the **WAIT** statement.

WAIT UNTIL Expression

The **WAIT UNTIL boolean_expression** clause suspends execution of the process until the expression returns a value of true. This statement effectively creates an implicit sensitivity list of the signals used in the expression. When any of the signals in the expression have events occur upon them, the expression is evaluated. The expression must return a boolean type or the compiler complains. When the expression returns a true value, execution continues with the statement following the **WAIT** statement. Otherwise, the process continues to be suspended. For example:

```
WAIT UNTIL (( x * 10 ) < 100 );
```

In this example, as long as the value of signal **x** is greater than or equal to 10, the **WAIT** statement suspends the process or subprogram. When the value of **x** is less than 10, execution continues with the statement following the **WAIT** statement.

WAIT FOR time_expression

The **WAIT FOR time_expression** clause suspends execution of the process for the time specified by the time expression. After the time specified in the time expression has elapsed, execution continues on the statement following the **WAIT** statement. A couple of examples are shown here:

```
WAIT FOR 10 ns;
WAIT FOR ( a * ( b + c ));
```

In the first example, the time expression is a simple constant value. The **WAIT** statement suspends execution for 10 nanoseconds. After 10 nanoseconds has elapsed, execution continues with the statement following the **WAIT** statement.

In the second example, the time expression is an expression that first must be evaluated to return a time value. After this value is calculated, the **WAIT** statement uses this value as the time value to wait for.

Multiple WAIT Conditions

The **WAIT** statement examples we have examined so far have shown the different options of the **WAIT** statement used separately. The different options can be used together. A single statement can include an **ON** signal, **UNTIL** expression, and **FOR** **time_expression** clauses. Following is an example:

```
WAIT ON nmi,interrupt UNTIL ((nmi = TRUE) or
            (interrupt = TRUE)) FOR 5 usec;
```

This statement waits for an event on signals **nmi** and **interrupt** and continues only if **interrupt** or **nmi** is true at the time of the event, or until 5 microseconds of time has elapsed. Only when one or more of these conditions are true does execution continue.

When using a statement such as this:

```
WAIT UNTIL (interrupt = TRUE) OR ( old_clk = '1');
```

be sure to have at least one of the values in the expression contain a signal. This is necessary to ensure that the **WAIT** statement does not wait forever. If both interrupt and **old_clk** are variables, the **WAIT** statement does not reevaluate when these two variables change value. (In fact, the variables cannot change value because they are declared in the suspended process.) Only signals have events on them, and only signals can cause a **WAIT** statement or concurrent signal assignment to reevaluate.

WAIT Time-Out

There are instances while designing a model when you are not sure that a condition will be met. To prevent the **WAIT** statement from waiting forever, add a time-out clause. The time-out clause allows execution to proceed whether or not the condition has been met. Be careful, though, because this method can cause erroneous behavior unless properly handled. The following example shows this problem:

```
ARCHITECTURE wait_example of wait_example IS
   SIGNAL sendB, sendA : std_logic;
BEGIN
   sendA <= '0';
   A : PROCESS
   BEGIN
    WAIT UNTIL sendB = '1';
    sendA <= '1' AFTER 10 ns;

    WAIT UNTIL sendB = '0';
    sendA <= '0' AFTER 10 ns;

   END PROCESS A;

   B : PROCESS
   BEGIN
    WAIT UNTIL sendA = '0';
    sendB <= '0' AFTER 10 ns;

    WAIT UNTIL sendA = '1';
    sendB <= '1' AFTER 10 ns;

   END PROCESS B;
END wait_example;
```

This architecture has two processes that communicate through two signals, **sendA** and **sendB**. This example does not do anything real but is a simple illustration of how **WAIT** statements can wait forever, a condition commonly referred to as *deadlock*.

During simulator initialization, all processes are executed exactly once. This allows the processes to always start at a known execution point at the start of simulation. In this example, the process labeled A executes at startup and stops at the following line:

```
WAIT UNTIL sendB = 1;
```

The process labeled **B** also executes at startup. Execution starts at the first line of the process and continues until this line:

```
WAIT UNTIL sendA = 1;
```

Execution stops at the first **WAIT** statement of the process even though the expression **sendA = 0** is satisfied by the first signal assignment of signal **sendA**. This is because the **WAIT** statement needs an event to occur on signal **sendA** to cause the expression to be evaluated. Both processes are now waiting on each other. Neither process can continue because they are both waiting for a signal **set** by the other process. If a time out interval is inserted on each **WAIT** statement, execution can be allowed to continue. There is one catch to this last statement. Execution continues when the condition is not met. An **ASSERT** statement can be added to check for continuation of the process without the condition being met. The following example shows the architecture **wait_example** rewritten to include time out clauses:

```
ARCHITECTURE wait_timeout OF wait_example IS
  SIGNAL sendA, sendB : std_logic;
BEGIN
  A : PROCESS
  BEGIN
   WAIT UNTIL (sendB = '1') FOR 1 us;

   ASSERT (sendB = '1')
     REPORT "sendB timed out at '1'"
     SEVERITY ERROR;

   sendA <= '1' AFTER 10 ns;

   WAIT UNTIL (sendB = '0') FOR 1 us;

   ASSERT (sendB = '0')
     REPORT "sendB timed out at '0'"
     SEVERITY ERROR;

   sendA <= '0' AFTER 10 ns;
  END PROCESS A;

  B : PROCESS
  BEGIN
   WAIT UNTIL (sendA = '0') FOR 1 us;

   ASSERT (sendA =  '0')
     REPORT "sendA timed out at '0'"
     SEVERITY ERROR;
```

```
    sendB <= '0' AFTER 10 ns;

    WAIT UNTIL (sendA = '1') FOR 1 us;

    ASSERT (sendA = '1')
      REPORT "sendA timed out at '1'"
      SEVERITY ERROR;

    sendB <= '1' AFTER 10 ns;

  end PROCESS B;
END wait_timeout;
```

Each of the **WAIT** statements now has a time-out expression specified as 1 usec. However, if the time out does happen, the **ASSERT** statement reports an error that the **WAIT** statement in question has timed out.

Sensitivity List Versus WAIT Statement

A process with a sensitivity list is an implicit **WAIT ON** the signals in the sensitivity list. This is shown by the following example:

```
PROCESS (clk)
  VARIABLE last_clk : std_logic := 'X';
BEGIN
  IF (clk /= last_clk ) AND (clk = '1') THEN
    q <= din AFTER 25 ns;
END IF;

last_clk := clk;

END PROCESS;
```

This example can be rewritten using a **WAIT** statement:

```
PROCESS
  VARIABLE last_clk : std_logic := 'X';
BEGIN
  IF (clk /= last_clk ) AND (clk = '1') THEN
    q <= din AFTER 25 ns;
  END IF;

  last_clk := clk;

  WAIT ON clk;
END PROCESS;
```

The **WAIT** statement at the end of the process is equivalent to the sensitivity list at the beginning of the process. But why is the **WAIT** statement at the end of the process and not at the beginning? During initialization of the simulator, all processes are executed once. To mimic the behavior of the sensitivity list, the **WAIT** statement must be at the end of the process to allow the **PROCESS** statement to execute once.

Concurrent Assignment Problem

One of the problems that most designers using sequential signal assignment statements encounter is that the value assigned in the last statement does not appear immediately. This can cause erroneous behavior in the model if the designer is depending on the new value. An example of this problem is shown here:

```
LIBRARY IEEE;
USE IEEE.std_logic_1164ALL;
ENTITY mux IS
  PORT (I0, I1, I2, I3, A, B : IN std_logic;
        Q : OUT std_logic);
END mux;

ARCHITECTURE mux_behave OF mux IS
  SIGNAL sel : INTEGER RANGE 0 TO 3;
BEGIN
  B : PROCESS(A, B, I0, I1, I2, I3)
  BEGIN

    sel <= 0;
    IF (A = '1') THEN sel <= sel + 1; END IF;
    IF (B = '1') THEN sel <= sel + 2; END IF;

    CASE sel IS
      WHEN 0 =>
        Q <= I0;
      WHEN 1 =>
        Q <= I1;
      WHEN 2 =>
        Q <= I2;
      WHEN 3 =>
        Q <= I3;
    END CASE;
  END PROCESS;
END mux_behave;
```

This model is for a 4 to 1 multiplexer. Depending on the values of **A** and **B**, one of the four inputs (**I0** to **I3**) is transferred to output **Q**.

The architecture starts processing by initializing internal signal **sel** to the value 0. Then, based on the values of **A** and **B**, the values 1 or 2 are added to **sel** to select the correct input. Finally, a **CASE** statement selected by the value of **sel** transfers the value of the input to output **Q**.

This architecture does not work as presently implemented. The value of signal **sel** will never be initialized by the first line in the architecture:

```
sel <= 0;
```

This statement inside of a process statement schedules an event for signal **sel** on the next delta time point, with the value 0. However, processing continues in the process statement with the next sequential statement. The value of **sel** remains at whatever value it had at the entry to the process. Only when the process has completed is this current delta finished and the next delta time point started. Only then is the new value of **sel** reflected. By this time, however, the rest of the process has already been processed using the wrong value of **sel**.

There are two ways to fix this problem. The first is to insert **WAIT** statements after each sequential signal assignment statement as shown here:

```
ARCHITECTURE mux_fix1 OF mux IS
  SIGNAL sel : INTEGER RANGE 0 TO 3;
BEGIN
  PROCESS
  BEGIN
   sel <= 0;
   WAIT FOR 0 ns;   -- or wait on sel

   IF (a = '1') THEN sel <= sel + 1; END IF;
   WAIT for 0 ns;

   IF (b = '1') THEN sel <= sel + 2; END IF;
   WAIT FOR 0 ns;

   CASE sel IS
     WHEN 0 =>
       Q <= I0;
     WHEN 1 =>
       Q <= I1;
     WHEN 2 =>
       Q <= I2;
     WHEN 3 =>
       Q <= I3;
   END CASE;

   WAIT ON A, B, I0, I1, I2, I3;
```

```
      END PROCESS;
    END mux_fix1;
```

The **WAIT** statements after each signal assignment cause the process to wait for one delta time point before continuing with the execution. By waiting for one delta time point, the new value has a chance to propagate. Therefore, when execution continues after the **WAIT** statement, signal **sel** has the new value.

One consequence of the **WAIT** statements, however, is that the process can no longer have a sensitivity list. A process with **WAIT** statements contained within it or within a subprogram called from within the process cannot have a sensitivity list. A sensitivity list implies that execution starts from the beginning of the procedure, while a **WAIT** statement allows suspending a process at a particular point. The two are mutually exclusive.

Because the process can no longer have a sensitivity list, a **WAIT** statement has been added to the end of the process that exactly imitates the behavior of the sensitivity list. This is the following statement:

```
WAIT ON A, B, I0, I1, I2, I3;
```

The **WAIT** statement proceeds whenever any of the signals on the right side of the keyword **ON** have an event upon them.

This method of solving the sequential signal assignment problem causes the process to work, but a better solution is to use an internal variable instead of the internal signal, as shown here:

```
ARCHITECTURE mux_fix2 OF mux IS
BEGIN
  PROCESS(A, B, I0, I1, I2, I3)
   VARIABLE sel : INTEGER RANGE 0 TO 3;
  BEGIN
   sel := 0;
   IF (A = '1') THEN sel := sel + 1; END IF;
   IF (B = '1') THEN sel := sel + 2; END IF;

   CASE sel IS
     WHEN 0 =>
       Q <= I0;
     WHEN 1 =>
       Q <= I1;
     WHEN 2 =>
       Q <= I2;
     WHEN 3 =>
       Q <= I3;
   END CASE;
  END PROCESS;
END mux_fix2;
```

The signal **sel** from the preceding example has been converted from an internal signal to an internal variable. This was accomplished by moving the declaration from the architecture declaration section to the process declaration section. Variables can only be declared in the process or subprogram declaration section.

Also, the signal assignments to **sel** have been changed to variable assignment statements. Now, when the first assignment to **sel** is executed, the value is updated immediately. Each successive assignment is also executed immediately so that the correct value of **sel** is available in each statement of the process.

Passive Processes

Passive processes are processes that exist in the entity statement part of an entity. They are different from a normal process in that no signal assignment is allowed. These processes are used to do all sorts of checking functions. For instance, one good use of a passive process is to check the data setup time on a flip-flop.

The advantage of the passive process over the example discussed in the **ASSERT** statement section is that, because the passive process exists in the entity, it can be applied to any architecture of the entity. Take a look at the following example:

```
LIBRARY IEEE;
USE IEEE.std_logic_1164ALL;
ENTITY dff IS
PORT( CLK, din : IN std_logic;
      Q, QB : OUT std_logic);
BEGIN
  PROCESS(CLK, din)
   VARIABLE last_d_change : TIME := 0 ns;
   VARIABLE last_clk, last_d_value : std_logic := 'X';
  BEGIN
   IF (din /= last_d_value) THEN
     last_d_change := now;
     last_d_value := din;
   END IF;

   IF (CLK /= last_clk) THEN
     IF  (CLK = '1') THEN
       ASSERT(now - last_d_change >= 15 ns)
         REPORT "setup error"
         SEVERITY ERROR;
     END IF;
```

```
        last_clk := CLK;
      END IF;
    END PROCESS;
  END dff;

ARCHITECTURE behave OF dff IS
BEGIN
  .
  .
  .
END behave;

ARCHITECTURE struct OF dff IS
BEGIN
  .
  .
  .
END struct;

ARCHITECTURE switch OF dff IS
BEGIN
  .
  .
  .
END switch;
```

This example shows the entity for a D flip-flop with a passive process included in the entity that performs a data setup check with respect to the clock. This setup check function was described in detail in the ASSERT statement description. What this example shows is that, when the setup check function is contained in the entity statement part, each of the architectures for the entity have the data setup check performed automatically. Without this functionality, each of the architectures would have to have the setup check code included. This introduces more code to maintain and can introduce inconsistencies between architectures.

The only restriction on these processes, as mentioned earlier, is that no signal assignment is allowed in a passive process. In the preceding example, a process statement was used to illustrate a passive process. A passive process can also exist as a concurrent statement that does not do any signal assignment. Examples of such statements are concurrent assert statements and concurrent subprogram invocations. An example of two concurrent assert statements as passive processes are shown here:

```
ENTITY adder IS
  PORT( A, B : IN INTEGER;
        X : OUT INTEGER);
```

```
BEGIN
  ASSERT (A < 256)
   REPORT "A out of range"
   SEVERITY ERROR;

  ASSERT  (B < 256)
   REPORT "B out of range"
   SEVERITY ERROR;

END adder;
```

The first **ASSERT** statement checks to make sure that input **A** is not out of range, and the second assertion checks that input **B** is not out of the range of the adder. Each of these statements acts as an individual process that is sensitive to the signal in its expression. For instance, the first assertion is sensitive to signal **A** because that signal is contained in its expression.

SUMMARY

In this chapter, we discussed the following:

- How process statements are concurrent statements that delineate areas of sequential statements.
- How process statements can be used to control when a process is activated.
- How signal assignments are scheduled and variable assignments happen immediately within a process statement.
- How **IF, CASE**, and **LOOP** statements can be used to control the flow of execution within a model.
- How **ASSERTION** statements can be used to check for error conditions or report information to the user.
- The three forms of the **WAIT** statement. How **WAIT UNTIL** is used for specifying clocks for synthesis, and how **WAIT ON** can be used to modify the sensitivity list.
- How passive processes can be used to perform error checking and other tasks across a number of architectures by existing in an **ENTITY** statement.

The next chapter focuses on all of the different data types of VHDL that can be used in models.

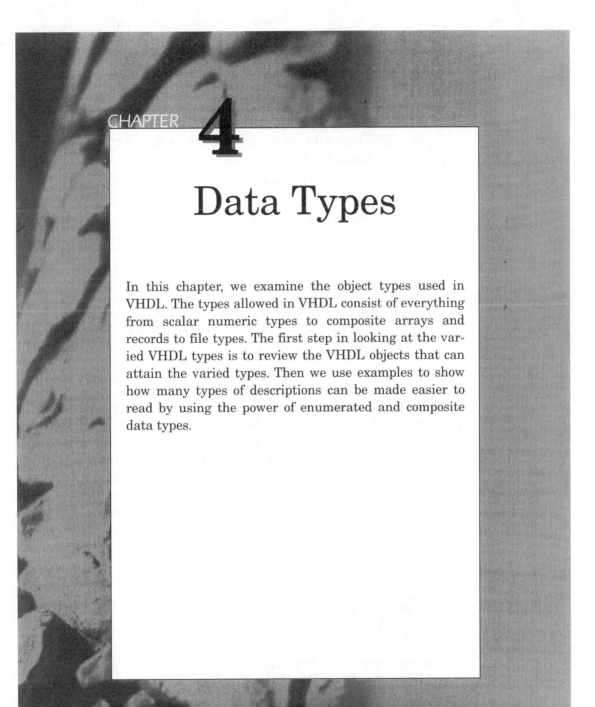

CHAPTER 4

Data Types

In this chapter, we examine the object types used in VHDL. The types allowed in VHDL consist of everything from scalar numeric types to composite arrays and records to file types. The first step in looking at the varied VHDL types is to review the VHDL objects that can attain the varied types. Then we use examples to show how many types of descriptions can be made easier to read by using the power of enumerated and composite data types.

Object Types

A VHDL object consists of one of the following:

- Signal, which represents interconnection wires that connect component instantiation ports together.
- Variable, which is used for local storage of temporary data, visible only inside a process.
- Constant, which names specific values.

Signal

Signal objects are used to connect entities together to form models. Signals are the means for communication of dynamic data between entities. A signal declaration looks like this:

```
SIGNAL signal_name : signal_type [:= initial_value];
```

The keyword **SIGNAL** is followed by one or more signal names. Each signal name creates a new signal. Separating the signal names from the signal type is a colon. The signal type specifies the data type of the information that the signal contains. Finally, the signal can contain an initial value specifier so that the signal value may be initialized.

Signals can be declared in entity declaration sections, architecture declarations, and package declarations. Signals in package declarations are also referred to as global signals because they can be shared among entities.

Following is an example of signal declarations:

```
LIBRARY IEEE;
USE IEEE.std_logic_1164.ALL;
PACKAGE sigdecl IS
  TYPE bus_type IS ARRAY(0 to 7) OF std_logic;

  SIGNAL vcc    : std_logic := '1';
  SIGNAL ground : std_logic := '0';

  FUNCTION magic_function( a : IN bus_type) RETURN
      bus_type;

END sigdecl;

USE WORK.sigdecl.ALL;
LIBRARY IEEE;
```

```
USE IEEE.std_logic_1164.ALL;
ENTITY board_design is
  PORT( data_in : IN bus_type;
        data_out : OUT bus_type);

  SIGNAL sys_clk : std_logic := '1';

END board_design;

ARCHITECTURE data_flow OF board_design IS
  SIGNAL int_bus : bus_type;
  CONSTANT disconnect_value : bus_type
   := ('X', 'X', 'X', 'X', 'X', 'X', 'X', 'X');
BEGIN
  int_bus <= data_in WHEN sys_clk = '1'
   ELSE int_bus;

  data_out <= magic_function(int_bus) WHEN sys_clk = '0'
   ELSE disconnect_value;

  sys_clk <= NOT(sys_clk) after 50 ns;
END data_flow;
```

Signals **vcc** and **ground** are declared in package **sigdecl**. Because these signals are declared in a package, they can be referenced by more than one entity and are therefore global signals. For an entity to reference these signals, the entity needs to use package **sigdecl**. To use the package requires a VHDL **USE** clause, as shown here:

```
USE work.sigdecl.vcc;
USE work.sigdecl.ground;
```

Or:

```
USE work.sigdecl.ALL;
```

In the first example, the objects are included in the entity by specific reference. In the second example, the entire package is included in the entity. In the second example, problems may arise because more than what is absolutely necessary is included. If more than one object of the same name results because of the **USE** clause, none of the objects is visible, and a compile operation that references the object fails.

SIGNALS GLOBAL TO ENTITIES   Inside the entity declaration section for entity **board_design** is a signal called **sys_clk**. This signal can be referenced in entity **board_design** and any architecture for entity **board_design**. In this example, there is only one architecture, **data_flow**, for `board_design`. The signal `sys_clk` can therefore be assigned to and read from in entity `board_design` and architecture `data_flow`.

ARCHITECTURE LOCAL SIGNALS Inside of architecture `data_flow` is a signal declaration for signal `int_bus`. Signal `int_bus` is of type `bus_type`, a type defined in package `sigdecl`. The `sigdecl` package is used in entity `board`; therefore, the type `bus_type` is available in architecture `data_flow`. Because the signal is declared in the architecture declaration section, the signal can only be referenced in architecture `data_flow` or in any process statements in the architecture.

Variables

Variables are used for local storage in process statements and subprograms. (Subprograms are discussed in Chapter 6, "Predefined Attributes.") As opposed to signals, which have their values scheduled in the future, all assignments to variables occur immediately. A variable declaration looks like this:

```
VARIABLE variable_name {,variable_name} : variable_type[:=
      value];
```

The keyword **VARIABLE** is followed by one or more variable names. Each name creates a new variable. The construct `variable_type` defines the data type of the variable, and an optional initial value can be specified.

Variables can be declared in the process declaration and subprogram declaration sections only. An example using two variables is shown here:

```
LIBRARY IEEE;
USE IEEE.std_logic_1164.ALL;
ENTITY and5 IS
  PORT ( a, b, c, d, e : IN  std_logic;
         q : OUT std_logic);
END and5;

ARCHITECTURE and5 OF and5 IS
BEGIN
PROCESS(a, b, c, d, e)
 VARIABLE state : std_logic;
 VARIABLE delay : time;
BEGIN

 state := a AND b AND c AND d AND e;

 IF state = '1' THEN
```

```
      delay := 4.5 ns;
 ELSIF state = '0' THEN
    delay := 3 ns;
 ELSE
    delay := 4 ns;
 END IF;

 q <= state AFTER delay;

END PROCESS;
END and5;
```

This example is the architecture for a five-input **AND** gate. There are two variable declarations in the process declaration section: one for variable **state** and one for variable **delay**. Variable **state** is used as a temporary storage area to hold the value of the **AND** function of the inputs. Temporary-storage value **delay** is used to hold the delay value that will be used when scheduling the output value. Both of these values cannot be static data because their values depend on the values of inputs **a**, **b**, **c**, **d**, and **e**. Signals could have been used to store the data, but there are several reasons why a signal was not used:

▪ Variables are inherently more efficient because assignments happen immediately, while signals must be scheduled to occur.

▪ Variables take less memory, while signals need more information to allow for scheduling and signal attributes.

▪ Using a signal would have required a **WAIT** statement to synchronize the signal assignment to the same execution iteration as the usage.

When any of the input signals **a**, **b**, **c**, **d**, or **e** change, the process is invoked. Variable **state** is assigned the **AND** of all of the inputs. Next, based on the value of variable **state**, variable **delay** is assigned a delay value. Based on the delay value assigned to variable **delay**, output signal **q** will have the value of variable state assigned to it.

Constants

Constant objects are names assigned to specific values of a type. Constants give the designer the ability to have a better-documented model, and a model that is easy to update. For instance, if a model requires a fixed value in a number of instances, a constant should be used. By using a constant, the designer can change the value of the constant and recompile, and all of the instances of the constant value are updated to reflect the new value of the constant.

A constant also provides a better-documented model by providing more meaning to the value being described. For instance, instead of using the value 3.1414 directly in the model, the designer should create a constant as in the following:

```
CONSTANT PI: REAL := 3.1414;
```

Even though the value is not going to change, the model becomes more readable.

A constant declaration looks like this:

```
CONSTANT  constant_name {,constant_name} : type_name[:=
    value];
```

The **value** specification is optional, because VHDL also supports deferred constants. These are constants declared in a package declaration whose value is specified in a package body.

A constant has the same scoping rules as signals. A constant declared in a package can be global if the package is used by a number of entities. A constant in an entity declaration section can be referenced by any architecture of that entity. A constant in an architecture can be used by any statement inside the architecture, including a process statement. A constant declared in a process declaration can be used only in a process.

Data Types

All of the objects we have been discussing until now—the signal, the variable, and the constant—can be declared using a type specification to specify the characteristics of the object. VHDL contains a wide range of types that can be used to create simple or complex objects.

To define a new type, you must create a type declaration. A type declaration defines the name of the type and the range of the type. Type declarations are allowed in package declaration sections, entity declaration sections, architecture declaration sections, subprogram declaration sections, and process declaration sections.

A type declaration looks like this:

```
TYPE type_name IS type_mark;
```

A **type_mark** construct encompasses a wide range of methods for specifying a type. It can be anything from an enumeration of all of the values of a type to a complex record structure. In the next few sections, type marks are examined. All of the scoping rules that were defined for signals and variables apply to type declarations also.

Figure 4-1 is a diagram showing the types available in VHDL. The four broad categories are scalar types, composite types, access types, and file types. Scalar types include all of the simple types such as integer and real. Composite types include arrays and records. Access types are the equivalent of pointers in typical programming languages. Finally, file types give the designer the ability to declare file objects with designer-defined file types.

Scalar Types

Scalar types describe objects that can hold, at most, one value at a time. The type itself can contain multiple values, but an object that is declared to be a scalar type can hold, at most, one of the scalar values at any point

Figure 4-1
VHDL Data Types
Diagram.

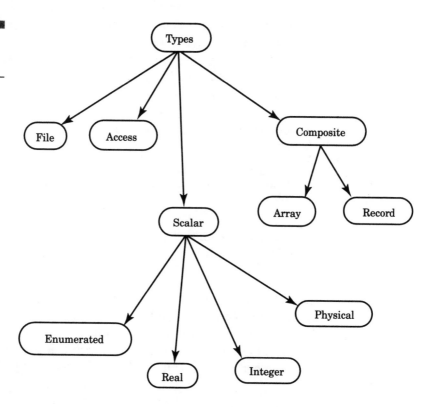

in time. Referencing the name of the object references the entire object. Scalar types encompass these five classes of types:

■ Integer types

■ Real types

■ Enumerated types

■ Physical types

■ Integer types

Integer types are exactly like mathematical integers. All of the normal predefined mathematical functions like add, subtract, multiply, and divide apply to integer types. The VHDL LRM does not specify a maximum range for integers, but does specify the minimum range: from -2,147,483,647 to +2,147,483,647. The minimum range is specified by the Standard package which is contained in the Standard Library.

The Standard package defines all of the predefined VHDL types provided with the language. The Standard library is used to hold any packages or entities provided as standard with the language. The Standard package is listed in the VHDL Language Reference Manual.

It may seem strange to some designers who are familiar with two's complement representations that the integer range is specified from −2,147,483,647 to +2,147,483,647 when two's complement integer representations usually allow one smaller negative number, −2,147,483,648. The language defines the integer range to be symmetric around 0.

Following are some examples of integer values:

```
ARCHITECTURE test OF test IS
BEGIN
PROCESS(X)
  VARIABLE a : INTEGER;
  VARIABLE b : int_type;
BEGIN
  a := 1;      --Ok     1
  a := -1;     --Ok     2
  a := 1.0;    --error  3
END PROCESS;
END test;
```

The first two statements (1 and 2) show examples of a positive integer assignment and a negative integer assignment. Line 3 shows a noninteger assignment to an integer variable. This line causes the compiler to issue an error message. Any numeric value with a decimal point is considered a real number value. Because VHDL is a strongly typed language, for the assignment to take place, either the base types must match or a type-casting operation must be performed.

REAL TYPES Real types are used to declare objects that emulate mathematical real numbers. They can be used to represent numbers out of the range of integer values as well as fractional values. The minimum range of real numbers is also specified by the Standard package in the Standard Library, and is from $-1.0E+38$ to $+1.0E+38$. These numbers are represented by the following notation:

```
+ or -number.number[E + or -number]
```

Following are a few examples of some real numbers:

```
ARCHITECTURE test OF test IS
  SIGNAL a : REAL;
BEGIN
    a <= 1.0;          --Ok   1
    a <= 1;            --error 2
    a <= -1.0E10;      --Ok     3
    a <= 1.5E-20;      --Ok     4
    a <= 5.3 ns;       --error 5
END test;
```

Line 1 shows how to assign a real number to a signal of type **REAL**. All real numbers have a decimal point to distinguish them from integer values. Line 2 is an example of an assignment that does not work. Signal **a** is of type **REAL**, and a real value must be assigned to signal **a**. The value 1 is of type **INTEGER**, so a type mismatch is generated by this line.

Line 3 shows a very large negative number. The numeric characters to the left of the character **E** represent the mantissa of the real number, while the numeric value to the right represents the exponent.

Line 4 shows how to create a very small number. In this example, the exponent is negative so the number is very small.

Line 5 shows how a type **TIME** cannot be assigned to a real signal. Even though the numeric part of the value looks like a real number, because of the units after the value, the value is considered to be of type **TIME**.

ENUMERATED TYPES An enumerated type is a very powerful tool for abstract modeling. A designer can use an enumerated type to represent exactly the values required for a specific operation. All of the values of an enumerated type are user-defined. These values can be identifiers or single-character literals. An identifier is like a name. Examples are x, abc, and black. Character literals are single characters enclosed in quotes, such as 'X', '1', and '0'.

A typical enumerated type for a four-state simulation value system looks like this:

```
TYPE fourval IS ( 'X', '0', '1', 'Z' );
```

This type contains four character literal values that each represent a unique state in the four-state value system. The values represent the following conditions:

- ■ `'X'`—An unknown value
- ■ `'0'`—A logical 0 or false value
- ■ `'1'`—A logical 1 or true value
- ■ `'Z'`—A tristate or open collector value

Character literals are needed for values `'1'` and `'0'` to separate these values from the integer values 1 and 0. It would be an error to use the values 1 and 0 in an enumerated type, because these are integer values. The characters **x** and **z** do not need quotes around them because they do not represent any other type, but the quotes were used for uniformity.

Another example of an enumerated type is shown here:

```
TYPE color IS ( red, yellow, blue, green, orange );
```

In this example, the type values are very abstract—that is, not representing physical values that a signal might attain. The type values in type **color** are also all identifiers. Each identifier represents a unique value of the type; therefore, all identifiers of the type must be unique.

Each identifier in the type has a specific position in the type, determined by the order in which the identifier appears in the type. The first identifier has a position number of 0, the next a position number of 1, and so on. (Chapter 5, "Subprograms and Packages" includes some examples using position numbers of a type.)

A typical use for an enumerated type would be representing all of the instructions for a microprocessor as an enumerated type. For instance, an enumerated type for a very simple microprocessor could look like this:

```
TYPE instruction IS ( add, sub, lda, ldb, sta, stb, outa,
    xfr );
```

The model that uses this type might look like this:

```
PACKAGE instr IS
  TYPE instruction IS ( add, sub, lda, ldb, sta, stb,
    outa, xfr );
```

```
END instr;

USE WORK.instr.ALL;
ENTITY mp IS
   PORT (instr : IN instruction;
         addr  : IN INTEGER;
         data  : INOUT INTEGER);
END mp;

ARCHITECTURE mp OF mp IS
BEGIN
   PROCESS(instr)
    TYPE regtype IS ARRAY(0 TO 255) OF INTEGER;
    VARIABLE a, b : INTEGER;
    VARIABLE reg : regtype;
    BEGIN
                           --select instruction to
      CASE instr is        --execute
        WHEN lda =>
           a := data;      --load a accumulator

        WHEN ldb =>
           b := data;      --load b accumulator

        WHEN add =>
           a := a 1 b;     --add accumulators

        WHEN sub =>
           a := a -b;      --subtract accumulators

        WHEN sta =>
           reg(addr) := a; --put a accum in reg array

        WHEN stb =>
           reg(addr) := b; --put b accum in reg array

        WHEN outa =>
           data <= a;      --output a accum

        WHEN xfr =>        --transfer b to a
           a := b;

      END CASE;
    END PROCESS;
END mp;
```

The model receives an instruction stream (**instr**), an address stream
(**addr**), and a data stream (**data**). Based on the value of the enumerated
value of **instr**, the appropriate instruction is executed. A **CASE** statement
is used to select the instruction to execute. The statement is executed and
the process then waits for the next instruction.

Another common example using enumerated types is a state machine.
State machines are commonly used in designing the control logic for ASIC

or FPGA devices. They represent a very easy and understandable method for specifying a sequence of actions over time, based on input signal values.

```
ENTITY traffic_light IS
  PORT(sensor : IN std_logic;
       clock : IN std_logic;
       red_light : OUT std_logic;
       green_light : OUT std_logic;
       yellow_light : OUT std_logic);
END traffic_light;

ARCHITECTURE simple OF traffic_light IS
TYPE t_state is (red, green, yellow);
Signal present_state, next_state : t_state;
BEGIN
  PROCESS(present_state, sensor)
  BEGIN
    CASE present_state IS
      WHEN green =>
        next_state <= yellow;
        red_light <= '0';
        green_light <= '1';
        yellow_light <= '0';
      WHEN red =>
        red_light <= '1';
        green_light <= '0';
        yellow_light <= '0';
        IF (sensor = '1') THEN
          next_state <= green;
        ELSE
          next_state <= red;
        END IF;
      WHEN yellow =>
        red_light <= '0';
        green_light <= '0';
        yellow_light <= '1';
        next_state <= red;
    END CASE;
  END PROCESS;

  PROCESS
  BEGIN
    WAIT UNTIL clock'EVENT and clock = '1';
    present_state <= next_state;
  END PROCESS;
END simple;
```

The state machine is described by two processes: the first calculates the next state logic, and the second latches the next state into the current state. Notice how the enumerated type makes the model much more readable because the state names represent the color of the light that is currently being displayed.

PHYSICAL TYPES Physical types are used to represent physical quantities such as distance, current, time, and so on. A physical type provides for a base unit, and successive units are then defined in terms of this unit. The smallest unit representable is one base unit; the largest is determined by the range specified in the physical type declaration. An example of a physical type for the physical quantity current is shown here:

```
TYPE current IS RANGE 0 to 1000000000

UNITS
  na;                  --nano amps
  ua = 1000 na;        --micro amps
  ma = 1000 ua;        --milli amps
  a  = 1000 ma;        --amps
END UNITS;
```

The type definition begins with a statement that declares the name of the type (**current**) and the range of the type (0 to 1,000,000,000). The first unit declared in the **UNITS** section is the base unit. In the preceding example, the base unit is **na**. After the base unit is defined, other units can be defined in terms of the base unit or other units already defined. In the preceding example, the unit **ua** is defined in terms of the base unit as 1000 base units. The next unit declaration is **ma**. This unit is declared as 1000 **ua**. The units declaration section is terminated by the **END UNITS** clause.

More than one unit can be declared in terms of the base unit. In the preceding example, the **ma** unit can be declared as 1000 **ma** or 1,000,000 **na**. The range constraint limits the minimum and maximum values that the physical type can represent in base units. The unit identifiers all must be unique within a single type. It is illegal to have two identifiers with the same name.

PREDEFINED PHYSICAL TYPES

The only predefined physical type in VHDL is the physical type **TIME**. This type is shown here:

```
TYPE TIME IS RANGE <implementation defined>
UNITS
  fs;                  --femtosecond
  ps  =  1000 fs;      --picosecond
  ns  =  1000 ps;      --nanosecond
  us  =  1000 ns;      --microsecond
  ms  =  1000 us;      --millisecond
  sec =  1000 ms;      --second
  min =  60 sec;       --minute
  hr  =  60 min;       --hour
END UNITS;
```

The range of time is implementation-defined but has to be at least the range of integer, in base units. This type is defined in the Standard package.

Following is an example using a physical type:

```
PACKAGE example IS
  TYPE current IS RANGE 0 TO 1000000000
  UNITS
  na;                  --nano amps
  ua = 1000 na;        --micro amps
  ma = 1000 ua;        --milli amps
  a  = 1000 ma;        --amps
  END UNITS;

TYPE load_factor IS (small, med, big );
END example;

USE WORK.example.ALL;
ENTITY delay_calc IS
  PORT ( out_current : OUT current;
         load : IN load_factor;
         delay : OUT time);
END delay_calc;

ARCHITECTURE delay_calc OF delay_calc IS
BEGIN
  delay <= 10 ns WHEN (load = small) ELSE
           20 ns WHEN (load = med) ELSE
           30 ns WHEN (load = big) ELSE
           10 ns;

  out_current <= 100 ua   WHEN (load = small)ELSE
                 1 ma      WHEN (load = med) ELSE
                 10 ma     WHEN (load = big) ELSE
                 100 ua;
END delay_calc;
```

In this example, two examples of physical types are represented. The first is of predefined physical type **TIME** and the second of user-specified physical type **current**. This example returns the current output and delay value for a device based on the output load factor.

Composite Types

Looking back at the VHDL types diagram in Figure 4-1, we see that composite types consist of array and record types. Array types are groups of elements of the same type, while record types allow the grouping of elements of different types. Arrays are useful for modeling linear structures such as RAMs and ROMs, while records are useful for modeling data packets, instructions, and so on.

Composite types are another tool in the VHDL toolbox that allow very abstract modeling of hardware. For instance, a single array type can represent the storage required for a ROM.

ARRAY TYPES   Array types group one or more elements of the same type together as a single object. Each element of the array can be accessed by one or more array indices. Elements can be of any VHDL type. For instance, an array can contain an array or a record as one of its elements.

In an array, all elements are of the same type. The following example shows a type declaration for a single dimensional array of bits:

```
TYPE data_bus IS ARRAY(0 TO 31) OF BIT;
```

This declaration declares a data type called **data_bus** that is an array of 32 bits. Each element of the array is the same as the next. Each element of the array can be accessed by an array index. Following is an example of how to access elements of the array:

```
VARIABLE X: data_bus;
VARIABLE Y: BIT;

Y := X(0);     --line 1
Y := X(15);    --line 2
```

This example represents a small VHDL code fragment, not a complete model. In line 1, the first element of array **x** is being accessed and assigned to variable **y**, which is of bit type. The type of **y** must match the base type of array **x** for the assignment to take place. If the types do not match, the compiler generates an error.

In line 2, the sixteenth element of array **x** is being assigned to variable **y**. Line 2 is accessing the sixteenth element of array **x** because the array index starts with 0. Element 0 is the first element, element 1 is the second, and so on.

Following is another more comprehensive example of array accessing:

```
PACKAGE array_example IS
   TYPE data_bus IS ARRAY(0 TO 31) OF BIT;
   TYPE small_bus IS ARRAY(0 TO 7) OF BIT;
END array_example;
```

```
USE WORK.array_example.ALL;
ENTITY extract IS
   PORT (data : IN data_bus;
         start : IN INTEGER;
         data_out : OUT small_bus);
END extract;

ARCHITECTURE test OF extract IS
BEGIN
   PROCESS(data, start)
   BEGIN
    FOR i IN 0 TO 7 LOOP
      data_out(i) <= data(i + start);
    END LOOP;
   END PROCESS;
END test;
```

This entity takes in a 32-bit array element as a port and returns 8 bits of the element. The 8 bits of the element returned depend on the value of index start. The 8 bits are returned through output port **data_out**. (There is a much easier method to accomplish this task, with functions, described in Chapter 5, "Subprograms and Packages.")

A change in value of **start** or **data** triggers the process to execute. The **FOR loop** loops 8 times, each time copying a single bit from port **data** to port **data_out**. The starting point of the copy takes place at the integer value of port **start**. Each time through the loop, the ith element of **data_out** is assigned the (**i** + start) element of data.

The examples shown so far have been simple arrays with scalar base types. In the next example, the base type of the array is another array:

```
LIBRARY IEEE;
USE IEEE.std_logic_1164.ALL;
PACKAGE memory IS
   CONSTANT width   : INTEGER := 3;
   CONSTANT memsize : INTEGER := 7;

   TYPE data_out IS ARRAY(0 TO width) OF std_logic;
   TYPE mem_data IS ARRAY(0 TO memsize) OF data_out;
END memory;

LIBRARY IEEE;
USE IEEE.std_logic_1164.ALL;
USE WORK.memory.ALL;
ENTITY rom IS
   PORT( addr : IN INTEGER;
         data : OUT data_out;
         cs   : IN std_logic);
END rom;
```

```
ARCHITECTURE basic OF rom IS
  CONSTANT z_state : data_out := ('Z', 'Z', 'Z', 'Z');
  CONSTANT x_state : data_out := ('X', 'X', 'X', 'X');
  CONSTANT rom_data : mem_data :=
   ( ( '0', '0', '0', '0'),
     ( '0', '0', '0', '1'),
     ( '0', '0', '1', '0'),
     ( '0', '0', '1', '1'),
     ( '0', '1', '0', '0'),
     ( '0', '1', '0', '1'),
     ( '0', '1', '1', '0'),
     ( '0', '1', '1', '1') );
BEGIN
  ASSERT addr <= memsize
   REPORT "addr out of range"
   SEVERITY ERROR;

  data <= rom_data(addr) AFTER 10 ns WHEN cs = '1' ELSE
          z_state AFTER 20 ns WHEN cs = '0' ELSE
          x_state AFTER 10 ns;
END basic;
```

Package memory uses two constants to define two data types that form the data structures for entity **rom**. By changing the constant width and recompiling, we can change the output width of the memory. The initialization data for the ROM would also have to change to reflect the new width.

The data types from package memory are also used to define the data types of the ports of the entity. In particular, the data port is defined to be of type **data_out**.

The architecture defines three constants used to determine the output value. The first defines the output value when the **cs** input is a '0'. The value output is consistent with the **rom** being unselected. The second constant defines the output value when **rom** has an unknown value on the **cs** input. The value output by **rom** is unknown as well. The last constant defines the data stored by **rom**. (This is a very efficient method to model the ROM, but if the ROM data changes, the model needs to be recompiled.) Depending on the address to **rom**, an appropriate entry from this third constant is output. This happens when the **cs** input is a '1' value.

The **rom** data type in this example is organized as eight rows (0 to 7) and four columns (0 to 3). It is a two-dimensional structure, as shown in Figure 4-2.

To initialize the constant for the **rom** data type, an aggregate initialization is required. The table after the **rom_data** constant declaration is an aggregate used to initialize the constant. The aggregate value is constructed as a table for readability; it could have been all on one line. The

Figure 4-2
Rom Data Represen-
tation.

Addr	Bit 3	Bit 2	Bit 1	Bit 0
0	0	0	0	0
1	0	0	0	1
2	0	0	1	0
3	0	0	1	1
4	0	1	0	0
5	0	1	0	1
6	0	1	1	0
7	0	1	1	1

structure of the aggregate must match the structure of the data type for the assignment to occur. Following is a simple example of an aggregate assignment:

```
PROCESS(X)
   TYPE bitvec IS ARRAY(0 TO 3) OF BIT;
   VARIABLE Y : bitvec;
BEGIN
   Y := ('1', '0', '1', '0');
     .
     .
     .
END PROCESS;
```

Variable **Y** has an element of type **BIT** in the aggregate for each element of its type. In this example, the variable **Y** is 4 bits wide, and the aggregate is 4 bits wide as well.

The constant **rom_data** from the **rom** example is an array of arrays. Each element of type **mem_data** is an array of type **data_out**. The aggregate assignment for an array of arrays can be represented by the form shown here:

```
value := ((e1, e2, . . . ,en), . . . ,(e1, e2, . . . ,en));

                E1                    . . .              En
```

This is acceptable, but a much more readable form is shown here:

```
value := ((e1, e2, . . . , en),      --E1
          (e1, e2, . . . , en),      --E2
             .    . . . .    .
             .    . . . .    .
          (e1, e2, . . . , en) )     --En
```

In the statement part of the **rom** example, there is one conditional signal assignment statement. The output port data is assigned a value based on the value of the **cs** input. The data type of the value assigned to port data must be of type **data_out** because port data has a type of **data_out**. By addressing the **rom_data** constant with an integer value, a data type of **data_out** is returned.

A single value can be returned from the array of arrays by using the following syntax:

```
bit_value := rom_data(addr) (bit_index);
```

The first index (**addr**) returns a value with a data type of **data_out**. The second index (**bit_index**) indexes the **data_out** type and returns a single element of the array.

MULTIDIMENSIONAL ARRAYS
The constant **rom_data** in the **rom** example was represented using an array of arrays. Following is another method for representing the data is with a multidimensional array:

```
TYPE mem_data_md IS ARRAY(0 TO memsize, 0 TO width) OF
     std_logic;

CONSTANT rom_data_md : mem_data :=
( ( '0', '0', '0', '0'),
  ( '0', '0', '0', '1'),
  ( '0', '0', '1', '0'),
  ( '0', '0', '1', '1'),
  ( '0', '1', '0', '0'),
  ( '0', '1', '0', '1'),
  ( '0', '1', '1', '0'),
  ( '0', '1', '1', '1') );
```

The declaration shown here declares a two-dimensional array type **mem_data_md**. When constant **rom_data_md** is declared using this type, the initialization syntax remains the same, but the method of accessing an element of the array is different. In the following example, a single element of the array is accessed:

```
X := rom_data_md(3, 3);
```

This access returns the fourth element of the fourth row, which, in this example, is a '1'.

UNCONSTRAINED ARRAY TYPES

An unconstrained array type is a type whose range or size is not completely specified when the type is declared. This allows multiple subtypes to share a common base type. Entities and subprograms can then operate on all of the different subtypes with a single subprogram, instead of a subprogram or entity per size.

Following is an example of an unconstrained type declaration:

```
TYPE BIT_VECTOR IS ARRAY(NATURAL RANGE <>) OF BIT;
```

This is the type declaration for type **BIT_VECTOR** from the Standard package. This type declaration declares a type that is an array of type **BIT**. However, the number of elements of the array is not specified. The notation that depicts this is:

```
RANGE <>
```

This notation specifies that the type being defined has an unconstrained range. The word **NATURAL** before the keyword **RANGE**, in the type declaration, specifies that the type is bounded only by the range of **NATURAL**. Type **NATURAL** is defined in the Standard package to have a range from 0 to **integer'high** (the largest integer value). Type **BIT_VECTOR**, then, can range in size from 0 elements to **integer'high** elements. Each element of the **BIT_VECTOR type** is of type **BIT**.

Unconstrained types are typically used as types of subprogram arguments, or entity ports. These entities or subprograms can be passed items of any size within the range of the unconstrained type.

For instance, let's assume that a designer wants a shift-right function for type **BIT_VECTOR**. The function uses the unconstrained type **BIT_VECTOR** as the type of its ports, but it can be passed any type that is a subtype of type **BIT_VECTOR**. Let's walk through an example to illustrate how this works. Following is an example of an unconstrained shift-right function:

```
PACKAGE mypack IS
  SUBTYPE eightbit IS BIT_VECTOR(0 TO 7);
  SUBTYPE fourbit IS BIT_VECTOR(0 TO 3);
  FUNCTION shift_right(val : BIT_VECTOR)
  RETURN BIT_VECTOR;
END mypack;

PACKAGE BODY mypack IS
  FUNCTION shift_right(val : BIT_VECTOR) RETURN BIT_VECTOR
```

```
        IS VARIABLE result : BIT_VECTOR(0 TO (val'LENGTH -1));
      BEGIN
        result := val;
        IF (val'LENGTH > 1) THEN
          FOR i IN 0 TO (val'LENGTH -2) LOOP
            result(i) := result(i + 1);
          END LOOP;
          result(val'LENGTH -1) := 0;
        ELSE
          result(0) := 0;
        END IF;
        RETURN result;
      END shift_right;
END mypack;
```

The package declaration (the first five lines of the model) declares two subtypes: **eightbit** and **fourbit**. These two subtypes are subtypes of the unconstrained base type **BIT_VECTOR**. These two types constrain the base type to range 0 to 7 for type **eightbit** and range 0 to 3 for type **fourbit**.

In a typical hardware description language without unconstrained types, two different shift-right functions would need to be written to handle the two different-sized subtypes. One function would work with type **eightbit**, and the other would work with type **fourbit**. With unconstrained types in VHDL, a single function can be written that will handle both input types and return the correct type.

Based on the size of input argument **val**, the internal variable result is created to be of the same size. Variable **result** is then initialized to the value of input argument **val**. This is necessary because the value of input argument **val** can only be read in the function; it cannot have a value assigned to it in the function. If the size of input argument **val** is greater than 1, then the shift-right function loops through the length of the subtype value passed into the function. Each loop shifts one of the bits of variable **result** one bit to the right. If the size of input argument **val** is less than 2, we treat this as a special case and return a single bit whose value is '0'.

RECORD TYPES   Record types group objects of many types together as a single object. Each element of the record can be accessed by its field name. Record elements can include elements of any type, including arrays and records. The elements of a record can be of the same type or different types. Like arrays, records are used to model abstract data elements.

Following is an example of a record type declaration:

```
TYPE optype IS ( add, sub, mpy, div, jmp );
```

```
TYPE instruction IS
RECORD
  opcode : optype;
  src    : INTEGER;
  dst    : INTEGER;
END RECORD;
```

The first line declares the enumerated type **optype**, which is used as one of the record field types. The second line starts the declaration of the record. The record type declaration begins with the keyword **RECORD** and ends with the clause **END RECORD**. All of the declarations between these two keywords are field declarations for the record.

Each field of the record represents a unique storage area that can be read from and assigned data of the appropriate type. This example declares three fields: **opcode** of type **optype**, and **src** and **dst** of type **INTEGER**. Each field can be referenced by using the name of the record, followed by a period and the field name. Following is an example of this type of access:

```
PROCESS(X)
  VARIABLE inst : instruction;
  VARIABLE source, dest : INTEGER;
  VARIABLE operator : optype;
BEGIN
  source := inst.src;            --Ok line 1
  dest    := inst.src;           --Ok line 2

  source := inst.opcode;         --error line 3
  operator := inst.opcode;       --Ok line 4

  inst.src  := dest;             --Ok line 5
  inst.dst := dest;              --Ok line 6

  inst := (add, dest, 2);        --Ok line 7
  inst := (source);              --error line 8
END PROCESS;
```

This example declares variable **inst**, which is of type instruction. Also, variables matching the record field types are declared. Lines 1 and 2 show fields of the record being assigned to local process variables. The assignments are legal because the types match. Notice the period after the name of the record to select the field.

Line 3 shows a case that is illegal. The type of field **opcode** does not match the type of variable source. The compiler will flag this statement as a type mismatch error. Line 4 shows the correct assignment occurring between the field **opcode** and a variable that matches its type.

Lines 5 and 6 show that not only can record fields be read from, but they can be assigned to as well. In these two lines, two of the fields of the record are assigned the values from variable **dest**.

Line 7 shows an example of an aggregate assignment. In this line, all of the fields of the record are being assigned at once. The aggregate assigned contains three entries: an **optype** value, an **INTEGER** variable value, and an **INTEGER** value. This is a legal assignment to variable record **inst**.

Line 8 shows an example of an illegal aggregate value for record **inst**. There is only one value present in the aggregate, which is an illegal type for the record.

In the examples so far, all of the elements of the records have been scalars. Let's examine some examples of records that have more complex field types. A record for a data packet is shown here:

```
TYPE word IS ARRAY(0 TO 3) OF std_logic;
TYPE t_word_array IS ARRAY(0 TO 15) OF word;
TYPE addr_type IS
RECORD
 source : INTEGER;
 key    : INTEGER;
END RECORD;

TYPE data_packet IS
RECORD
 addr : addr_type;
 data : t_word_array;
 checksum : INTEGER;
 parity : BOOLEAN;
END RECORD;
```

The first two type declarations define type **word** and **addr_type**, which are used in the record **data_packet**. Type **word** is a simple array and type **addr_type** is a simple record. Record type **data_packet** contains four fields using these two types in combination with two VHDL predefined types.

The following example shows how a variable of type **data_packet** would be accessed:

```
PROCESS(X)
  VARIABLE packet : data_packet;
BEGIN

  packet.addr.key := 5;                    --Ok line 1
  packet.addr := (10, 20);                 --Ok line 2

  packet.data(0) := ('0', '0', '0', '0');  --Ok line 3
```

```
packet.data(10)(4) := '1';          --error line 4
packet.data(10)(0) := '1';          --Ok line 5

END PROCESS;
```

This example shows how complex record types are accessed. In line 1, a record field of a record is accessed. Field **key** is a record field of record **addr_type**, which is a field of record **data_packet**. This line assigns the value 5 to that field. Line 2 assigns an aggregate to the whole field called **addr** in record **data_packet**.

In line 3, the data field is assigned an aggregate for the 0th element of the array. Line 4 tries to assign to only one bit of the eleventh element of the data array field in record **data_packet**, but the second index value is out of range. Finally, line 5 shows how to assign to a single bit of the array correctly.

Composite types are very powerful tools for modeling complex and abstract data types. By using the right combination of records and arrays, you can make models easy to understand and efficient.

ACCESS TYPES   Most hardware design engineers using VHDL probably never use access types directly (a hardware designer may use the TextIO package, which uses access types, thereby an indirect use of access types), but access types provide very powerful programming language type operations. An access type in VHDL is very similar to a pointer in a language like Pascal or C. It is an address, or a handle, to a specific object.

Access types allow the designer to model objects of a dynamic nature. For instance, dynamic queues, fifos, and so on can be modeled easily using access types. Probably the most common operation using an access type is creating and maintaining a linked list.

Only variables can be declared as access types. By the nature of access types, they can only be used in sequential processing. Access types are currently not synthesizable because they are usually used to model the behavior of dynamically sized structures such as a linked list.

When an object is declared to be of an access type, two predefined functions are automatically available to manipulate the object. These functions are named **NEW** and **DEALLOCATE**. Function **NEW** allocates memory of the size of the object in bytes and returns the access value. Function **DEALLOCATE** takes in the access value and returns the memory back to the system. Following is an example that shows how this all works:

```
PROCESS(X)
  TYPE fifo_element_t IS ARRAY(0 TO 3)
OF std_logic; --line 1

TYPE fifo_el_access IS
 ACCESS fifo_element_t; --line 2

 VARIABLE fifo_ptr : fifo_el_access := NULL; --line 3
 VARIABLE temp_ptr : fifo_el_access := NULL; --line 4
BEGIN

  temp_ptr := new fifo_element_t;              --Ok line 5
  temp_ptr.ALL := ('0', '1', '0', '1');        --Ok line 6

  temp_ptr.ALL := ('0', '0', '0', '0');        --Ok line 7
  temp_ptr.ALL(0) := '0';                      --Ok line 8

  fifo_ptr := temp_ptr;                        --Ok line 9
  fifo_ptr.ALL := temp_ptr.ALL;                --Ok line 10
END PROCESS;
```

In line 2, an access type is declared using the type declared in line 1. Lines 3 and 4 declare two access type variables of **fifo_el_access** type from line 2. This process now has two access variable objects that can be used to access objects of type **fifo_element_t**.

Line 5 calls the predefined function **NEW**, which allocates enough memory for a variable of type **fifo_element_t** and returns an access value to the memory allocated. The access value returned is then assigned to variable **temp_ptr**. Variable **temp_ptr** is now pointing to an object of type **fifo_element_t**. This value can be read from or assigned to using variable assignment statements.

In line 6, a value is assigned to the object pointed to by **temp_ptr**. Line 7 shows another way to assign a value using an access value. The keyword **.ALL** specifies that the entire object is being accessed. Subelements of the object can be assigned by using a subelement name after the access variable name. Line 8 shows how to reference a subelement of an array pointed to by an access value. In this example, the first element of the array will have a value assigned to it.

In the next few statements, we examine how access values can be copied among different objects. In line 9, the access value of **temp_ptr** is assigned to **fifo_ptr**. Now both **temp_ptr** and **fifo_ptr** are pointing to the same object. This is shown in Figure 4-3.

Both **temp_ptr** and **fifo_ptr** can be used to read from and assign to the object being accessed.

Line 10 shows how one object value can be assigned to another using access types. The value of the object pointed to by **temp_ptr** is assigned to the value pointed to by **fifo_ptr**.

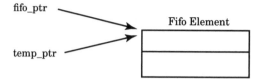

Figure 4-3
Multiple Access Type
References.

Incomplete Types

When implementing recursive structures such as linked lists, you need another VHDL language feature to complete the declarations. This feature is called the incomplete type. The incomplete type allows the declaration of a type to be defined later.

Following is an example that demonstrates why this would be useful:

```
PACKAGE stack_types IS
   TYPE data_type IS ARRAY(0 TO 7) OF std_logic; --line 1

   TYPE element_rec;   --incomplete type         line 2

   TYPE element_ptr IS ACCESS element_rec;       --line 3
   TYPE element_rec IS                           --line 4
    RECORD                                        --line 5
       data : data_type;                          --line 6
       nxt  : element_ptr;                        --line 7
    END RECORD;                                   --line 8

END stack_types;

USE WORK.stack_types.ALL;
ENTITY stack IS
   PORT(din : IN data_type;
        clk : IN std_logic;
        dout : OUT data_type;
        r_wb : IN std_logic);
END stack;

ARCHITECTURE stack OF stack IS
BEGIN
   PROCESS(clk)

      VARIABLE list_head : element_ptr := NULL;   --line 9
      VARIABLE temp_elem : element_ptr := NULL;   --line 10
      VARIABLE last_clk : std_logic     := U;     --line 11
   BEGIN
    IF (clk = `1') AND (last_clk = `0') THEN      --line 12
       IF (r_wb = `0') THEN                        --line 13
          temp_elem := NEW element_rec;            --line 14
          temp_elem.data := din;                   --line 15
```

```
        temp_elem.nxt := list_head;              --line 16
        list_head := temp_elem;                  --line 17
        --read mode                              line 18
    ELSIF (r_wb = '1') THEN
        dout <= list_head.data;                  --line 19
        temp_elem := list_head;                  --line 20
        list_head := temp_elem.nxt;              --line 21
        DEALLOCATE (temp_elem);                  --line 22
    ELSE
        ASSERT FALSE
          REPORT "read/write unknown while clock active"
          SEVERITY WARNING;                      --line 23
    END IF;
  END IF;
  last_clk := clk;                               --line 24
END PROCESS;
END stack;
```

This example implements a stack using access types. The package **stack_types** declares all of the types needed for the stack. In line 2, there is a declaration of the incomplete type **element_rec**. The name of the type is specified, but no specification of the type is present. The purpose of this declaration is to reserve the name of the type and allow other types to gain access to the type when it is fully specified. The full specification for this incomplete type appears in lines 4 through 8.

The fundamental reason for the incomplete type is to allow self-referencing structures as linked lists. Notice that type **element_ptr** is used in type **element_rec** in line 6. To use a type, it must first be defined. Notice also that, in the declaration for type **element_ptr** in line 3, type **element_rec** is used. Because each type uses the other in its respective declarations, neither type can be declared first without a special way of handling this case. The incomplete type allows this scenario to exist.

Lines 4 through 8 declare the record type **element_rec**. This record type is used to store the data for the stack. The first field of the record is the data field, and the second is an access type that points to the next record in the stack.

The entity for stack declares port **din** for data input to the stack, a **clk** input on which all operations are triggered, a **dout** port which transfers data out of the stack, and, finally, a **r_wb** input which causes a read operation when high and a write operation when low. The process for the stack is only triggered when the **clk** input has an event occur. It is not affected by changes in **r_wb**.

Lines 9 through 11 declare some variables used to keep track of the data for the stack. Variable **list_head** is the head of the linked list of data. It always points to the first element of the list of items in the stack. Variable `temp_elem` is used to hold a newly allocated element until it is connected into the stack list. Variable `last_clk` is used to hold the previous value of `clk` to enable transitions on the clock to be detected. (This behavior can be duplicated with attributes, which are discussed in Chapter 7, "Configurations.")

Line 12 checks to see if a 0 to 1 transition has occurred on the `clk` input. If so, then the stack needs to do a read or write depending on the `r_wb` input. Line 13 checks to see if `r_wb` is set up for a write to the stack. If so, lines 14 through 17 create a new data storage element and connect this element to the list.

Line 14 uses the predefined function **NEW** to allocate a record of type `element_rec` and return an access value to be assigned to variable `temp_elem`. This creates a structure that is shown in Figure 4-4.

Lines 15 and 16 fill in the newly allocated object with the data from input `din` and the access value to the head of the list. After line 16, the data structures look like Figure 4-5.

Finally, in line 17, the new element is added to the head of the list. This is shown in Figure 4-6.

Lines 18 through 22 of the model provide the behavior of the stack when an element is read from the stack. Line 19 copies the data from the stack element to the output port. Lines 20 through 22 disconnect the element from the stack list and return the memory to the system.

Line 20 assigns the `temp_elem` access variable to point to the head of the list. This is shown in Figure 4-7.

Line 21 moves the head of the list to the next element in the list. This is shown in Figure 4-8.

Figure 4-4
Allocate New Stack Element.

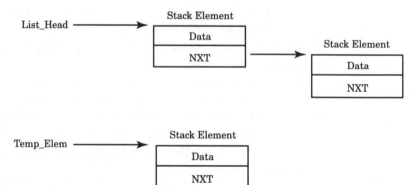

Figure 4-5
Point New Element
to Head of List.

Figure 4-6
Point **List_Head** to
New Element.

Figure 4-7
Point **Temp_Elem** to
List_Head.

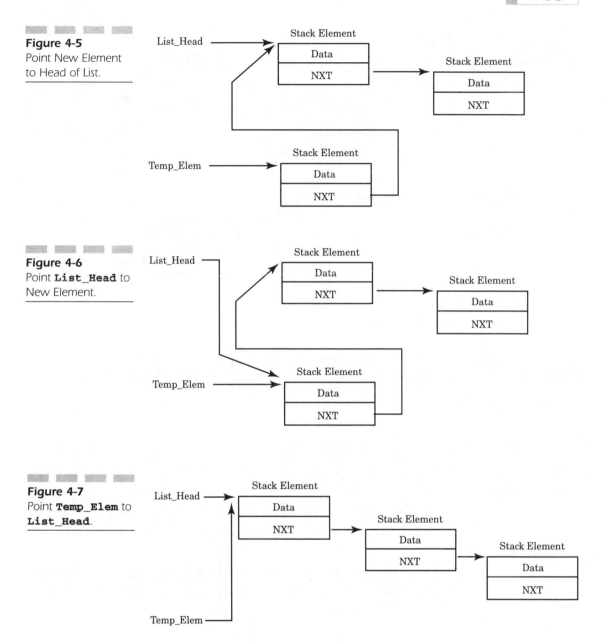

Finally, in line 22, the element that had its data transferred out is deallocated, and the memory is returned to the memory pool. This is shown in Figure 4-9.

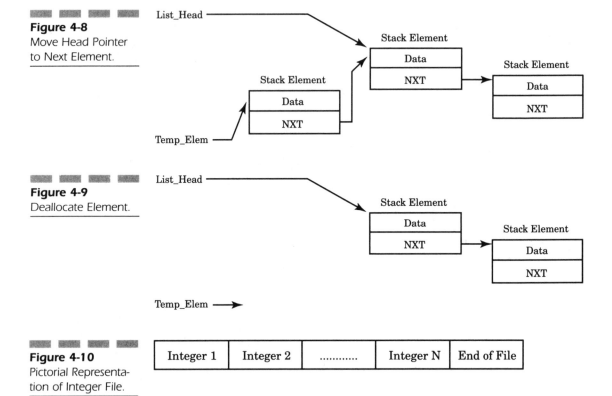

Figure 4-8
Move Head Pointer
to Next Element.

Figure 4-9
Deallocate Element.

Figure 4-10
Pictorial Representa-
tion of Integer File.

Figure 4-11
Pictorial Representa-
tion of Complex File.

OPCODE	OPCODE	....	OPCODE	END
ADDRMODE	ADDRMODE	....	ADDRMODE	OF
SRC	SRC	....	SRC	FILE
DST	DST	....	DST	MARK

Access types are very powerful tools for modeling complex and abstract types of systems. Access types bring programming language types of operations to VHDL processes.

File Types

A file type allows declarations of objects that have a type **FILE**. A file object type is actually a subset of the variable object type. A variable object can be assigned with a variable assignment statement, while a file object cannot be assigned. A file object can be read from, written to, and checked for end of file only with special procedures and functions.

Files consist of sequential streams of a particular type. A file whose base object type is integer consists of a sequential stream of integers. This is shown in Figure 4-10.

A file whose object type is a complex record type consists of a sequential stream of complex records. An example of how this might look is shown in Figure 4-11.

At the end of the stream of data is an end-of-file mark. Two procedures and one function allow operations on file objects:

- **READ** (file, data)Procedure

- **WRITE** (file, data)Procedure

- **ENDFILE** (file)Function, returns boolean

Procedure **READ** reads an object from the file and returns the object in argument data. Procedure **WRITE** writes argument data to the file specified by the file argument. Finally, function **ENDFILE** returns true when the file is currently at the end-of-file mark.

Using these procedures and functions requires a file type declaration and a file object declaration.

FILE TYPE DECLARATION   A file type declaration specifies the name of the file type and the base type of the file. Following is an example of a file type declaration:

```
TYPE integer_file IS FILE OF INTEGER;
```

This declaration specifies a file type whose name is **integer_file** and is of type **INTEGER**. This declaration corresponds to the file in Figure 4-10.

FILE OBJECT DECLARATION   A file object makes use of a file type and declares an object of type **FILE**. The file object declaration specifies the name of the file object, the mode of the file, and the physical disk path name. The file mode can be **IN** or **OUT**. If the mode is **IN**, then the file can be read with the **READ** procedure. If the mode is **OUT**, then the file can be written with the **WRITE** procedure. Here is an example:

```
FILE myfile : integer_file IS IN
        "/doug/test/examples/data_file";
```

This declaration declares a file object called **myfile** that is an input file of type **integer_file**. The last argument is the path name on the physical disk where the file is located. (In most implementations this is true, but it is not necessarily true.)

FILE TYPE EXAMPLES To read the contents of a file, you can call the READ procedure within a loop statement. The loop statement can perform read operations until an end of file is reached, at which time the loop is terminated. Following is an example of a file read operation:

```
LIBRARY IEEE;
USE IEEE.std_logic_1164.ALL;
ENTITY rom IS
  PORT(addr : IN INTEGER;
       cs   : IN std_logic;
       data : OUT INTEGER);
END rom;

ARCHITECTURE rom OF rom IS
BEGIN
  PROCESS(addr, cs)
    VARIABLE rom_init : BOOLEAN := FALSE;      --line 1
    TYPE rom_data_file_t IS FILE OF INTEGER;   --line 2

    FILE rom_data_file : rom_data_file_t IS IN
         "/doug/dlp/test1.dat";                --line 3

    TYPE dtype IS ARRAY(0 TO 63) OF INTEGER;

    VARIABLE rom_data : dtype;                  --line 4
    VARIABLE i : INTEGER := 0;                  --line 5
  BEGIN
    IF (rom_init = false) THEN                  --line 6
      WHILE NOT ENDFILE(rom_data_file)          --line 7
                AND (i < 64) LOOP
        READ(rom_data_file, rom_data(i));       --line 8
        i := i + 1;                             --line 9
      END LOOP;
      rom_init := true;                         --line 10
    END IF;
    IF (cs = '1') THEN                          --line 11
      data <= rom_data(addr);                   --line 12
    ELSE
      data <= -1;                               --line 13
    END IF;
  END PROCESS;
END rom;
```

This example shows how a rom can be initialized from a file the first time the model is executed and never again. A variable called rom_init is used to keep track of whether the rom has been initialized or not. If false, the rom has not been initialized; if true, the rom has already been initialized.

Line 2 of the example declares a file type **rom_data_file_t** that is used to declare a file object. In line 3, a **rom_data_file** object is declared. In this example, the physical disk path name was hard-coded into the model, but a generic could have been used to pass a different path name for each instance of the **rom**.

Line 6 of the example tests variable **rom_init** for true or false. If false, the initialization loop is executed. Line 7 is the start of the initialization loop. The loop test makes use of the predefined function **ENDFILE**. The loop executes until there is no more data in the file or when the **rom** storage area has been filled.

Each pass through the loop calls the predefined procedure **READ**. This procedure reads one integer at a time and places it in the element of **rom_data** that is currently being accessed. Each time through the loop, the index **i** is incremented to the next element position.

Finally, when the loop finishes, the variable **rom_init** is set to true. The next time the process is invoked, variable **rom_init** will be true, so the initialization loop will not be invoked again.

Writing a file is analogous to reading, except that the loop does not test every time through for an end-of-file condition. Each time a loop writing data is executed, the new object is appended to the end of the file. When the model is writing to a file, the file must have been declared with mode **OUT**.

File Type Caveats

In general, the file operations allowed are limited. Files cannot be opened, closed, or accessed in a random sequence. All that VHDL provides is a simple sequential capability. See Appendix D for a description of VHDHL93 file access. For textual input and output, there is another facility that VHDL provides called TextIO. This facility provides for formatted textual input and output and is discussed in Chapter 8, "Advanced Topics."

Subtypes

Subtype declarations are used to define subsets of a type. The subset can contain the entire range of the base type but does not necessarily need to. A typical subtype adds a constraint or constraints to an existing type.

The type integer encompasses the minimum range -2,147,483,647 to +2,147,483,647. In the Standard package (a designer should never redefine any of the types used in the Standard package; this can result in incompatible VHDL, because of type mismatches), there is a subtype called **NATURAL** whose range is from 0 to +2,147,483,647. This subtype is defined as shown here:

```
TYPE INTEGER IS -2,147,483,647 TO 12,147,483,647;
SUBTYPE NATURAL IS INTEGER RANGE 0 TO 12,147,483,647;
```

After the keyword **SUBTYPE** is the name of the new subtype being created. The keyword **IS** is followed by the base type of the subtype. In this example, the base type is **INTEGER**. An optional constraint on the base type is also specified.

So why would a designer want to create a subtype? There are two main reasons for doing so:

- To add constraints for selected signal assignment statements or case statements.
- To create a resolved subtype. (Resolved types are discussed along with resolution functions in Chapter 5.)

When a subtype of the base type is used, the range of the base type can be constrained to be what is needed for a particular operation. Any functions that work with the base type also work with the subtype.

Subtypes and base types also allow assignment between the two types. A subtype can always be assigned to the base type because the range of the subtype is always less than or equal to the range of the base type. The base type may or may not be able to be assigned to the subtype, depending on the value of the object of the base type. If the value is within the value of the subtype, then the assignment succeeds; otherwise, a range constraint error results.

A typical example where a subtype is useful is adding a constraint to a numeric base type. In the previous example, the **NATURAL** subtype constrained the integer base type to the positive values and zero. But what if this range is still too large? The constraint specified can be a user-defined expression that matches the type of the base type. In the following example, an 8-bit multiplexer is modeled with a much smaller constraint on the integer type:

```
PACKAGE mux_types IS
   SUBTYPE eightval IS INTEGER RANGE 0 TO 7; --line 1
END mux_types;
```

```
USE WORK.mux_types.ALL;
LIBRARY IEEE;
USE IEEE.std_logic_1164.ALL;
ENTITY mux8 IS
PORT(I0, I1, I2, I3, I4, I5,
     I6, I7: IN std_logic;
     sel : IN eightval; --line 2
     q   : OUT std_logic);
END mux8;

ARCHITECTURE mux8 OF mux8 IS
BEGIN
WITH sel SELECT    --line 3
    Q <= I0 AFTER 10 ns WHEN 0, --line 4
         I1 AFTER 10 ns WHEN 1, --line 5
         I2 AFTER 10 ns WHEN 2, --line 6
         I3 AFTER 10 ns WHEN 3, --line 7
         I4 AFTER 10 ns WHEN 4, --line 8
         I5 AFTER 10 ns WHEN 5, --line 9
         I6 AFTER 10 ns WHEN 6, --line 10
         I7 AFTER 10 ns WHEN 7; --line 11
END mux8;
```

The package **mux_types** declares a subtype **eightval**, which adds a constraint to base type **INTEGER**. The constraint allows an object of **eightval** to take on values from 0 to 7.

The package is included in entity **mux8**, which has one of its input ports **sel** declared using type **eightval**. In the architecture at line 3, a selected signal assignment statement uses the value of **sel** to determine which output is transferred to the output **Q**. If **sel** was not of the subtype **eightval**, but was strictly an integer type, then the selected signal assignment would need a value to assign for each value of the type, or an **OTHERS** clause. By adding the constraint to the integer type, all values of the type can be directly specified.

SUMMARY ▪ ▪ ▪ ▪ ▪ ▪ ▪ ▪ ▪

In this chapter, we have examined the different types available in VHDL to the designer. We discussed the following:

■ How types can be used by three different types of objects: the signal, variable, and constant.

■ How signals are the main mechanism for the connection of entities, and how signals are used to pass information between entities.

- How variables are local to processes and subprograms and are used mainly as scratch pad areas for local calculations.

- How constants name a particular value of a type.

- How integers behave like mathematical integers, and real numbers behave like mathematical real numbers.

- How enumerated types can be used to describe user-defined operations and make a model much more readable.

- How physical types represent physical quantities such as distance, current, time, and so on.

- The composite type, arrays and records. Arrays are a group of elements of the same type, and records are a group of elements of any type(s).

- How access types are like pointers in typical programming languages.

- How file types are linear streams of data of a particular type that can be read and written from a model.

- How subtypes can add constraints to a type.

In the next chapter, we focus on another method of sequential statement modeling: the subprogram.

Subprograms and Packages

In this chapter, subprograms and packages are discussed. Subprograms consist of procedures and functions used to perform common operations. Packages are mechanisms that allow sharing data among entities. Subprograms, types, and component declarations are the tools to build designs with, and packages are the toolboxes.

Subprograms

Subprograms consist of procedures and functions. A procedure can return more than one argument; a function always returns just one. In a function, all parameters are input parameters; a procedure can have input parameters, output parameters, and inout parameters.

There are two versions of procedures and functions: a concurrent procedure and concurrent function, and a sequential procedure and sequential function. The concurrent procedure and function exist outside of a process statement or another subprogram; the sequential function and procedure exist only in a process statement or another subprogram statement.

All statements inside of a subprogram are sequential. The same statements that exist in a process statement can be used in a subprogram, including **WAIT** statements.

A procedure exists as a separate statement in an architecture or process; a function is usually used in an assignment statement or expression.

Function

The following example is a function that takes in an array of the **std_logic** type (described in Chapter 9, "Synthesis" and Appendix A, "Standard Logic Package") and returns an integer value. The integer value represents the numeric value of all of the bits treated as a binary number:

```
USE LIBRARY IEEE;
USE IEEE.std_logic_1164.ALL;
PACKAGE num_types IS
  TYPE log8 IS ARRAY(0 TO 7) OF std_logic;   --line 1
END num_types;

USE LIBRARY IEEE; USE IEEE.std_logic_1164.ALL;
USE WORK.num_types.ALL;
ENTITY convert IS
  PORT(I1 : IN log8;          --line 2
     O1 : OUT INTEGER);    --line 3
END convert;

ARCHITECTURE behave OF convert IS
  FUNCTION vector_to_int(S : log8)    --line 4
   RETURN INTEGER is                  --line 5
   VARIABLE result : INTEGER := 0;    --line 6
  BEGIN
```

```
FOR i IN 0 TO 7 LOOP                --line 7
   result := result * 2;            --line 8
   IF S(i) = '1' THEN               --line 9
      result := result + 1;         --line 10
   END IF;
END LOOP;
RETURN result;                      --line 11
END vector_to_int;

BEGIN
 O1 <= vector_to_int(I1);           --line 12
END behave;
```

Line 1 of the example declares the array type used throughout the example. Lines 2 and 3 show the input and output ports of the convert entity and their types. Lines 4 through 11 describe a function that is declared in the declaration region of the architecture **behave**. By declaring the function in the declaration region of the architecture, the function is visible to any region of the architecture.

Lines 4 and 5 declare the name of the function, the arguments to the function, and the type that the function returns. In line 6, a variable local to the function is declared. Functions have declaration regions very similar to process statements. Variables, constants, and types can be declared, but no signals.

Lines 7 through 10 declare a loop statement that loops once for each value in the array type. The basic algorithm of the function is to do a shift and add for each bit position in the array. The result is first shifted (by multiplying by 2), and then, if the bit position is a logical 1, a 1 value is added to the result.

At the end of the loop statement, variable **result** contains the integer value of the array passed in. The value of the function is passed back via the **RETURN** statement. An example **RETURN** statement is shown in line 11.

Finally, line 12 shows how a function is called. The name of the function is followed by its arguments enclosed in parentheses. The function always returns a value; therefore, the calling process, concurrent statement, and so on must have a place for the function to return the value to. In this example, the output of the function is assigned to an output port.

Parameters to a function are always input only. No assignment can be done to any of the parameters of the function. In the preceding example, the parameters were of a constant kind because no explicit kind was specified and the default is constant. The arguments are treated as if they were constants declared in the declaration area of the function.

The other kind of parameter that a function can have is a signal parameter. With a signal parameter, the attributes (which discussed in Chapter 6, "Predefined Attributes") of the signal are passed in and are available for use in the function. The exception to this statement are at-tributes `STABLE, `QUIET, `TRANSACTION, and `DELAYED, which create spe-cial signals.

Following is an example showing a function that contains signal para-meters:

```
USE LIBRARY IEEE;
USE IEEE.std_logic_1164.ALL;
ENTITY dff IS
  PORT(d, clk : IN std_logic;
       q : OUT std_logic);

  FUNCTION rising_edge(SIGNAL S : std_logic)      --line 1
    RETURN BOOLEAN IS                             --line 2
  BEGIN
    --this function makes use of attributes
    --`event and `last_value discussed
    --in Chapter 6
    IF (S'EVENT) AND (S = `1') AND                --line 3
        (S'LAST_VALUE = `0') THEN                 --line 4
      RETURN TRUE;                                --line 5
    ELSE
      RETURN FALSE;                               --line 6
    END IF;
  END rising_edge;
END dff;

ARCHITECTURE behave OF dff IS
BEGIN
  PROCESS( clk)
  BEGIN
    IF rising_edge(clk) THEN                      --line 7
     q <= d;                                      --line 8
    END IF;
  END PROCESS;
END behave;
```

This example provides a rising edge detection facility for the D flip-flop being modeled. The function is declared in the entity declaration section and is available to any architecture of the entity.

Lines 1 and 2 show the function declaration. There is only one para-meter (**s**) to the function, and it is of a signal type. Lines 3 and 4 show an **IF** statement that determines whether the signal has just changed or not, if the current value is a `1', and whether the previous value was a `0'. If all of these conditions are true, then the **IF** statement returns a true value, signifying that a rising edge was found on the signal.

If any one of the conditions is not true, the value returned is false, as shown in line 6. Line 7 shows an invocation of the function using the signal created by port **clk** of entity **dff**. If there is a rising edge on the signal **clk**, then the **d** value is transferred to the output **q**.

The most common use for a function is to return a value in an expression; however, there are two more classes of use available in VHDL. The first is a conversion function, and the second is a resolution function. Conversion functions are used to convert from one type to another. Resolution functions are used to resolve bus contention on a multiply-driven signal.

Conversion Functions

Conversion functions are used to convert an object of one type to another. They are used in component instantiation statements to allow mapping of signals and ports of different types. This type of situation usually arises when a designer wants to make use of an entity from another design that uses a different data type.

Assume that designer A was using a data type that had the following four values:

```
TYPE fourval IS (X, L, H, Z);
```

Designer B was using a data type that also contained four values, but the value identifiers were different, as shown here:

```
TYPE fourvalue IS ('X', '0', '1', 'Z');
```

Both of these types can be used to represent the states of a four-state value system for a VHDL model. If designer A wanted to use a model from designer B, but designer B used the values from type **fourvalue** as the interface ports to the model, then designer A cannot use the model without converting the types of the ports to the value system used by designer B. This problem can be solved through the use of conversion functions.

First, let's write the function that converts between these two value systems. The values from the first type represent these distinct states:

- **x**—Unknown value
- **L**—Logical 0 value
- **H**—Logical 1 value
- **z**—High-impedance or open-collector value

The values from the second type represent these states:

- **'x'**—Unknown value

▓ `'0'`—Logical 0 value

▓ `'1'`—Logical 1 value

▓ `'z'`—High-impedance or open-collector value

From the description of the two value systems, the conversion function is trivial. Following is an example of one:

```
FUNCTION convert4val(S : fourval) RETURN fourvalue IS
BEGIN
 CASE S IS
   WHEN X =>
     RETURN 'X';
   WHEN L =>
     RETURN '0';
   WHEN H =>
     RETURN '1';
   WHEN Z =>
     RETURN 'Z';
 END CASE;
END convert4val;
```

This function accepts a value of type **fourval** and returns a value of type **fourvalue**. The next example shows where such a function might be used:

```
PACKAGE my_std IS
 TYPE fourval IS (X, L, H, Z);
 TYPE fourvalue IS ('X', '0', '1', 'Z');

 TYPE fvector4 IS ARRAY(0 TO 3) OF fourval;
END my_std;

USE WORK.my_std.ALL;
ENTITY reg IS
 PORT(a : IN fvector4;
    clr : IN fourval;
    clk : IN fourval;
   q : OUT fvector4);

 FUNCTION convert4val(S : fourval)
  RETURN fourvalue IS
 BEGIN
   CASE S IS
     WHEN X =>
       RETURN 'X';
     WHEN L =>
       RETURN '0';
     WHEN H =>
       RETURN '1';
     WHEN Z =>
```

```
          RETURN 'Z';
      END CASE;
    END convert4val;

    FUNCTION convert4value(S : fourvalue)
      RETURN fourval IS
    BEGIN
     CASE S IS
       WHEN 'X' =>
          RETURN X;
       WHEN '0' =>
          RETURN L;
       WHEN '1' =>
          RETURN H;
       WHEN 'Z' =>
          RETURN Z;
     END CASE;
    END convert4value;
  END reg;

  ARCHITECTURE structure OF reg IS
   COMPONENT dff
     PORT(d, clk, clr : IN fourvalue;
         q : OUT fourvalue);
   END COMPONENT;
  BEGIN
   U1 : dff PORT MAP(convert4val(a(0)),
          convert4val(clk),
          convert4val(clr),
          convert4value(q) => q(0));

   U2 : dff PORT MAP(convert4val(a(1)),
          convert4val(clk),
          convert4val(clr),
          convert4value(q) => q(1));

   U3 : dff PORT MAP(convert4val(a(2)),
          convert4val(clk),
          convert4val(clr),
          convert4value(q) => q(2));

   U4 : dff PORT MAP(convert4val(a(3)),
          convert4val(clk),
          convert4val(clr),
          convert4value(q) => q(3));

  END structure;
```

This example is a 4-bit register built out of flip-flops. The type used in the entity declaration for the register is a vector of type **fourval**. However, the flip-flops being instantiated have ports that are of type **fourvalue**. A type mismatch error is generated if the ports of entity register are mapped directly to the component ports. A conversion function is needed to convert between the two value systems.

If the ports are all of mode **IN**, then only one conversion is needed to map from the containing entity type to the contained entity type. In this example, if all of the ports were of mode input, then only function **convert4val** would be required.

If the component has output ports as well, then the output values of the contained entity need to be converted back to the containing entity type. In this example, the **q** port of component **dff** is an output port. The type of the output values is fourvalue. These values cannot be mapped to the type **fourval** ports of entity **xregister**. Function **convert4value** converts from a **fourvalue** type to a **fourval** type. Applying this function on the output ports allows the port mapping to occur.

There are four component instantiations that use these conversion functions: components U1 through U4. Notice that the input ports use the **convert4val** conversion function; the output ports use the **convert4value** conversion function.

Using the named association form of mapping for component instantiation, U1 would look like this:

```
U1: dff PORT MAP (
  d   => convert4val( a(0) ),
  clk => convert4val( clk ),
  clr => convert4val( clr ),
  convert4value(q) => q(0) );
```

What this notation shows is that, for the input ports, the conversion functions are applied to the appropriate input signals (ports) before being mapped to the **dff** ports, and the output port value is converted with the conversion function before being mapped to the output port **q(0)**.

Conversion functions free the designer from generating a lot of temporary signals or variables to perform the conversion. The following example shows another method for performing conversion functions:

```
  temp1 <= convert4val( a(0) );
  temp2 <= convert4val( clk );
  temp3 <= convert4val( clr );

U1: dff PORT MAP (
  d   => temp1,
  clk => temp2,
  clr => temp3,
  q   => temp4);

  q(0) <=  convert4value(temp4);
```

This method is much more verbose, requiring an intermediate temporary signal for each port of the component being mapped. This clearly is not the preferred method.

If a port is of mode **INOUT**, conversion functions cannot be used with positional notation. The ports must use named association because two conversion functions must be associated with each inout port. One conversion function is used for the input part of the inout port, and the other is used for the output part of the inout port.

In the following example, two bidirectional transfer devices are contained in an entity called **trans2**:

```
PACKAGE my_pack IS
  TYPE nineval IS (Z0, Z1, ZX,
                   R0, R1, RX,
                   F0, F1, FX);

  TYPE nvector2 IS ARRAY(0 TO 1) OF nineval;
  TYPE fourstate IS (X, L, H, Z);

  FUNCTION convert4state(a : fourstate)
    RETURN nineval;

  FUNCTION convert9val(a : nineval)
    RETURN fourstate;

END my_pack;

PACKAGE body my_pack IS
  FUNCTION convert4state(a : fourstate)
    RETURN nineval IS
  BEGIN
    CASE a IS
      WHEN X =>
        RETURN FX;
      WHEN L =>
        RETURN F0;
      WHEN H =>
        RETURN F1;
      WHEN Z =>
        RETURN ZX;
    END CASE;
  END convert4state;

  FUNCTION convert9val(a : nineval)
    RETURN fourstate IS
  BEGIN
    CASE a IS
      WHEN Z0 =>
        RETURN Z;
      WHEN Z1 =>
```

```
          RETURN Z;
      WHEN ZX =>
          RETURN Z;
      WHEN R0 =>
          RETURN L;
      WHEN R1 =>
          RETURN H;
      WHEN RX =>
          RETURN X;
      WHEN F0 =>
          RETURN L;
      WHEN F1 =>
          RETURN H;
      WHEN FX =>
          RETURN X;
    END CASE;
  END convert9val;
END my_pack;

USE WORK.my_pack.ALL;
ENTITY trans2 IS
  PORT( a, b : INOUT nvector2;
        enable : IN nineval);
END trans2;

ARCHITECTURE struct OF trans2 IS
  COMPONENT trans
    PORT( x1, x2 : INOUT fourstate;
          en : IN fourstate);
END COMPONENT;
BEGIN
  U1 : trans PORT MAP(
    convert4state(x1) => convert9val(a(0)),
    convert4state(x2) => convert9val(b(0)),
    en => convert9val(enable) );

  U2 : trans PORT MAP(
    convert4state(x1) => convert9val(a(1)),
    convert4state(x2) => convert9val(b(1)),
    en => convert9val(enable) );
END struct;
```

Each component is a bidirectional transfer device called **trans**. The **trans** device contains three ports. Ports **x1** and **x2** are inout ports, and port **en** is an input port. When port **en** is an **H** value, **x1** is transferred to **x2**; and when port **en** is an **L** value, **x2** is transferred to **x1**.

The **trans** components use type **fourstate** for the port types; the containing entity uses type **nineval**. Conversion functions are required to allow the instantiation of the **trans** components in architecture **struct** of entity **trans2**.

The first component instantiation statement for the **trans** component labeled **U1** shows how conversion functions are used for inout ports. The first port mapping maps port**x1** to **a(0)**. Port **a(0)** is a **nineval** type; therefore, the signal created by the port is a **nineval** type. When this signal is mapped to port **x1** of component **trans**, it must be converted to a **fourstate** type. Conversion function **convert9val** must be called to complete the conversion. When data is transferred out to port **x1** for the out portion of the inout port, conversion function **convert4state** must be called.

The conversion functions are organized such that the side of the port mapping clause that changes contains the conversion function that must be called. When **x1** changes, function **convert4state** is called to convert the **fourstate** value to a **nineval** value before it is passed to the containing entity **trans2**. Conversely, when port **a(0)** changes, function **convert9val** is called to convert the **nineval** value to a **fourstate** value that can be used within the **trans** model.

Conversion functions are used to convert a value of one type to a value of another type. They can be called explicitly as part of execution or implicitly from a mapping in a component instantiation.

Resolution Functions

A resolution function is used to return the value of a signal when the signal is driven by multiple drivers. It is illegal in VHDL to have a signal with multiple drivers without a resolution function attached to that signal.

A resolution function consists of a function that is called whenever one of the drivers for the signal has an event occur on it. The resolution function is executed and returns a single value from all of the driver values; this value is the new value of the signal.

In typical simulators, resolution functions are built in, or fixed. With VHDL, the designer has the ability to define any type of resolution function desired, wired-or, wired-and, average signal value, and so on.

A resolution function has a single-argument input and returns a single value. The single-input argument consists of an unconstrained array of driver values for the signal that the resolution function is attached to. If the signal has two drivers, the unconstrained array is two elements long; if the signal has three drivers, the unconstrained array is three elements long. The resolution function examines the values of all of the drivers and returns a single value called the resolved value of the signal.

Let's examine a resolution function for the type **fourval** that was used in the conversion function examples. The type declaration for **fourval** is shown here:

```
TYPE fourval IS (X, L, H, Z);
```

Four distinct values are declared that represent all of the possible values that the signal can obtain. The value **L** represents a logical 0, the value **H** represents a logical 1, the value **Z** represents a high-impedance or open-collector condition, and, finally, the value **X** represents an unknown condition in which the value can represent an **L** or an **H**, but we're not sure which. This condition can occur when two drivers are driving a signal, one driver driving with an **H**, and the other driving with an **L**.

Listed by order of strength, with the weakest at the top, the values are as follows:

- **Z**—Weakest, **H**, **L**, or **X** can override
- **H,L**—Medium strength, only **X** can override
- **X**—Strong, no override

Using this information, a truth table for two inputs can be developed, as shown in Figure 5-1.

This truth table is for two input values. It can be expanded to more inputs by successively applying it to two values at a time. This can be done because the table is commutative and associative. An **L** and a **Z**, or a **Z** and an **L**, gives the same results. An (**L**, **Z**) with **H** gives the same results as an (**H**, **Z**) with an **L**. These principles are very important, because the order of driver values within the input argument to the resolution function is nondeterministic from the designer's point of view. Any dependence on order can cause nondeterministic results from the resolution function.

Figure 5-1
Four State Truth Table.

	Z	L	H	X
Z	Z	L	H	X
L	L	L	X	X
H	H	X	H	X
X	X	X	X	X

Using all of this information, a designer can write a resolution function for this type. The resolution function maintains the highest strength seen so far and compares this value with new values a single element at a time, until all values have been exhausted. This algorithm returns the highest-strength value.

Following is an example of such a resolution function:

```
PACKAGE fourpack IS
  TYPE fourval IS (X, L, H, Z);
  TYPE fourval_vector IS ARRAY (natural RANGE <> ) OF
    fourval;

  FUNCTION resolve( s: fourval_vector) RETURN fourval;
END fourpack;

PACKAGE BODY fourpack IS
  FUNCTION resolve( s: fourval_vector) RETURN fourval IS
   VARIABLE result : fourval := Z;
  BEGIN
   FOR i IN s'RANGE LOOP
    CASE result IS
      WHEN Z =>
        CASE s(i) IS
          WHEN H =>
            result := H;
          WHEN L =>
            result := L;
          WHEN X =>
            result := X;
          WHEN OTHERS =>
            NULL;
        END CASE;

      WHEN L =>
        CASE s(i) IS
          WHEN H =>
            result := X;
          WHEN X =>
            result := X;
          WHEN OTHERS =>
            NULL;
        END CASE;

      WHEN H =>
        CASE s(i) IS
          WHEN L =>
            result := X;
          WHEN X =>
            result := X;
          WHEN OTHERS =>
```

```
              NULL;
          END CASE;

          WHEN X =>
            result := X;

        END CASE;
      END LOOP;
     RETURN result;
   END resolve;
 END fourpack;
```

The input argument is an unconstrained array of the driver-base type, **fourval**. The resolution function examines all of the values of the drivers passed in argument **s** one at a time and returns a single value of **fourval** type to be scheduled as the signal value.

Variable **result** is initialized to a **z** value to take care of the case of zero drivers for the signal. In this case, the loop is never executed, and the result value returned is the initialization value. It is also a good idea to initialize the result value to the weakest value of the value system to allow overwriting by stronger values.

If a nonzero number of drivers exists for the signal being resolved, then the loop is executed once for each driver value passed in argument **s**. Each driver value is compared with the current value stored in variable **result**. If the new value is stronger according to the rules outlined earlier, then the current result is updated with the new value.

Let's look at some example driver values to see how this works. Assuming that argument **s** contained the driver values shown in Figure 5-2, what would the result be?

Figure 5-2

Four State Resolution with Two Values.

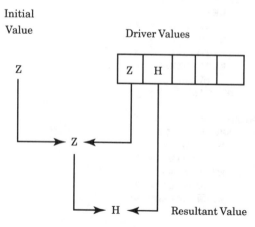

Initial Value

Driver Values

Resultant Value

Because there are two drivers, the loop is executed twice. The first time through, the loop variable result contains the initial value **z**. The first driver value is also a **z** value. Value **z** compared with value **z** produces a resulting value **z**.

The next iteration through the loop retrieves the next driver value, which is **H**. The value **H** compared with value **z** returns value **H**. The function therefore returns the value **H** as the resolved value of the signal.

Another case is shown in Figure 5-3. In this example, there are three drivers, and the resolution function executes the loop three times. In the first iteration of the loop, the initial value of result (**z**) is compared with the first driver value (**H**). The value **H** is assigned to `result`. In the next iteration, result (**H**) is compared with the second driver (**z**). The value **H** remains in `result` because the value **z** is weaker. Finally, the last iteration result (**H**) is compared with the last driver value (**L**). Beccause these values are of the same strength, the value **x** is assigned to `result`. The value **x** is returned from the function as the resolved value for the signal.

NINE-VALUE RESOLUTION FUNCTION Some simulators use more complex types to represent the value of a signal. For instance, what might a resolution function look like for a nine-value system, typical of most workstation-based simulators in use currently? Following are the nine values in the value system:

Figure 5-3
Four State Resolution with Three Values

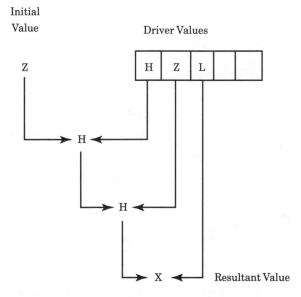

```
Z0, Z1, ZX, R0, R1, RX, F0, F1, FX

weakest----------------------------strongest
```

The system consists of three strengths and three logic values. The three strengths represent the following:

- **z**—High impedance strength, few hundred k of resistance
- **R**—Resistive, few k of resistance
- **F**—Forcing, few ohms of resistance

The three logic levels are represented as follows:

- **0**—Logical 0 or false
- **1**—Logical 1 or true
- **x**—Logical unknown

The nine states are described as follows:

- **Z0**—High-impedance 0
- **Z1**—High-impedance 1
- **ZX**—High-impedance unknown
- **R0**—Resistive 0
- **R1**—Resistive 1
- **RX**—Resistive unknown
- **F0**—Forcing 0
- **F1**—Forcing 1
- **FX**—Forcing unknown

A few simple rules can be used to define how the resolution function should work:

—Strongest strength always wins.

—If strengths are the same and values are different, return same strength but **x** value.

Following are the type declarations needed for the value system:

```
PACKAGE ninepack IS
  TYPE strength IS (Z, R, F);
  TYPE nineval IS ( Z0, Z1, ZX,
                    R0, R1, RX,
                    F0, F1, FX );

  TYPE ninevalvec IS ARRAY(natural RANGE <>) OF nineval;
```

```
TYPE ninevaltab IS ARRAY(nineval'LOW TO
  nineval'HIGH) OF nineval;

TYPE strengthtab IS ARRAY(strength'LOW TO
  strength'HIGH) OF nineval;

FUNCTION resolve9( s: ninevalvec) RETURN nineval;

END ninepack;
```

The package body contains the resolution function (package bodies are discussed near the end of this chapter).

```
PACKAGE BODY ninepack IS
  FUNCTION resolve9( s: ninevalvec) RETURN nineval IS
    VARIABLE result: nineval;
    CONSTANT get_strength : ninevaltab :=
      (Z,      --Z0
       Z,      --Z1
       Z,      --ZX
       R,      --R0
       R,      --R1
       R,      --RX
       F,      --F0
       F,      --F1
       F);     --FX

    CONSTANT x_tab :   strengthtab :=
      (ZX,     --Z
       RX,     --R
       FX);    --F
  BEGIN
    IF s'LENGTH = 0 THEN RETURN ZX; END IF;
      result := s(0);

    FOR i IN s'RANGE LOOP
      IF get_strength(result) < get_strength(s(i)) THEN
        result := s(i);

      ELSIF get_strength(result) = get_strength(s(i)) THEN
        IF result /= s(i) THEN
          result := x_tab(get_strength(result));
        END IF;

      END IF;
    END LOOP;
      RETURN result;

  END resolve9;
END ninepack;
```

The package **ninepack** declares a number of types used in this example, including some array types to make the resolution function easier to implement. The basic algorithm of the function is the same as the **fourval** resolution function; however, the operations with nine values are a little more complex. Function **resolve9** still does a pairwise comparison of the input values to determine the resultant value. With a nine-value system, the comparison operation is more complicated, and therefore some constant arrays were declared to make the job easier.

The constant **get_strength** returns the driving strength of the driver value. The constant **x_tab** returns the appropriate unknown nine-state value, given the strength of the input. These constants could have been implemented as **IF** statements or **CASE** statements, but constant arrays are much more efficient.

In the nine-value system, there are three values at the lowest strength level, so the variable result has to be initialized more carefully to predict correct results. If there are no drivers, the range attribute of argument **s** returns 0, and the default value (**zx**) is returned.

Let's look at a few examples of driver-input arguments and see what the resolution function predicts. An example of two drivers is shown in Figure 5-4.

This example contains two driver values, **z1** and **R0**. Variable **result** is initialized to the first driver value, and the loop executes as many times as there are drivers. The first time through the loop, **result** equals **z1** and the first driver equals **z1**. Variable **result** remains at **z1** because the values are equal. The next time through the loop, variable **result** contains **z1**, and the second driver contains **R0**. The constant **get_strength** returns strength **R**. The constant **get_strength** for variable **result** returns strength **z**. Strength **R** is lexically greater than strength **z**. This is because value **R** has a higher position number than **z**, because **R** is listed after **z** in the type declaration for type strength. The fact that the new driver has

Figure 5-4

Nine State Resolution with Two Values.

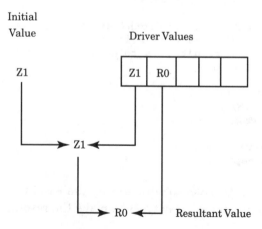

Initial Value

Driver Values

Resultant Value

a stronger strength value than variable **result** causes variable **result** to be updated with the stronger value, **R0**.

Another example shows how the constant **x_tab** is used to predict the correct value for conflicting inputs. The driver values are shown in the array in Figure 5-5.

In this example, variable **result** is initialized to **F0**. The first iteration of the loop does nothing because the first driver and the result-initialization value are the same value. The next iteration starts with variable **result** containing the value **F0**, and the next driver value as **R0**. Because the value in variable **result** is greater in strength than the value of the new driver, no action is implemented, except to advance the loop to the next driver.

The last driver contains the value **F1**. The strength of the value contained in variable **result** and the new driver value are the same. Therefore, the **IF** statement checking this condition is executed and succeeds. The next **IF** statement checks to see if the logical values are the same for both variable **result** and the new driver. Variable **result** contains an **F0**, and the new driver value contains an **F1**. The values are not the same, and the **x_tab** table is used to return the correct unknown value for the strength of the driver values. The **x_tab** table returns the value **FX**, which is returned as the resolved value.

A more efficient method to implement the loop would be to skip the first iteration where the first driver is compared to itself, because the value in variable **result** is initialized to the first driver value. It is left as an exercise to the reader to write this new loop iteration mechanism.

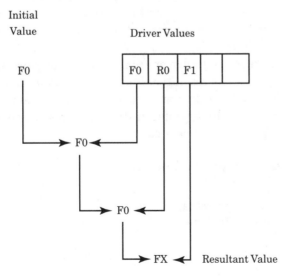

Figure 5-5
Nine State Resolution with Three Values.

Initial Value

Driver Values

FX ◄— Resultant Value

Although VHDL simulators can support any type of resolution that can be legally written in the language, synthesis tools can only support a subset. The reason stems from the fact that the synthesis tools must build actual hardware from the VHDL description. If the Resolution Function maps into a common hardware behavior such as wired-or or wired-and, then most synthesis tools allow the user the ability to tag the resolution function appropriately. For instance, a Resolution Function that performs a wired-or function is tagged with an attribute that tells the synthesis tools to connect the outputs together.

COMPOSITE TYPE RESOLUTION   For simple signal values such as the **nineval** and **fourval** types, it is easy to see how to create the resolution function. But for signals of composite types, it is not so obvious. How can one value of a composite type be stronger than another?

The answer is that one value must be designated as weaker than all of the other values. Then the principle is the same as any other type being resolved. In the **fourval** type, the value **z** was considered the weakest state, and any of the other values could overwrite this value. In the **nineval** type, all values with a strength of **z** could be overridden by values with a strength of **R** or **F**, and all values with strength **R** could be overridden by strength **F**.

To resolve a composite type, designate one value of the composite type as unusable except to indicate that the signal is not currently being driven. The resolution function checks how many drivers have this value and how many drivers have a driving value. If only one driving value exists, then the resolution function can return this value as the resolved value. If more than one driving value is present, then an error condition probably exists and the resolution function can announce the error.

A typical application for a Composite Type Resolution Function is shown in Figure 5-6.

Signal **XBUS** can be driven from a number of sources, but hopefully only one at a time. The resolution function must determine how many drivers are trying to drive **XBUS** and return the correct value for the signal.

Following is the type declarations and resolution function for a composite type used in such a circuit:

```
PACKAGE composite_res IS
  TYPE xtype IS
  RECORD
    addr : INTEGER;
    data : INTEGER;
```

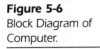

Figure 5-6
Block Diagram of
Computer.

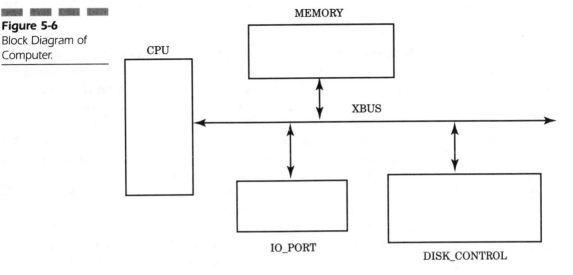

```
END RECORD;

TYPE xtypevector IS ARRAY( natural RANGE <>) OF xtype;
CONSTANT notdriven : xtype := (-1,-1);

FUNCTION cresolve( t : xtypevector) RETURN xtype;
END composite_res;

PACKAGE BODY composite_res IS
 FUNCTION cresolve( t : xtypevector) RETURN xtype IS
  VARIABLE result : xtype := notdriven;
  VARIABLE drive_count : INTEGER := 0;
 BEGIN
  IF t'LENGTH = 0 THEN RETURN notdriven;
  END IF;

  FOR i IN t'RANGE LOOP
   IF t(i) /= notdriven THEN
     drive_count := drive_count + 1;
     IF drive_count = 1 THEN
       result := t(i);
     ELSE
       result := notdriven;
       ASSERT FALSE
         REPORT "multiple drivers detected"
         SEVERITY ERROR;
     END IF;
   END IF;
  END LOOP;
  RETURN result;
```

```
END cresolve;
END composite_res;
```

Type **xtype** declares the record type for signal **xbus**. Type **xtypevector** is an unconstrained array type of **xtype** values used for the resolution function input argument **t**. Constant **notdriven** declares the value of the record that is used to signify that a signal driver is not driving. Negative number values were used to represent the notdriven state because, in this example, only positive values are used in the **addr** and **data** fields. But what happens if all of the values must be used for a particular type? The easiest solution is probably to declare a new type which is a record, containing the original type as one field of the record, and a new field which is a boolean that determines whether the driver is driving or not driving.

In this example, resolution function **cresolve** first checks to make certain that at least one driver value is passed in argument **t** (drivers can be turned off using guarded signal assignment). If at least one driver is driving, the loop statement loops through all driver values, looking for driving values. If a driving value is detected, and it is the first, then this value is assumed to be the output resolved value, until proven otherwise. If only one driving value occurs, that value is returned as the resolved value.

If a second driving value appears, the output is set to the nondriven value, signifying that the outcome is uncertain, and the **ASSERT** statement writes out an error message to that effect.

In this example, the negative numbers of the integer type were not used except to indicate whether the signal was driving or not. We reserved one value to indicate this condition. Another value could be reserved to indicate the multiply-driven case such that when multiple drivers are detected on the signal, this value would be returned as the resolved value. An example might look like this:

```
CONSTANT multiple_drive : xtype := (-2,-2);
```

This constant provides the capability of distinguishing between a nondriven signal and a multiply-driven signal.

RESOLVED SIGNALS So far we have discussed how to write resolution functions that can resolve signals of multiple drivers, but we have not discussed how all of the appropriate declarations are structured to accomplish this.

Resolved signals are created using one of two methods. The first is to create a resolved subtype and declare a signal using this type. The second is to declare a signal specifying a resolution function as part of the signal declaration.

Let's discuss the resolved subtype method first. To create a resolved sub-type, the designer declares the base type, then declares the subtype speci-fying the resolution function to use for this type. An example looks like this:

```
TYPE fourval IS (X, L, H, Z);        -- won't compile
SUBTYPE resfour IS resolve fourval; -- as is
```

The first declaration declares the enumerated type **fourval**. The sec-ond declaration is used to declare a subtype named **resfour**, which uses a resolution function named **resolve** to resolve the base type **fourval**. This syntax does not compile as is because the function **resolve** is not vis-ible. To declare a resolved subtype requires a very specific combination of statements, in a very specific ordering.

Following is a correct example of the resolved type:

```
PACKAGE fourpack IS
 TYPE fourval IS (X, L, H, Z); -- line 1
 TYPE fourvalvector IS ARRAY(natural RANGE <>)
   OF fourval;       -- line 2

 FUNCTION resolve( s: fourvalvector) RETURN fourval;
      -- line 3

 SUBTYPE resfour IS resolve fourval; -- line 4
END fourpack;
```

The statement in line 2 declares an unconstrained array of the base type that is used to contain the driver values passed to the resolution function. The statement in line 3 declares the definition of the resolution function resolve so that the subtype declaration can make use of it. The body of the resolution function is implemented in the package body. Fi-nally, the statement in line 4 declares the resolved subtype using the base type and the resolution function declaration.

The order of the statements is important, because each statement de-clares something that is used in the next statement. If the unconstrained array declaration is left out, the resolution function could not be declared, and if the resolution function was not declared, the subtype could not be declared.

The second method of obtaining a resolved signal is to specify the res-olution function in the signal declaration. In the following example, a sig-nal is declared using the resolution function **resolve**:

```
PACKAGE fourpack IS
  TYPE fourval IS (X, L, H, Z);
```

```
    TYPE fourvalvector IS ARRAY(natural RANGE <>) OF fourval;

    FUNCTION resolve( s: fourvalvector) RETURN fourval;
    SUBTYPE resfour IS resolve fourval;
END fourpack;

USE WORK.fourpack.ALL;
ENTITY mux2 IS
  PORT( i1, i2, a : IN fourval;
     q : OUT fourval);
END mux2;

ARCHITECTURE different OF mux2 IS
  COMPONENT and2
    PORT( a, b : IN fourval;
       c : OUT fourval);
  END COMPONENT;

  COMPONENT inv
    PORT( a : IN fourval;
       b : OUT fourval);
  END COMPONENT;

  SIGNAL nota : fourval;

  -- resolved signal
  SIGNAL intq : resolve fourval := X;

BEGIN

  U1: inv PORT MAP(a, nota);

  U2: and2 PORT MAP(i1, a, intq);

  U3: and2 PORT MAP(i2, nota, intq);

  q <= intq;

END different;
```

The package **fourpack** declares all of the appropriate types and function declarations so that the resolution function **resolve** is visible in the entity. In the architecture declaration section, signal **intq** is declared of type **fourval**, using the resolution function **resolve**. This signal is also given an initial value of **x**.

Signal **intq** is required to have a resolution function because it is the output signal for components **U2** and **U3**. Each component provides a driver to signal **intq**. Resolution function **resolve** is used to determine the end result of the two driver values. Signal **nota** is not required to have a resolution function because it only has one driver, component **U1**.

Procedures

In the earlier section describing functions, we discussed how functions can have a number of input parameters and always return one value. In contrast, procedures can have any number of in, out, and inout parameters. A procedure call is considered a statement of its own; a function usually exists as part of an expression. The most usual case of using a procedure is when more than one value is returned.

Procedures have basically the same syntax and rules as functions. A procedure declaration begins with the keyword **PROCEDURE**, followed by the procedure name, and then an argument list. The main difference between a function and a procedure is that the procedure argument list most likely has a direction associated with each parameter; the function argument list does not. In a procedure, some of the arguments can be mode **IN**, **OUT**, or **INOUT**; in a function, all arguments are of mode **IN** by default and can only be of mode **IN**.

A typical example where a procedure is very useful is during the conversion from an array of a multivalued type to an integer. A procedure showing an example of how to accomplish this is shown here:

```
USE LIBRARY IEEE;
USE IEEE.std_logic_1164.ALL;
PROCEDURE vector_to_int (z : IN std_logic_vector;
      x_flag : OUT BOOLEAN; q : INOUT INTEGER) IS
BEGIN
  q := 0;
  x_flag := false;

  FOR i IN z'RANGE LOOP
    q := q * 2;

  IF z(i) = '1' THEN
    q := q + 1;
  ELSIF z(i) /= F0 THEN
    x_flag := TRUE;
  END IF;
 END LOOP;
END vector_to_int;
```

The behavior of this procedure is to convert the input argument **z** from an array of a type to an integer. However, if the input array has unknown values contained in it, an integer value cannot be generated from the array. When this condition occurs, output argument **x_flag** is set to true, indicating that the output integer value is unknown. A procedure was required to implement this behavior because more than one output value results from the procedure. Let's examine what the result from the procedure is from the input array value shown here:

```
`0'  `0'  `1'  `1'
```

The first step for the procedure is to initialize the output values to known conditions, in case a zero length input argument is passed in. Output argument **x_flag** is initialized to false and stays false until proven otherwise.

The loop statement loops through the input vector **z** and progressively adds each value of the vector until all values have been added. If the value is a `1', then it is added to the result. If the value is a `0', then no addition is done. If any other value is found in the vector, the **x_flag** result is set true, indicating that an unknown condition was found on one of the inputs. (Notice that parameter **q** is defined as an inout parameter. This is needed because the value is read in the procedure.)

PROCEDURE WITH INOUT PARAMETERS   The examples we have discussed so far have dealt mostly with in and out parameters, but procedures can have inout parameters also. The next example shows a procedure that has an inout argument that is a record type. The record contains an array of eight integers, along with a field used to hold the average of all of the integers. The procedure calculates the average of the integer values, writes the average in the average field of the record, and returns the updated record:

```
PACKAGE intpack IS
  TYPE bus_stat_vec IS ARRAY(0 to 7) OF INTEGER;
  TYPE bus_stat_t IS
   RECORD
     bus_val: bus_stat_vec;
     average_val : INTEGER;
   END RECORD;

  PROCEDURE bus_average( x : inout bus_stat_t );

END intpack;

PACKAGE BODY intpack IS
  PROCEDURE bus_average( x : inout bus_stat_t ) IS
   VARIABLE total : INTEGER := 0;
  BEGIN
   FOR i IN 0 TO 7 LOOP
     total := total + x.bus_val(i);
   END LOOP;
   x.average_val := total / 8;
```

```
    END bus_average;
END intpack;
```

A process calling the procedure might look as shown below:

```
PROCESS( mem_update )
  VARIABLE bus_statistics : bus_stat_t;
BEGIN
  bus_statistics.bus_val :=
    (50, 40, 30, 35, 45, 55, 65, 85 );

  bus_average(bus_statistics);
  average <= bus_statistics.average_val;

END PROCESS;
```

The variable assignment to **bus_statistics.bus_val** fills in the appropriate bus utilization values to be used for the calculation. The next line is the call to the **bus_average** procedure, which performs the averaging calculation. Initially, the argument to the **bus_average** procedure is an input value, but after the procedure has finished, the argument becomes an output value that can be used inside the calling process. The output value from the procedure is assigned to an output signal in the last line of the process.

SIDE EFFECTS Procedures have an interesting problem that is not shared by their function counterparts. Procedures can cause side effects to occur. A side effect is the result of changing the value of an object inside a procedure when that object was not an argument to the procedure. For instance, a signal of an architecture can be assigned a value from within a procedure, without that signal being an argument passed into the procedure. For instance, if two signals are not declared in the argument list of a procedure, but are assigned from within a procedure called from the current procedure, any assignments to these signals are side effects.

This is not a recommended method for writing a model. The debugging and maintenance of a model of this type can be very difficult. This feature was presented so the reader would understand the behavior if such a model were examined.

Packages

The primary purpose of a package is to encapsulate elements that can be shared (globally) among two or more design units. A package is a common storage area used to hold data to be shared among a number of entities.

Declaring data inside of a package allows the data to be referenced by other entities; thus, the data can be shared.

A package consists of two parts: a package declaration section and a package body. The package declaration defines the interface for the package, much the same way that the entity defines the interface for a model. The package body specifies the actual behavior of the package in the same method that the architecture statement does for a model.

Package Declaration

The package declaration section can contain the following declarations:

- Subprogram declaration
- Type, subtype declaration
- Constant, deferred constant declaration
- Signal declaration creates a global signal
- File declaration
- Alias declaration
- Component declaration
- Attribute declaration, a user-defined attribute (Chapter 8, "Advanced Topics")
- Attribute specification
- Disconnection specification
- Use clause

All of the items declared in the package declaration section are visible to any design unit that uses the package with a **USE** clause. The interface to a package consists of any subprograms or deferred constants declared in the package declaration. The subprogram and deferred constant declarations must have a corresponding subprogram body and deferred constant value in the package body or an error results.

Deferred Constants

Deferred constants are constants that have their name and type declared in the package declaration section but have the actual value specified in the package body section. Following is an example of a deferred constant in the package declaration:

```
PACKAGE tpack IS
  CONSTANT timing_mode : t_mode;
END tpack;
```

This example shows a deferred constant called **timing_mode** being defined as type **t_mode**. The actual value of the constant is specified when the package body for package **tpack** is compiled. This feature allows late binding of the value of a constant so that the value of the constant can be specified at the last possible moment and can be changed easily. Any design unit that uses a deferred constant from the package declaration need not be recompiled if the value of the constant is changed in the package body. Only the package body needs to be recompiled.

Subprogram Declaration

The other item that forms the interface to the package is the subprogram declaration. A subprogram declaration allows the designer to specify the interface to a subprogram separately from the subprogram body. This functionality allows any designers using the subprogram to start or continue with the design, while the specification of the internals of the subprograms are detailed. It also gives the designer of the subprogram bodies freedom to change the internal workings of the subprograms, without affecting any designs that use the subprograms. Following is an example of a subprogram declaration:

```
PACKAGE cluspack IS
  TYPE nineval IS (Z0, Z1, ZX,
                   R0, R1, RX,
                   F0, F1, FX );
  TYPE t_cluster IS ARRAY(0 to 15) OF nineval;
  TYPE t_clus_vec IS ARRAY(natural range <>) OF t_cluster;

  FUNCTION resolve_cluster( s: t_clus_vec )
     RETURN t_cluster;

  SUBTYPE t_wclus IS resolve_cluster t_cluster;
  CONSTANT undriven : t_wclus;

END cluspack;
```

The subprogram declaration for **resolve_cluster** specifies the name of the subprogram, any arguments to the subprogram, their types and modes, and the return type if the subprogram is a function. This declaration can be used to compile any models that intend to use it, without the actual subprogram body specified yet. The subprogram body must exist before the simulator is built, during elaboration.

Package Body

The main purpose of the package body is to define the values for deferred constants and specify the subprogram bodies for any subprogram declarations from the package declaration. However, the package body can also contain the following declarations:

- Subprogram declaration
- Subprogram body
- Type, subtype declaration
- Constant declaration, which fills in the value for the deferred constant
- File declaration
- Alias declaration
- Use clause

All of the declarations in the package body, except for the constant declaration that is specifying the value of a deferred constant and the subprogram body declaration, are local to the package body.

Let's examine a package body for the package declaration that was discussed in the last section:

```
PACKAGE BODY cluspack IS
  CONSTANT undriven : t_wclus :=
        (ZX,  ZX,  ZX,  ZX,
         ZX,  ZX,  ZX,  ZX,
         ZX,  ZX,  ZX,  ZX,
         ZX,  ZX,  ZX,  ZX);

  FUNCTION resolve_cluster ( s: t_clus_vec )
     return t_cluster IS
    VARIABLE result : t_cluster;
    VARIABLE drive_count : INTEGER;
  BEGIN
    IF s'LENGTH = 0 THEN RETURN undriven;
    END IF;
    FOR i in s'RANGE LOOP
      IF s(i) /= undriven THEN
        drive_count := drive_count + 1;
      IF drive_count = 1 THEN
```

```
                result := s(i);
            ELSE
                result := undriven;
            ASSERT FALSE
                REPORT "multiple drivers detected"
                SEVERITY ERROR;
            END IF;
         END IF;
      END LOOP;
   RETURN result;
  END resolve_cluster;
END cluspack;
```

The package body statement is very similar to the package declaration, except for the keyword **BODY** after package. The contents of the two design units are very different, however. This package body example contains only two items: the deferred constant value for deferred constant **undriven** and the subprogram body for subprogram **resolve_cluster**. Notice how the deferred constant value specification matches the deferred constant declaration in the package declaration, and the subprogram body matches the subprogram declaration in the package declaration. The subprogram body must match the subprogram declaration exactly in the number of parameters, the type of parameters, and the return type.

A package body can also contain local declarations that are used only within the package body to create other subprogram bodies, or deferred constant values. These declarations are not visible outside of the package body but can be very useful within the package body. Following is an example of a complete package making use of this feature:

```
USE LIBRARY IEEE;
USE IEEE.std_logic_1164.ALL;
PACKAGE math IS
   TYPE st16 IS ARRAY(0 TO 15) OF std_logic;

   FUNCTION add(a, b: IN st16) RETURN st16;
   FUNCTION sub(a, b: IN st16) RETURN st16;

END math;

PACKAGE BODY math IS

   FUNCTION vect_to_int(S : st16) RETURN INTEGER IS
     VARIABLE result : INTEGER := 0;
   BEGIN
     FOR i IN 0 TO 7 LOOP
       result := result * 2;

       IF S(i) = '1' THEN
```

```
              result := result + 1;
          END IF;
      END LOOP;

    RETURN result;
  END vect_to_int;

  FUNCTION int_to_st16(s : INTEGER) RETURN st16 IS
    VARIABLE result : st16;
    VARIABLE digit : INTEGER := 2**15;
    VARIABLE local : INTEGER;
  BEGIN
    local : = s;
    FOR i IN 15 DOWNTO 0 LOOP
      IF local/digit >>= 1 THEN
        result(i) := '1';
        local := local - digit;
      ELSE
        result(i) := '0';
      END IF;

      digit := digit/2;

    END LOOP;
    RETURN result;
  END int_to_st16;

  FUNCTION add(a, b: IN st16) RETURN st16 IS
    VARIABLE result : INTEGER;
  BEGIN
    result := vect_to_int(a) + vect_to_int(b);
    RETURN int_to_st16(result);
  END add;

  FUNCTION sub(a, b: IN st16) RETURN st16 IS
    VARIABLE result : INTEGER;
  BEGIN
    result := vect_to_int(a) - vect_to_int(b);
    RETURN int_to_st16(result);
  END sub;

END math;
```

The package declaration declares a type **st16** and two functions, **add** and **sub**, that work with this type. The package body has function bodies for function declarations **add** and **sub** and also includes two functions that are only used in the package body. These functions are **int_to_st16** and **vect_to_int**. These functions are not visible outside of the package body. To make these functions visible, a function declaration would need to be added to the package declaration, for each function.

Functions `vect_to_int` and `int_to_st16` must be declared ahead of function **add** to compile correctly. All functions must be declared before they are used to compile correctly.

SUMMARY

In this chapter, we discussed the different kinds of subprograms and some of the uses for them. Specifically, we covered the following:

- How subprograms consist of functions and procedures. Functions have only input parameters and a single return value; procedures can have any number of in, out, and inout parameters.
- How functions can be used as conversion functions to convert from one type to another.
- How functions can be used as resolution functions to calculate the proper value on a multiply driven network.
- How procedures are considered statements; functions are usually part of an expression. Procedures can exist alone; functions are usually called as part of a statement.
- How packages are used to encapsulate information that is to be shared among multiple design units.
- How packages consist of a package declaration in which all of the type, subprogram, and other declarations exist and a package body in which subprogram bodies and deferred constants exist.

In the next chapter, we discuss how attributes can make some descriptions easier to read and more compact.

Predefined Attributes

This chapter discusses VHDL predefined attributes and the way that concise readable models can be written using attributes. Predefined attributes are data that can be obtained from blocks, signals, and types or subtypes. The data obtained falls into one of the following categories shown:

- *Value kind*—A simple value is returned.
- *Function kind*—A function call is performed to return a value.
- *Signal kind*—A new signal is created whose value is derived from another signal.
- *Type kind*—A type mark is returned.
- *Range kind*—A range value is returned.

Predefined attributes have a number of very important applications. Attributes can be used to detect clock edges, perform timing checks in concert with **ASSERT** statements, return range information about unconstrained types, and much more. All of these applications are examined in this chapter. First, we discuss each of the predefined attribute kinds and the ways that these attributes can be applied to modeling.

Value Kind Attributes

Value attributes are used to return a particular value about an array of a type, a block, or a type in general. Value attributes can be used to return the length of an array or the lowest bound of a type. Value attributes can be further broken down into three subclasses:

- Value type attributes, which return the bounds of a type
- Value array attributes, which return the length of an array
- Value block attributes, which return block information

Value Type Attributes

Value type attributes are used to return the bounds of a type. For instance, a type defined as shown in the following would have a low bound of 0 and a high bound of 7:

```
TYPE state IS (0 TO 7);
```

There are four predefined attributes in the value type attribute category:

- **T'LEFT**, which returns the left bound of a type or subtype
- **T'RIGHT**, which returns the right bound of a type or subtype
- **T'HIGH**, which returns the upper bound of a type or subtype
- **T'LOW**, which returns the lower bound of a type or subtype

Attributes are specified by the character ′ and then the attribute name. The object preceding the ′ is the object that the attribute is attached to. The capital **T** in the preceding descriptions means that the object that the attribute is attached to is a type. The ′ character is pronounced "tick" among VHDL hackers. Therefore, the first attribute in the preceding list is specified "T tick left."

The left bound of a type or subtype is the leftmost entry of the range constraint. The right bound is the rightmost entry of the type or subtype. In the following example, the left bound is -32,767, and the right bound is 32,767:

```
TYPE smallint IS -32767 TO 32767;
```

The upper bound of a type or subtype is the bound with the largest value, and the lower bound is the bound with the lowest value. In the preceding example, for the type **smallint**, the upper bound is 32,767, and the lower bound is -32,767.

To use one of these value attributes, the type mark name is followed by the attribute desired. For example, following is the syntax to return the left bound of a type:

```
PROCESS(x)
 SUBTYPE smallreal IS REAL RANGE -1.0E6 TO 1.0E6;
 VARIABLE q : real;
BEGIN
 q := smallreal'LEFT;
 -- use of 'left returns
 -- -1.0E6
END test;
```

In this example, variable **q** is assigned the left bound of type **smallreal**. Variable **q** must have the same type as the bounds of the type for the assignment to occur. (The assignment could also occur if variable **q** was cast into the appropriate type.) After the assignment has occurred, variable **q** contains -1.0E6, which is the left bound of type **smallreal**.

In the next example, all of the attributes are used to show what happens when a **DOWNTO** range is used for a type:

```
PROCESS(a)
 TYPE bit_range IS ARRAY(31 DOWNTO 0) OF BIT;
 VARIABLE left_range, right_range, uprange, lowrange :
     integer;
BEGIN
 left_range  := bit_range'LEFT;
 -- returns 31

 right_range := bit_range'RIGHT;
 -- returns 0

 uprange     := bit_range'HIGH;
 -- returns 31
```

```
    lowrange     := bit_range'LOW;
    -- returns 0
END PROCESS;
```

This example shows how the different attributes can be used to return information about a type. When ranges of a type are defined using **(a TO b)** where b > a, the **'LEFT** attribute will always equal the **'LOW** attribute; but when a range specification using **(b DOWNTO a)** where b > a is used, the **'HIGH** and **'LOW** can be used to determine the upper and lower bounds of the type.

Value type attributes are not restricted to numeric types. These attributes can also be used with any scalar type. Following is an example using enumerated types:

```
ARCHITECTURE b OF a IS
  TYPE color IS (blue, cyan, green, yellow, red, magenta);
  SUBTYPE reverse_color IS color RANGE red DOWNTO green;
  SIGNAL color1, color2, color3,
      color4, color5, color6,
      color7, color8 : color;
BEGIN

  color1 <= color'LEFT;     -- returns blue
  color2 <= color'RIGHT;    -- returns magenta

  color3 <= color'HIGH;     -- returns magenta
  color4 <= color'LOW;      -- returns blue

  color5 <= reverse_color'LEFT;
  -- returns red

  color6 <= reverse_color'RIGHT;
  -- returns green

· color7 <= reverse_color'HIGH;
  -- returns red

  color8 <= reverse_color'LOW;
  -- returns green
END b;
```

This example illustrates how value type attributes can be used with enumerated types to return information about the type. Signals **color1** and **color2** are assigned **blue** and **magenta**, respectively, the left and right bounds of the type. It is easy to see how these values are obtained by examining the declaration of the type. The left bound of the type is **blue** and the right bound is **magenta**. What is returned for the **'HIGH** and **'LOW** attributes of an enumerated type? The answer relates to the position numbers of the type. For an integer and real type, the position numbers of a value are equal to the value itself; but for an enumerated type, the position numbers of a value are determined by the declaration of the type. Values declared earlier have lower position numbers than values declared later. Value **blue** from the preceding example has a position number of 0, because it is the first value of the type. Value **cyan** has a position number 1, **green** has 2, and so on. From these position numbers, the high and low bounds of the type can be found.

Signals **color5** through **color8** are assigned attributes of the type **reverse_color**. This type has a DOWNTO range specification. Attributes **'HIGH** and **'RIGHT** do not return the same value because the range is reversed. Value **red** has a higher position number than value **green**, and therefore a DOWNTO is needed for the range specification.

Value Array Attributes

There is only one value array attribute: **'LENGTH**. Given an array type, this attribute returns the total length of the array range specified. This attribute works with array ranges of any scalar type and with multi-dimensional arrays of scalar-type ranges. Following is a simple example:

```
PROCESS(a)
  TYPE bit4 IS ARRAY(0 TO 3) of BIT;
  TYPE bit_strange IS ARRAY(10 TO 20) OF BIT;
  VARIABLE len1, len2 : INTEGER;
BEGIN
  len1 := bit4'LENGTH;          -- returns 4
  len2 := bit_strange'LENGTH; -- returns 11
END PROCESS;
```

The assignment to **len1** assigns the value of the number of elements in array type **bit4**. The assignment to **len2** assigns the value of the number of elements of type **bit_strange**.

This attribute also works with enumerated-type ranges, as shown by the following example:

```
PACKAGE p_4val IS
  TYPE t_4val IS ('x', '0', '1', 'z');
  TYPE t_4valX1 IS ARRAY(t_4val'LOW TO t_4val'HIGH) OF
      t_4val;

  TYPE t_4valX2 IS ARRAY(t_4val'LOW TO t_4val'HIGH) OF
      t_4valX1;
```

```
TYPE t_4valmd IS ARRAY(t_4val'LOW TO t_4val'HIGH,
               t_4val'LOW TO t_4val'HIGH) OF t_4val;

CONSTANT andsd : t_4valX2 :=
    (('x',      -- xx
     '0',       -- x0
     'x',       -- x1
     'x'),      -- xz        (Notice this is an
    ('0',       -- 0x         array of arrays.)
     '0',       -- 00
     '0',       -- 01
     '0'),      -- 0z
    ('x',       -- 1x
     '0',       -- 10
     '1',       -- 11
     'x'),      -- 1z
    ('x',       -- zx
     '0',       -- z0
     'x',       -- z1
     'x'));     -- zz

CONSTANT andmd : t_4valmd :=
    (('x',      -- xx
     '0',       -- x0
     'x',       -- x1
     'x'),      -- xz
    ('0',       -- 0x        (Notice this example
     '0',       -- 00         is a multidimensional
     '0',       -- 01         array.)
     '0'),      -- 0z
    ('x',       -- 1x
     '0',       -- 10
     '1',       -- 11
     'x'),      -- 1z
    ('x',       -- zx
     '0',       -- z0
     'x',       -- z1
     'x'));     -- zz
END p_4val;
```

The two composite type constants, **andsd** and **andmd**, provide a lookup table for an **AND** function of type **t_4val**. The first constant **andsd** uses an array of array values, while the second constant **andmd** uses a multi-dimensional array to store the values. The initialization of both constants is specified by the same syntax. If the **'LENGTH** attribute is applied to these types as shown in the following, the results shown in the VHDL comments are obtained:

```
PROCESS(a)
 VARIABLE len1, len2, len3, len4 : INTEGER;
BEGIN
```

```
len1 := t_4valX1'LENGTH;        -- returns 4
len2 := t_4valX2'LENGTH;        -- returns 4

len3 := t_4valmd'LENGTH(1);     -- returns 4
len4 := t_4valmd'LENGTH(2);     -- returns 4
END PROCESS;
```

Type t_4valX1 is a four-element array of type t_4val. The range of the array is specified using the predefined attributes 'LOW and 'HIGH of the t_4val type. Assigning the length of type t_4valX1 to len1 returns the value 4, the number of elements in array type t_4valX1. The assignment to len2 also returns the value 4, because the range of type t_valX2 is from 'LOW to 'HIGH of element type t_4valX1.

The assignments to len3 and len4 make use of a multidimensional array type t_4valmd. Because a multidimensional array has more than one range, an argument is used to specify a particular range. The range defaults to the first range, if none is specified. In the type t_4valmd example, the designer can pick the first or second range, because there are only two to choose from. To pick a range, the argument passed to the attribute specifies the number of the range starting at 1. An argument value of 1 picks the first range, an argument value of 2 picks the second range, and so on.

The assignment to len3 in the previous example passed in the value 1 to pick the first range. The first range is from t_4val'LOW to t_4val'HIGH, or four entries. The second range is exactly the same as the first; therefore, both assignments return 4 as the length of the array.

If the argument to 'LENGTH is not specified, it defaults to 1. This was the case in the first examples of 'LENGTH, when no argument was specified. There was only one range, so the correct range was selected.

Value Block Attributes

There are two attributes that form the set of attributes that work with blocks and architectures. Attributes 'STRUCTURE and 'BEHAVIOR return information about how a block in a design is modeled. Attribute 'BEHAVIOR returns true if the block specified by the block label, or architecture specified by the architecture name, contains no component instantiation statements. Attribute 'STRUCTURE returns true if the block or architecture contains only component instantiation statements and/or passive processes.

The following two examples illustrate how these attributes work. The first example contains only structural VHDL:

```
LIBRARY IEEE;
USE IEEE.std_logic_1164.ALL;
ENTITY shifter IS
 PORT( clk, left : IN std_logic;
     right : OUT std_logic);
END shifter;

ARCHITECTURE structural OF shifter IS
 COMPONENT dff
  PORT( d, clk : IN std_logic;
    q : OUT std_logic);
 END COMPONENT;

 SIGNAL i1, i2, i3: std_logic;

BEGIN

 u1: dff PORT MAP(d => left, clk => clk, q => i1);

 u2: dff PORT MAP(d => i1, clk => clk, q => i2);

 u3: dff PORT MAP(d => i2, clk => clk, q => i3);

 u4: dff PORT MAP(d => i3, clk => clk, q => right);

 checktime: PROCESS(clk)
  VARIABLE last_time : time := time'left;
 BEGIN
  ASSERT (NOW - last_time = 20 ns)
   REPORT "spike on clock"
   SEVERITY WARNING;
  last_time := now;
 END PROCESS checktime;
END structural;
```

The preceding example is a shift register modeled using four **dff** components connected in series. A passive process statement exists in the architecture for entity **shifter**, used to detect spikes on the **clk** input. The following example shows the results of the attributes for the architecture **structural**:

```
structural'BEHAVIOR: returns false

structural'STRUCTURE: returns true
```

The passive process **checktime** has no effect on the fact that the architecture is structural. If the process contained signal assignment statements, then the process would no longer be considered passive, and attribute **'STRUCTURE** would also return false.

For any block or architecture that does not contain any component instantiation statements, attribute **'BEHAVIOR** is true, and attribute **'STRUCTURE** is false. For blocks or architectures that mix structure and behavior, both attributes return false.

Function Kind Attributes

Function attributes return information to the designer about types, arrays, and signals. When a function kind attribute is used in an expression, a function call occurs that uses the value of the input argument to return a value. The value returned can be a position number of an enumerated value, an indication of whether a signal has changed this delta, or one of the bounds of an array.

Function attributes can be subdivided into three general classifications:

- Function type attributes, which return type values
- Function array attributes, which return array bounds
- Function signal attributes, which return signal history information

Function Type Attributes

Function type attributes return particular information about a type. Given the position number of a value within a type, the value can be returned. Also values to the left or right of an input value of a particular type can be returned.

Function type attributes are one of the following:

- **'POS** (value), which returns position number of value passed in
- **'VAL** (value), which returns value from position number passed in
- **'SUCC** (value), which returns next value in type after input value
- **'PRED** (value), which returns previous value in type before input value
- **'LEFTOF** (value), which returns value immediately to the left of the input value

■ `'RIGHTOF` (value), which returns value immediately to the right of the input value

A typical use of a function type attribute is to convert from an enumerated or physical type to an integer type. Following is an example of conversion from a physical type to an integer type:

```
PACKAGE ohms_law IS
  TYPE current IS RANGE 0 TO 1000000
    UNITS
     ua;                    -- micro amps
     ma = 1000 ua;      -- milli amps
     a  = 1000 ma;      -- amps
    END UNITS;

  TYPE voltage IS RANGE 0 TO 1000000
    UNITS
     uv;                    -- micro volts
     mv = 1000 uv;      -- milli volts
     v  = 1000 mv;      -- volts
    END UNITS;

  TYPE resistance IS RANGE 0 TO 100000000
    UNITS
     ohm;                  -- ohms
     Kohm = 1000 ohm;  -- kilo ohms
     Mohm = 1000 Kohm;-- mega ohms
    END UNITS;
END ohms_law;

use work.ohms_law.all;
ENTITY calc_resistance IS
  PORT( i : IN current; e : IN voltage;
      r : OUT resistance);
END calc_resistance;

ARCHITECTURE  behave OF calc_resistance IS
BEGIN
  ohm_proc: PROCESS( i, e )
   VARIABLE convi, conve, int_r : integer;
  BEGIN
    convi := current'POS(i); -- current in ua
    conve := voltage'POS(e); -- voltage in uv

    -- resistance in ohms
    int_r := conve / convi;

    r <= resistance'VAL(int_r);

    -- another way to write this example
    -- is shown below
    -- r <=resistance'VAL(current'POS(i)
```

```
          -- / voltage'POS(e));

     END PROCESS;
     END behave;
```

Package `ohms_law` declares three physical types used in this example. Types `current`, `voltage`, and `resistance` are used to show how physical types can be converted to type `INTEGER` and back to a physical type.

Whenever ports `i` or `e` have an event occur on them, process `ohm_proc` is invoked and calculates a new value of resistance (`r`) from the current (`i`) and the voltage (`e`). Variables `conve`, `convi`, and `int_r` were not necessary in this example but were added for ease of understanding. The commented-out assignment to output `r` shows an example where the internal variables are not needed.

The first statement of the process assigns the position number of the input value to variable `convi`. If the input value is 10 ua, then 10 is assigned to variable `convi`.

The second statement assigns the position number of the value of input `e` to variable `conve`. The base unit of type voltage is uv (microvolts); therefore, the position number of any voltage value is determined based on how many uv the input value is equal to.

The last line in the process converts the resistance value calculated from the previous line to the appropriate ohms value in type `resistance`. The `'VAL` attribute is used to convert a position number to a physical type value of type `resistance`.

The preceding example illustrates how `'POS` and `'VAL` work, but not `'SUCC`, `'PRED`, `'RIGHTOF`, and `'LEFTOF`. Following is a very simple example using these attributes:

```
PACKAGE p_color IS
  TYPE color IS ( red, yellow, green, blue, purple,
                  orange );

  SUBTYPE reverse_color is color RANGE orange downto red ;

END p_color;
```

Assuming the preceding types, the following results are obtained:

- `color'SUCC` (blue) returns purple.
- `color'PRED` (green) returns yellow.
- `reverse_color'SUCC` (blue) returns green.
- `reverse_color'PRED` (green) returns blue.

- `color'RIGHTOF` (blue) returns purple.
- `color'LEFTOF` (green) returns yellow.
- `reverse_color'RIGHTOF` (blue) returns green.
- `reverse_color'LEFTOF` (green) returns blue.

For ascending ranges, the following is true:

```
'SUCC(x) = 'RIGHTOF(x);
'PRED(x) = 'LEFTOF(x);
```

For descending ranges, the opposite is true:

```
'SUCC(x) = 'LEFTOF(x);
'PRED(x) = 'RIGHTOF(x);
```

What happens if the value passed to `'SUCC`, `'PRED`, and so on is at the limit of the type? For instance, for type `color`, what is the value of the expression shown here:

```
y := red;
x := color'PRED(y);
```

The second expression causes a runtime error to be reported, because a range constraint has been violated.

Function Array Attributes

Function array attributes return the bounds of array types. An operation that requires accessing every location of an array can use these attributes to find the bounds of the array.

The four kinds of function array attributes are:

- `array'LEFT` (n), which returns the left bound of index range n
- `array'RIGHT` (n), which returns the right bound of index range n
- `array'HIGH` (n), which returns the upper bound of index range n
- `array'LOW` (n), which returns the lower bound of index range n

These attributes are exactly like the value type attributes that were discussed earlier, except that these attributes work with arrays.

For ascending ranges, the following is true:

```
array'LEFT = array'LOW
array'RIGHT = array'HIGH
```

For descending ranges, the opposite is true:

```
array'LEFT = array'HIGH
array'RIGHT = array'LOW
```

Following is an example where these attributes are very useful:

```
PACKAGE p_ram IS
 TYPE t_ram_data IS ARRAY(0 TO 511) OF INTEGER;

 CONSTANT x_val : INTEGER := -1;
 CONSTANT z_val : INTEGER := -2;
END p_ram;

USE WORK.p_ram.ALL;
LIBRARY IEEE; USE IEEE.std_logic_1164.ALL;
ENTITY ram IS
 PORT( data_in : IN INTEGER;
       addr : IN INTEGER;
       data : OUT INTEGER;
       cs : IN std_logic;
       r_wb: in std_logic);
END ram;

ARCHITECTURE behave_ram OF ram IS
BEGIN
 main_proc: PROCESS( cs, addr, r_wb )
  VARIABLE ram_data : t_ram_data;
  VARIABLE ram_init : boolean := false;
 BEGIN
  IF NOT(ram_init) THEN
   FOR i IN ram_data'LOW TO ram_data'HIGH LOOP
    ram_data(i) := 0;
   END LOOP;

   ram_init := TRUE;
  END IF;

  IF (cs = 'X') OR (r_wb = 'X')THEN
   data <= x_val;

  ELSIF ( cs = '0' ) THEN
   data <=z_val;

  ELSIF (r_wb = '1') THEN
   IF (addr = x_val) OR (addr = z_val) THEN
    data <=x_val;
   ELSE
    data <= ram_data(addr);
   END IF;

  ELSE
```

```
IF (addr = x_val) OR (addr = z_val) THEN
  ASSERT FALSE
   REPORT " writing to unknown address"
   SEVERITY ERROR;
 data <= x_val;
ELSE
 ram_data(addr) :=data_in;
 data <= ram_data(addr);
END IF;

 END IF;
 END PROCESS;
END behave_ram;
```

This example implements an integer-based RAM device. There are 512 integer locations in the RAM, which is controlled by two control lines. The first is **cs** (chip select), and the second is **r_wb** (read/write bar). The model contains an **IF** statement that initializes the contents of the RAM to a known value. A boolean variable (**ram_init**) is declared to keep track of whether the RAM has been initialized or not. If this variable is false, the RAM has not yet been initialized. If true, initialization has been performed.

The first time the process is executed, variable **ram_init** is false, and the **IF** statement is executed. Inside the **IF** statement is a loop statement that loops through every location of the RAM and sets the location to a known value. This process is necessary because the starting value of type **INTEGER** is the value **integer'LEFT**, or -2,147,483,647. Notice the use of function array attributes **'LOW** and **'HIGH** to control the range of the initialization loop.

After the loop has been executed and all RAM locations have been initialized, the **ram_init** variable is set to true. Setting the variable **ram_init** to true prevents the initialization loop from executing again.

The rest of the model implements the read and write functions based on the values of **addr**, **data_in**, **r_wb**, and **cs**. This model performs a lot of error checking for unknown values on input ports. The model tries to intelligently handle these unknown input values.

Function Signal Attributes

Function signal attributes are used to return information about the behavior of signals. These attributes can be used to report whether a signal has just changed value, how much time has passed since the last event transition, or what the previous value of the signal was. There are five attributes that fall into this category. Following is a brief description of each:

- **s'EVENT**, which returns true if an event occurred during the current delta; otherwise, returns false

- **s'ACTIVE**, which returns true if a transaction occurred during the current delta; otherwise, returns false

- **s'LAST_EVENT**, which returns time elapsed since the previous event transition of signal

- **s'LAST_VALUE**, which returns previous value of **s** before the last event

- **s'LAST_ACTIVE**, which returns time elapsed since the previous transaction of signal

Attributes 'EVENT and 'LAST_VALUE

Attribute **'EVENT** is very useful for determining clock edges. By checking if a signal is at a particular value, and if the signal has just changed, it can be deduced that an edge has occurred on the signal. Following is an example of a rising edge detector:

```
LIBRARY IEEE;
USE IEEE.std_logic_1164.ALL;
ENTITY dff IS
  PORT( d, clk : IN std_logic;
        q : OUT std_logic);
END dff;

ARCHITECTURE dff OF dff IS
BEGIN
  PROCESS(clk)
  BEGIN
    IF ( clk = '1') AND ( clk'EVENT ) THEN
     q <= d;
    END IF;
  END PROCESS;
END dff;
```

This example shows a very simple **dff** model. The **clk** input is used to transfer the **d** input to the **q** output, on a rising edge of the **clk**. To detect the rising edge of the **clk** input, this model makes use of the **'EVENT** attribute. If the value of the **clk** input is a **'1'**, and the value has just changed, then a rising edge must have occurred. (When a synthesis tool is applied to the preceding example, a flip flop results.)

What the preceding example ignores is the fact that an 'x' value to a '1' value also looks like a rising edge when it is not. The next example shows how to correct this problem using the 'LAST_VALUE attribute. The IF statement from the preceding example is rewritten here:

```
IF ( clk = '1' ) AND ( clk'EVENT )
       and ( clk'LAST_VALUE = '0') THEN
  q <= d;
END IF;
```

In this example, one more check is made to make certain that the last value of the clk input was a '0' before the new event occurred.

In both examples, the 'EVENT attribute was not really needed, because the process statement had only clk as its sensitivity list. The only way that the process statement could be executed would be because of an event on signal clk. This is a true statement, but it is a good modeling practice to check for the event anyway. Some time in the future, the model may be modified to include an asynchronous preset or clear, and these signals will be added to the sensitivity list for the process statement. Now, when an event occurs on any of the inputs, the process is invoked. Using the 'EVENT attribute, the process can determine which input caused the process to be invoked.

Attribute 'LAST_EVENT

Attribute 'LAST_EVENT returns the time since the previous event occurred on the signal. This attribute is very useful for implementing timing checks, such as setup checks, hold checks, and pulse width checks. An example of a setup time and a hold time are shown in Figure 6-1.

The rising edge of signal clk is the reference edge to which all checks are performed. A setup time check guarantees that the data input does not change during the setup time, and the hold time check guarantees that the data input does not change during the time equal to the hold time after the reference edge. This ensures correct operation of the device.

Following is an example of the setup time check using the 'LAST_EVENT attribute:

```
LIBRARY IEEE;
USE IEEE.std_logic_1164.ALL;
ENTITY dff IS
```

Figure 6-1
Setup and Hold Time
Waveform Descrip-
tion.

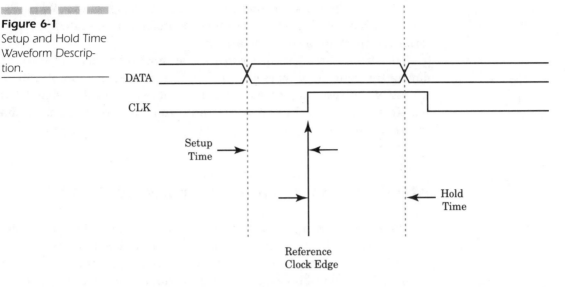

```
GENERIC ( setup_time, hold_time : TIME );
PORT( d, clk : IN std_logic;
   q : OUT std_logic);
BEGIN
 setup_check : PROCESS ( clk )
 BEGIN
  IF ( clk = '1' ) and ( clk'EVENT ) THEN
   ASSERT ( d'LAST_EVENT >= setup_time )
    REPORT "setup violation"
    SEVERITY ERROR;
  END IF;
 END PROCESS setup_check;
END dff;

ARCHITECTURE dff_behave OF dff IS
BEGIN
 dff_process : PROCESS ( clk )
 BEGIN
  IF ( clk = '1' ) AND ( clk'EVENT ) THEN
   q <= d;
  END IF;
 END PROCESS dff_process;
END dff_behave;
```

The setup_check procedure is contained in a passive process in the
entity for the **dff** model. The check could have been included in the
architecture for the **dff** model, but having the check in the entity allows
the timing check to be shared among any architecture of the entity.

The passive process executes for each event on signal **clk**. When the **clk** input has a rising edge, the **ASSERT** statement is executed and performs the check for a setup violation.

The **ASSERT** statement checks to see that input **d** has not had an event during the setup time passed in by the generic **setup_time**. Attribute **d'LAST_EVENT** returns the time since the most recent event on signal **d**. If the time returned is less than the setup time, the assertion fails and reports a violation.

Attribute 'ACTIVE and 'LAST_ACTIVE

Attributes **'ACTIVE** and **'LAST_ACTIVE** trigger on transactions of the signal attached to AND events. A transaction on a signal occurs when a model in or inout port has an event occur that triggers the execution of the model. The model is executed, but the result of the execution produces the same output values. For instance, if an AND gate has a **'1'** value on one input and a **'0'** on the other, the output value is **'0'**. If the input with a **'1'** value changes to a **'0'** value, the output remains **'0'**; no event is generated, but a transaction will have been generated on the output of the AND gate.

Attribute **'ACTIVE** returns true when a transaction or event occurs on a signal, and attribute **'LAST_ACTIVE** returns the time since a previous transaction or event occurred on the signal it is attached to. Both of these attributes are counterparts for attributes **'EVENT** and **'LAST_EVENT**, which provide the same behavior for events.

Signal Kind Attributes

Signal kind attributes are used to create special signals, based on other signals. These special signals return information to the designer about the signal that the attribute is attached to. The information returned is very similar to some of the functionality provided by some of the function attributes. The difference is that these special signals can be used anywhere that a normal signal can be used, including sensitivity lists.

Signal attributes return information such as whether a signal has been stable for a specified amount of time, when a transaction has occurred on a signal, and a delayed version of the signal can be created.

One restriction on the use of these attributes is that they cannot be used within a subprogram. A compiler error message results if a signal kind attribute is used within a subprogram.

There are four attributes in the signal kind category:

- **s'DELAYED [(time)]**, which creates a signal of the same type as the reference signal that follows the reference signal, delayed by the time of the optional time expression

- **s'STABLE [(time)]**, which creates a boolean signal that is true whenever the reference signal has had no events for the time specified by the optional time expression

- **s'QUIET [(time)]**, which creates a boolean signal that is true whenever the reference signal has had no transactions or events for the time specified by the optional time expression

- **s'TRANSACTION**, which creates a signal of type **BIT** that toggles its value for every transaction or event that occurs on **s**

Attribute **'DELAYED**

Attribute **'DELAYED** creates a delayed version of the signal that it is attached to. The same functionality can be obtained using a transport-delayed signal assignment. The difference between a transport delay assignment and the **'DELAYED** attribute is that the designer has to do more bookkeeping with the transport signal assignment method. With a transport signal assignment, a new signal must be declared.

Let's look at one use for the **'DELAYED** attribute. One method for modeling ASIC devices is to place path-related delays on the input pins of the ASIC library part. An example of this method is shown in Figure 6-2.

Typically, before the layout process, educated guesses are made for the delays of each input. After layout, the real delay values are back-annotated to the model, and the simulation is run again with the real delays. One method to provide for back annotation of the delay values is to use generic values specified in the configuration for the device. (Configurations are discussed in Chapter 7, "Configurations.") A typical model for one of the **and2** gates shown in Figure 6-2 might look like this:

```
LIBRARY IEEE;
USE IEEE.std_logic_1164.ALL;
ENTITY and2 IS
  GENERIC ( a_ipd, b_ipd, c_opd : TIME );
```

Figure 6-2
Gate Array Logic
with Input and Out-
put Delays.

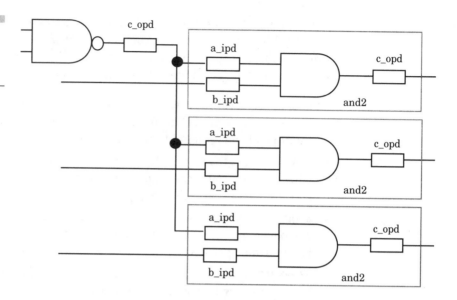

```
PORT ( a, b : IN std_logic;
    c: OUT std_logic);
END and2;

ARCHITECTURE int_signals OF and2 IS
  SIGNAL inta, intb : std_logic;
BEGIN
  inta <= TRANSPORT a AFTER a_ipd;
  intb <= TRANSPORT b AFTER b_ipd;

  c <= inta AND intb AFTER c_opd;
END int_signals;

ARCHITECTURE attr OF and2 IS
BEGIN
  c <= a'DELAYED(a_ipd) AND b'DELAYED(b_ipd) AFTER c_opd;
END attr;
```

In the preceding example, two architectures for entity **and2** show two different methods of delaying the input signals by the path delay. The first method uses transport-delayed internal signals to delay the input signals. These delayed signals are then **AND**ed together and assigned to output port **c**.

The second method makes use of the predefined signal attribute **'DELAYED**. Input signals **a** and **b** are delayed by the path delay generic value **a_ipd** (a input path delay) and **b_ipd** (b input path delay). The values of the delayed signals are **AND**ed together and assigned to output port **c**.

If the optional time expression for attribute 'DELAYED is not specified, 0 ns is assumed. A signal delayed by 0 ns is delayed by one delta. (Delta delay is discussed in Chapter 2.)

Another application for the 'DELAYED attribute is to perform a hold-check. Earlier in this chapter, we discussed what setup and hold times were and how to implement the setup check using 'LAST_EVENT. Implementing the hold-check requires the use of a delayed version of the clk signal. The example shown earlier has been modified to include the hold-check function as shown here:

```
LIBRARY IEEE;
USE IEEE.std_logic_1164.ALL;
ENTITY dff IS
 GENERIC ( setup_time, hold_time : TIME );
 PORT( d, clk : IN std_logic;
         q : OUT std_logic);
BEGIN
 setup_check : PROCESS ( clk )
 BEGIN
  IF ( clk = '1' ) and ( clk'EVENT ) THEN
   ASSERT ( d'LAST_EVENT >= setup_time )
     REPORT "setup violation"
     SEVERITY ERROR;
  END IF;
 END PROCESS setup_check;

 hold_check : PROCESS (clk'DELAYED(hold_time))
 BEGIN
  IF ( clk'DELAYED(hold_time) = '1' ) and
          ( clk'DELAYED(hold_time)'EVENT ) THEN

   ASSERT ( d'LAST_EVENT = 0 ns ) OR ( d'LAST_EVENT >
     hold_time )
    REPORT "hold violation"
    SEVERITY ERROR;

  END IF;
 END PROCESS hold_check;
END dff;

ARCHITECTURE dff_behave OF dff IS
BEGIN
 dff_process : PROCESS ( clk )
 BEGIN
  IF ( clk = '1' ) AND ( clk'EVENT ) THEN
   q <= d;
  END IF;

 END PROCESS dff_process;
END dff_behave;
```

A delayed version of the `clk` input is used to trigger the hold-check. The `clk` input is delayed by the amount of the hold-check. If the data input changes within the hold time, `d'LAST_EVENT` returns a value that is less than the hold time. When `d` changes exactly at the same time as the delayed `clk` input, `d'LAST_EVENT` returns 0 ns. This is a special case and is legal so it must be handled specially.

An alternative method for checking the hold time of a device is to trigger the hold-check process when the `d` input changes and then look back at the last change on the `clk` input. However, this is more complicated and requires the designer to manually keep track of the last reference edge on the `clk` input.

Another interesting feature of attributes that this model pointed out is the cascading of attributes. In the preceding example, the delayed version of the `clk` signal was checked for an event. This necessitated the use of `clk'DELAYED (hold_time) 'EVENT`. The return value from this attribute is true whenever the signal created by the `'DELAYED` attribute has an event during the current delta time point. In general, attributes can be cascaded any level if the values returned from the previous attribute are appropriate for the next attribute.

Attribute 'STABLE

Attribute `'STABLE` is used to determine the relative activity level of a signal. It can be used to determine if the signal just changed or has not changed in a specified period of time. The resulting value output is itself a signal that can be used to trigger other processes.

Following is an example of how attribute `'STABLE` works:

```
LIBRARY IEEE;
USE IEEE.std_logic_1164.ALL;
ENTITY pulse_gen IS
  PORT( a : IN std_logic;
        b : OUT BOOLEAN);
END pulse_gen;

ARCHITECTURE pulse_gen OF pulse_gen IS
BEGIN
  b <= a'STABLE( 10 ns );
END pulse_gen;
```

Figure 6-3 shows the resulting waveform `b` when waveform `a` is presented to the model.

Figure 6-3
Example Showing
'DELAYED(10ns).

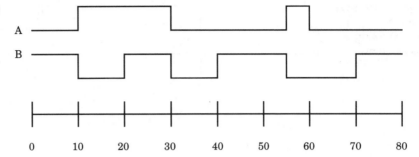

At the first two changes in signal **a** (10 ns and 30 ns), signal **b** immediately changes to false (actually at the next delta). Then when signal **a** has been stable for 10 ns, signal **b** changes to true. At time 55 ns, signal **a** changes value again, so signal **b** changes to false. Because signal **a** changes 5 ns later (60 ns), signal **a** has not been stable long enough to allow output **b** to go to a true value. Only at 10 ns after the last change on signal **a** (60 ns) is the input signal **a** stable long enough to allow signal **b** to change to true.

If the time value specified for the **'STABLE** attribute is 0 ns, or not specified, then the **'STABLE** attribute is false for 1 delta whenever the signal that the attribute is attached to changes. An example of this scenario is shown in Figure 6-4.

When used in this method, the resulting signal value has the same timing but opposite value as function attribute **'EVENT**. A statement to detect the rising edge of a clock could be written in two ways, as shown here:

```
IF (( clk'EVENT ) AND ( clk = '1' ) AND
         ( clk'LAST_VALUE = '0' )) THEN
   .
   .   -- DO PROCESSING
   .
END IF;


IF (( NOT( clk'STABLE) ) AND ( clk = '1' ) AND
      (
                clk'LAST_VALUE = '0' )) THEN
   .
   .   --- DO PROCESSING
   .
END IF;
```

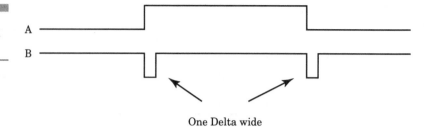

Figure 6-4
Example Showing
'DELAYED(0 ns).

One Delta wide

In both cases, the **IF** statement detects the rising edge; but the **IF** statement using **'EVENT** is more efficient in memory space and speed. The reason for this is that attribute **'STABLE** creates an extra signal in the design that uses more memory to store, and whenever the value for the new signal needs to be updated, it must be scheduled. Keeping track of signal events costs memory and time.

Attribute **'QUIET**

Attribute **'QUIET** has the same functionality as **'STABLE**, except that **'QUIET** is triggered by transactions on the signal that it is attached to in addition to events. Attribute **'QUIET** creates a **BOOLEAN** signal that is true whenever the signal it is attached to has not had a transaction or event for the time expression specified.

Typically, models that deal with transactions involve complex models of devices at the switch level or the resolution of driver values. Following is an interesting application using the attribute **'QUIET**:

```
ARCHITECTURE test OF test IS
  TYPE t_int is (int1, int2, int3, int4, int5 );

  SIGNAL int, intsig1, intsig2, intsig3 : t_int;

 SIGNAL lock_out : BOOLEAN;
BEGIN
 int1_proc: PROCESS
 BEGIN
   .
   .
   .
   WAIT ON trigger1; -- outside trigger signal
   WAIT UNTIL clk = '1';
    IF NOT(lock_out) THEN
      intsig1 <= int1;
    END IF;
```

```
      END PROCESS int1_proc;

   int2_proc: PROCESS
   BEGIN
     .
     .
     .
    WAIT ON trigger2;-- outside trigger signal
    WAIT UNTIL clk = '1';
    IF NOT(lock_out) THEN
       intsig2 <=int2;
     END IF;
   END PROCESS int2_proc;

   int3_proc: PROCESS
   BEGIN
     .
     .
     .
    WAIT ON trigger3;-- outside trigger signal
    WAIT UNTIL clk = '1';
     IF NOT(lock_out) THEN
       intsig3 <=int3;
     END IF;
   END PROCESS int3_proc;

   int <=intsig1 WHEN NOT(intsig1'QUIET) ELSE
       intsig2 WHEN NOT(intsig2'QUIET) ELSE
       intsig3 WHEN NOT(intsig3'QUIET) ELSE
       int;

   int_handle : PROCESS
   BEGIN
    WAIT ON int'TRANSACTION;-- described next
    lock_out <= TRUE;
    WAIT FOR 10 ns;
    CASE int IS
     WHEN int1 =>
      .
      .
     WHEN int2 =>
      .
      .
     WHEN int3 =>
      .
      .
     WHEN int4 =>
      .
      .
     WHEN int5 =>
      .
      .
     END CASE;
    lock_out <= false;
```

```
END PROCESS;
END test;
```

This example shows how a priority mechanism could be modeled for an interrupt handler. Process **int1_proc** has the highest priority, and process **int3_proc** has the lowest. Whenever one of the processes is triggered, the appropriate interrupt handler is placed on signal **int**, and the interrupt handler for that interrupt is called.

The model consists of three processes that drive the interrupt signal **int**, and another process to call the appropriate interrupt handling function. Signal **int** is not a resolved signal and therefore cannot have multiple drivers. If a resolution function is written for signal **int**, the order of the drivers cannot be used to determine priority. Therefore, the approach shown in the preceding was taken.

In this approach, three internal signals **intsig1**, **intsig2**, and **intsig3** are driven by each of the processes, respectively. These signals are then combined, using a conditional signal assignment statement. The conditional signal assignment statement makes use of the predefined attribute **'QUIET** to determine when a transaction has been assigned to a driver of a signal. It is required that transactions are detected on the internal signals, because the process always assigns the same value so an event only occurs on the first assignment.

The priority mechanism is controlled by the conditional signal assignment statement. When a transaction occurs on **intsig1**, **intsig2**, or **intsig3**, the assignment statement evaluates and assigns the appropriate value to signal **int** based on the signal(s) that had a transaction. If a transaction occurred only on **intsig2**, **intsig2'QUIET** would be false, causing the conditional signal assignment statement to place the value of **intsig2** on signal **int**. But what happens if **intsig3** and **intsig2** occur at the same time? The conditional signal assignment statement evaluates, and the first clause that has a **WHEN** expression return true does the assignment and then exits the rest of the statement. For this example, the value for **intsig2** is returned, because it is first in the conditional signal assignment statement. The priority of the inputs is determined by the order of the **WHEN** clauses in the conditional signal assignment statement.

Attribute 'TRANSACTION

The process that implemented the interrupt handling for the previous example uses the **'TRANSACTION** attribute in a **WAIT** statement. This attribute is another of the attributes that creates a signal where it is used. Attribute **'TRANSACTION** creates a signal of type **BIT** that toggles from '1' or '0' for every transaction of the signal that it is attached to. This attribute is useful for invoking processes when transactions occur on signals.

In the preceding example, the interrupt handler process needs to be executed whenever a transaction occurs on signal **int**. This is true because the same interrupt could happen twice or more in sequence. If this occurred, a transaction, not an event would be generated on signal **int**. Without the attribute **'TRANSACTION**, **WAIT** statements are sensitive to events. By using the attribute **'TRANSACTION**, the value of **int'TRANSACTION** toggles for every transaction causing an event to occur, thus activating the **WAIT** statement.

Type Kind Attributes

Type attributes return values of kind type. There is only one type attribute, and it must be used with another value or function type attribute. The only type attribute available in VHDL is the attribute **t'BASE**.

This attribute returns the base type of a type or subtype. This attribute can only be used as the prefix of another attribute, as shown in the following example:

```
do_nothing : PROCESS(x)
  TYPE color IS (red, blue, green, yellow, brown, black);
  SUBTYPE color_gun IS color RANGE red TO green;

  VARIABLE a : color;
BEGIN
  a := color_gun'BASE'RIGHT;        -- a = black
  a := color'BASE'LEFT;             -- a = red

  -- a = yellow
  a := color_gun'BASE'SUCC(green);

END PROCESS do_nothing;
```

In the first assignment to variable **a**, **color_gun'BASE** returns type **color**, the base type of **color_gun**. The statement **color'RIGHT** then returns the value **black**. In the second assignment statement, the base type of type `color` is type `color`. The statement `color'LEFT` returns the value `red`. In the last assignment, `color_gun'BASE` returns type `color`, and `color'SUCC(green)` returns yellow.

Range Kind Attributes

The last two predefined attributes in VHDL return a value kind of range. These attributes work only with constrained array types and return the index range specified by the optional input parameter. Following are the attribute notations:

■ `a'RANGE[(n)]`

■ `a'REVERSE_RANGE[(n)]`

Attributes `'RANGE` return the nth range denoted by the value of parameter `n`. Attribute `'RANGE` returns the range in the order specified, and `'REVERSE_RANGE` returns the range in reverse order.

Attributes `'RANGE` and `'REVERSE_RANGE` can be used to control the number of times that a loop statement loops. Following is an example:

```
FUNCTION vector_to_int(vect: std_logic_vector) RETURN
      INTEGER IS
 VARIABLE result : INTEGER := 0;
BEGIN
 FOR i IN vect'RANGE LOOP

   result := result * 2;

   IF vect(i) = '1' THEN
     result := result + 1;
   END IF;

 END LOOP;

   RETURN result;
END vector_to_int;
```

This function converts an array of bits into an integer value. The number of times that the loop needs to be executed is determined by the number of bits in the input argument `vect`. When the function call is made, the input argument cannot be an unconstrained value; therefore, the attribute `'RANGE` can be used to determine the range of the input `vector`. The range can then be used in the loop statement to determine the number of times to execute the loop and finish the conversion.

The 'REVERSE_RANGE attribute works similar to the 'RANGE attribute, except that the range is returned in the reverse order. For a type shown in the following, the 'RANGE attribute returns 0 TO 15, and the 'REVERSE_RANGE attribute returns 15 DOWNTO 0:

```
TYPE array16 IS ARRAY(0 TO 15) OF BIT;
```

VHDL attributes extend the language to provide some very useful functionality. They make models much easier to read and maintain.

SUMMARY

In this chapter, we discussed the following:

- The different kinds of attributes and how some just return values, while others create new signals.
- How 'LEFT, 'RIGHT, 'LENGTH, 'HIGH, and 'LOW can be used to get the bounds of a type or array.
- How 'POS, 'VAL, 'SUCC, 'PRED, 'LEFTOF, and 'RIGHTOF can be used to manipulate enumerated types.
- How 'ACTIVE, 'EVENT, 'LAST_ACTIVE, 'LAST_EVENT, and 'LAST_VALUE can be used to return information about when events occur.
- How 'DELAYED, 'STABLE, 'QUIET, and 'TRANSACTION create new signals that return information about other signals.
- How range attributes 'RANGE and 'REVERSE_RANGE can be used to control statements over the exact range of a type.

In the next chapter, we examine configurations, the method of binding architectures to entities.

7

Configurations

Configurations are a primary design unit used to bind component instances to entities. For structural models, configurations can be thought of as the parts list for the model. For component instances, the configuration specifies from many architectures for an entity which architecture to use for a specific instance. When the configuration for an entity-architecture combination is compiled into the library, a simulatable object is created.

Configurations can also be used to specify generic values for components instantiated in the architecture configured by the configuration. This mechanism, for example, provides a late-binding capability for delay values. Delay values calculated from a physical layout tool, such as a printed circuit board design system or a gate array layout system, can be inserted in a configuration to provide a simulation model with actual delays in the design.

If the designer wants to use a component in an architecture that has different port names from the architecture component declaration, the new component can have its ports mapped to the appropriate signals. With this functionality, libraries of components can be mixed and matched easily.

The configuration can also be used to provide a very fast substitution capability. Multiple architectures can exist for a single entity. One architecture might be a behavioral model for the entity, while another architecture might be a structural model for the entity. The architecture used in the containing model can be selected by specifying which architecture to use in the configuration, and recompiling only the configuration. After compilation, the simulatable model uses the specified architecture.

Default Configurations

The simplest form of explicit configuration is the default configuration. (The simplest configuration is none at all in which the last architecture compiled is used for an entity.) This configuration can be used for models that do not contain any blocks or components to configure. The default configuration specifies the configuration name, the entity being configured, and the architecture to be used for the entity. Following is an example of two default configurations shown by configurations **big_count** and **small_count**:

```
LIBRARY IEEE;
USE IEEE.std_logic_1164.ALL;
ENTITY counter IS
 PORT(load, clear, clk : IN std_logic;
      data_in : IN INTEGER;
      data_out : OUT INTEGER);
END counter;

ARCHITECTURE count_255 OF counter IS
BEGIN
 PROCESS(clk)
  VARIABLE count : INTEGER := 0;
 BEGIN
  IF clear = '1' THEN
   count := 0;
  ELSIF load = '1' THEN
   count := data_in;
  ELSE
   IF (clk'EVENT) AND (clk = '1') AND
     (clk'LAST_VALUE = '0') THEN
    IF (count = 255) THEN
     count := 0;
    ELSE
     count := count + 1;
    END IF;
```

```
      END IF;
     END IF;
    data_out <= count;
   END PROCESS;
  END count_255;

  ARCHITECTURE count_64k OF counter IS
  BEGIN
   PROCESS(clk)
    VARIABLE count : INTEGER := 0;
   BEGIN
    IF clear = '1' THEN
     count := 0;
    ELSIF load = '1' THEN
     count := data_in;
    ELSE
      IF (clk'EVENT) AND (clk = '1') AND
        (clk'LAST_VALUE = '0') THEN
        IF (count = 65535) THEN
         count := 0;
        ELSE
         count := count + 1;
        END IF;
      END IF;
    END IF;
    data_out <= count;
   END PROCESS;
  END count_64k;

  CONFIGURATION small_count OF counter IS
   FOR count_255
   END FOR;
  END small_count;

  CONFIGURATION big_count OF counter IS
   FOR count_64k
   END FOR;
  END big_count;
```

This example shows how two different architectures for a counter entity can be configured using two default configurations. The entity for the counter does not specify any bit width for the data to be loaded into the counter or data from the counter. The data type for the input and output data is INTEGER. With a data type of integer, multiple types of counters can be supported up to the integer representation limit of the host computer for the VHDL simulator.

The two architectures of entity counter specify two different-sized counters that can be used for the entity. The first architecture, count_255, specifies an 8-bit counter. The second architecture, count_64k, specifies a

16-bit counter. The architectures specify a synchronous counter with a synchronous **load** and **clear**. All operations for the device occur with respect to the clock.

Each of the two configurations for the entity specifies a different architecture for the counter entity. Let's examine the first configuration in more detail. The configuration design unit begins with the keyword **CONFIGURATION** and is followed by the name of the configuration. In this example, the name of the configuration is **small_count**. The keyword **OF** precedes the name of the entity begin configured (counter). The next line of the configuration starts the block configuration section. The keyword **FOR** is followed by a name of the architecture to use for the entity being configured or the name of the block of the architecture that will be configured. Any component or block configuration information then exists between the **FOR ARCHITECTURE** clause and the matching **END FOR**.

In this architecture, there are no blocks or components to configure; therefore, the block configuration area from the **FOR** clause to the **END FOR** clause is empty, and the default is used. The configuration is called the default configuration, because the default is used for all objects in the configuration.

The first configuration is called **small_count** and binds architecture **count_255** with entity **counter** to form a simulatable object. The second configuration binds architecture **count_64k** with entity **counter** and forms a simulatable object called **big_count**.

Component Configurations

In this section, we discuss how architectures that contain instantiated components can be configured. Architectures that contain other components are called *structural architectures*. These components are configured through component configuration statements.

Let's first look at some very simple examples of component configurations, and then at some progressively more complex examples. The first example is a simple 2 to 4 decoder device. Figure 7-1 shows the symbol for the decoder, and Figure 7-2 shows the schematic.

The components used in the design are defined using the VHDL description shown here:

```
LIBRARY IEEE;
USE IEEE.std_logic_1164.ALL;
```

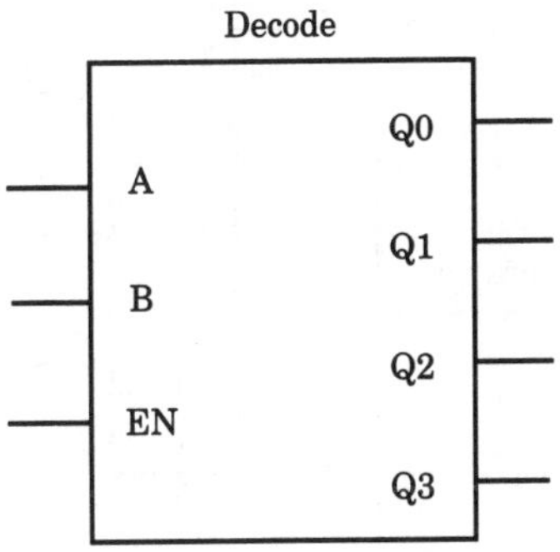

Figure 7-1
Symbol for Decoder
Example.

```
ENTITY inv IS
 PORT( a : IN std_logic;
       b : OUT std_logic);
END inv;

ARCHITECTURE behave OF inv IS
BEGIN
 b <= NOT(a) AFTER 5 ns;
END behave;

CONFIGURATION invcon OF inv IS
 FOR behave
 END FOR;
END invcon;

LIBRARY IEEE; USE IEEE.std_logic_1164.ALL;
ENTITY and3 IS
 PORT( a1, a2, a3 : IN std_logic;
       o1 : OUT std_logic);
END and3;

ARCHITECTURE behave OF and3 IS
BEGIN
 o1 <= a1 AND a2 AND a3 AFTER 5 ns;
END behave;

CONFIGURATION and3con OF and3 IS
 FOR behave
 END FOR;
END and3con;
```

Next, the entity and architecture for decode are shown:

```
LIBRARY IEEE; USE IEEE.std_logic_1164.ALL;
ENTITY decode IS
 PORT( a, b, en : IN std_logic;
       q0, q1, q2, q3 : OUT std_logic);
END decode;
```

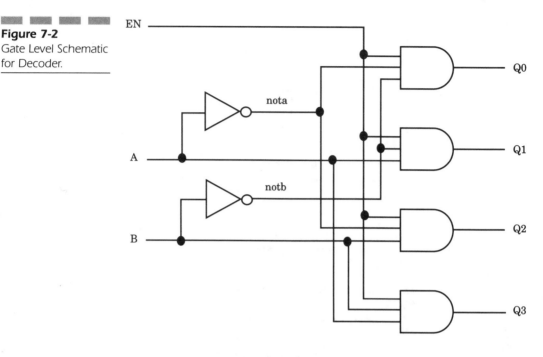

Figure 7-2
Gate Level Schematic
for Decoder.

```
ARCHITECTURE structural OF decode IS
  COMPONENT inv
    PORT( a : IN std_logic;
          b : OUT std_logic);
  END COMPONENT;

  COMPONENT and3
    PORT( a1, a2, a3 : IN std_logic;
          o1 : OUT std_logic);
  END COMPONENT;

  SIGNAL nota, notb : std_logic;
BEGIN
  I1 : inv
    PORT MAP(a, nota);

  I2 : inv
    PORT MAP(b, notb);

  A1 : and3
    PORT MAP(nota, en, notb, Q0);

  A2 : and3
    PORT MAP(a, en, notb, Q1);

  A3 : and3
    PORT MAP(nota, en, b, Q2);
```

```
A4 : and3
  PORT MAP(a, en, b, Q3);

END structural;
```

When all of the entities and architectures have been compiled into the working library, the circuit can be simulated. The simulator uses the last compiled architecture to build the executable design for the simulator because it is the default. Using the last compiled architecture for an entity to build the simulator works fine in a typical system, until more than one architecture exists for an entity. Then it can become confusing as to which architecture was compiled last. A better method is to specify exactly which architecture to use for each entity. The component configuration binds architectures to entities.

Two different styles can be used for writing a component configuration for an entity. The lower-level configuration style specifies lower-level configurations for each component, and the entity-architecture style specifies entity-architecture pairs for each component. The word *style* is used to describe these two different configurations because there is no hard-and-fast rule about how to use them. Lower-level configurations can be mixed with entity-architecture pairs, creating a mixed-style configuration.

Lower-Level Configurations

Let's examine the configuration for the lower-level configuration style first. Following is an example of such a configuration for the decode entity:

```
CONFIGURATION decode_11con OF decode IS
  FOR structural
    FOR I1 : inv USE CONFIGURATION WORK.invcon;
    END FOR;

    FOR I2 : inv USE CONFIGURATION WORK.invcon;
    END FOR;

    FOR ALL : and3 USE CONFIGURATION WORK.and3con;
    END FOR;
  END FOR;
END decode_11con;
```

This configuration specifies which configuration to use for each component in architecture **structural** of entity **decode**. The specified lower-level configuration must already exist in the library for the current configuration to compile. Each component being configured has a **FOR** clause to begin the configuration and an **END FOR** clause to end the configuration specification for the component. Each component can be specified with the component instantiation label directly, as shown for component **I1**, or with an **ALL** or **OTHERS** clause as shown by the **and3** components.

After the component is uniquely specified by label or otherwise, the **USE CONFIGURATION** clause specifies which configuration to use for this instance of the component. In the preceding example, the configuration specification for component **I1** uses the configuration called **invcon**, from the working library. For configuration **decode_11con** to compile, configuration **invcon** must have been already compiled into library **WORK**.

Notice that the names of the entities, architectures, and configurations reflect a naming convention. In general, this is a good practice. It helps distinguish the different types of design units from one another when they all exist in a library.

The advantage of this style of configurations is that most configurations are easy to write and understand. The disadvantage is not being able to change the configuration of a lower-level component, without implementing a two-step or more process of recompilation when hierarchy levels increase.

Entity-Architecture Pair Configuration

The other style of component configurations is the entity-architecture pair style. Following is an example of a configuration that uses the same entity and architectures as the previous example:

```
CONFIGURATION decode_eacon OF decode IS
  FOR structural
    FOR I1 : inv USE ENTITY WORK.inv(behave);
    END FOR;

    FOR OTHERS : inv USE ENTITY WORK.inv(behave);
    END FOR;

    FOR A1 : and3 USE ENTITY WORK.and3(behave);
    END FOR;

    FOR OTHERS : and3 USE ENTITY WORK.and3(behave);
```

```
      END FOR;

      END FOR;
   END decode_eacon;
```

This configuration looks very similar to the lower-level configuration style except for the USE clause in the component specification. In the previous example, a configuration was specified, but in this style, an entity-architecture pair is specified. The architecture is actually optional. If no architecture is specified, the last compiled architecture for the entity is used.

Let's take another look at the FOR clause for the first inverter, I1. In the preceding example, the component is still specified by the label or by an ALL or OTHERS clause. In this example, a USE ENTITY clause follows. This clause specifies the name of the entity to use for this component. The entity can have a completely different name than the component being specified. The component name comes from the component declaration in the architecture, while the entity name comes from the actual entity that has been compiled in the library specified. Following the entity is an optional architecture name that specifies which architecture to use for the entity.

Notice that the OTHERS clause is used for the second inverter in this example. The first inverter is configured from its label I1, and all components that have not yet been configured are configured by the OTHERS clause. This capability allows component I1 to use an architecture that is different from the other components to describe its behavior. This concept allows mixed-level modeling to exist. One component can be modeled at the switch or gate level, and the other can be modeled at the behavior level.

To change the architecture used for a component with the first configuration, decode_11con requires modifying the lower-level configuration and recompiling, then recompiling any higher-level configurations that depend on it. With the second configuration decode_eacon, changing the architecture for a component involves modifying configuration decode_eacon and recompiling. No other configurations need be recompiled.

Port Maps

In the last two examples of component configurations, default mapping of entity ports and component ports was used. When the port names for an entity being configured to a component match the component port names, no other mapping needs to take place. The default mapping causes the ports to match. What happens when the component ports do not match the entity being mapped to the component instance? Without any further information, the compiler cannot figure out which ports to map to which and produces an error. However, more information can be passed to the compiler with the configuration port map clause.

The configuration port map clause looks exactly like the component instantiation port map clause used in an architecture. The configuration port map clause specifies which of the component ports map to the actual ports of the entity. If the port names are different, then the port map clause specifies the mapping.

Let's change the port names of the **inv** component used in the previous example and see what the effect is in the configuration:

```
LIBRARY IEEE;
USE IEEE.std_logic_1164.ALL;
ENTITY inv IS
  PORT( x : IN std_logic;
        y : OUT std_logic);
END inv;

ARCHITECTURE behave OF inv IS
BEGIN
  y <= NOT(x) AFTER 5 ns;
END behave;

CONFIGURATION invcon OF inv IS
  FOR behave
  END FOR;
END invcon;
```

The entity and architecture for **decode** stays exactly the same, including the component declaration. The configuration, however, needs to add the port map clause, as shown in the following example:

```
CONFIGURATION decode_map_con OF decode IS
  FOR structural
    FOR I1 : inv USE ENTITY WORK.inv(behave);
    PORT MAP( x => a, y => b );
    END FOR;

    FOR I2 : inv USE ENTITY WORK.inv(behave);
    PORT MAP( x => a, y => b );
    END FOR;

    FOR ALL : and3 USE ENTITY WORK.and3(behave);
    END FOR;
```

```
END FOR;
END decode_map_con;
```

The port map clause maps the port names of the component declarations, called the formal ports, to the port names of the entities from the library. The term used for the ports of the entities from the library being mapped are *actuals*. The ports are mapped using named association. The rules for mapping ports using named association in the configuration port map clause are the same rules as used in the component instantiation port map clause.

In the preceding example, component declaration **inv**, port **a**, is mapped to entity **inv**, port **x**, of the actual entity. Component declaration **inv**, port **b**, is mapped to entity **inv**, port **y**, of the actual entity. Using the configuration port map clause can allow entities with completely different port names to be mapped into existing architectures.

Mapping Library Entities

Not only can the ports be mapped with the configuration statement, but entities from libraries can be mapped to components as well. This capability allows the names of components to differ from the actual entities being mapped to them. The designer can easily switch the entity used for each component in the architecture from one entity to another. This feature allows the designer to map component instances to different entities.

Let's look again at the decode example from the beginning of this chapter. The inverter **inv** for the decoder could be modeled using NMOS transistors as shown in Figure 7-3.

Component X1 is an NMOS depletion transistor, which for all intents and purposes in digital simulation acts like a resistor. Component X2 is a unidirectional pass transistor that transfers ground from the triangular-shaped component X3 to signal O1. Components X3 and X4 are ground and VCC, respectively. Following is the model that represents this circuit:

```
LIBRARY IEEE;
USE IEEE.std_logic_1164.ALL;
ENTITY ground IS
  PORT( x : OUT std_logic );
END ground;

ARCHITECTURE dirt OF ground IS
BEGIN
  X <= '0';
```

Figure 7-3
Transistor Level
Schematic for
Inverter.

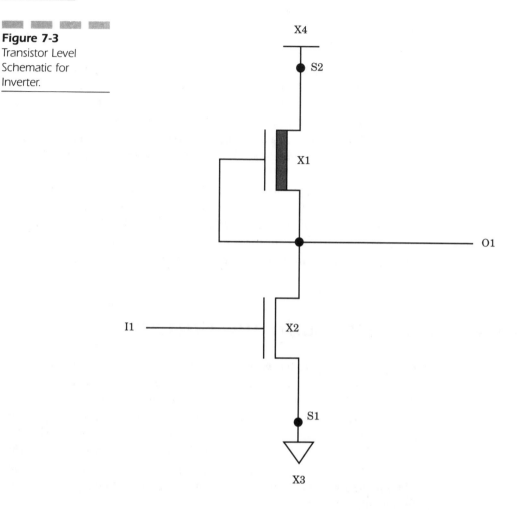

```
END dirt;

CONFIGURATION groundcon OF ground IS
 FOR dirt
 END FOR;
END groundcon;

LIBRARY IEEE; USE IEEE.std_logic_1164.ALL;
ENTITY vcc IS
 PORT( x : OUT std_logic);
END vcc;

ARCHITECTURE plus5 OF vcc IS
BEGIN
 x <= '1';
END plus5;
```

```
CONFIGURATION vcccon OF vcc IS
 FOR plus5
 END FOR;
END vcccon;
```

Following are the two transistor devices:

```
LIBRARY IEEE;
USE IEEE.std_logic_1164.ALL;
ENTITY dep IS
 PORT( top : IN std_logic;
       bottom : OUT std_logic);
END dep;

ARCHITECTURE behave OF dep IS
BEGIN
 bottom <= R1 WHEN top = '1' ELSE
           R0 WHEN top = '0' ELSE
           RX WHEN top = 'X' ELSE
           top;
END behave;

CONFIGURATION depcon OF dep IS
 FOR behave
 END FOR;
END depcon;

LIBRARY IEEE; USE IEEE.std_logic_1164.ALL;
ENTITY uxfr IS
 PORT( left, sw : IN std_logic;
       right : inout std_logic);
END uxfr;

ARCHITECTURE behave OF uxfr IS
BEGIN
 uxfr_proc: PROCESS(left, sw, right)
 BEGIN
  CASE f_state(sw) is
   WHEN '1' =>
    right <= left;
   WHEN '0' =>
    right <= f_convz(right);
   WHEN 'X' =>
    right <= 'X';
  END CASE;
 END PROCESS uxfr_proc;
END behave;

CONFIGURATION uxfrcon OF uxfr IS
 FOR behave
 END FOR;
END uxfrcon;
```

These parts can be connected together, with the following entity-architecture, to form a structural representation of the inverter:

```
LIBRARY IEEE;
USE IEEE.std_logic_1164.ALL;
ENTITY struc_inv IS
 PORT( I1 : IN std_logic;
       O1 : OUT std_logic);
END struc_inv;

ARCHITECTURE structural OF struc_inv IS
 COMPONENT ground
  PORT( X : OUT std_logic);
 END COMPONENT;

 COMPONENT vcc
  PORT( X : OUT std_logic);
 END COMPONENT;

 COMPONENT dep
  PORT( top : IN std_logic;
        bottom : OUT std_logic);
 END COMPONENT;

 COMPONENT uxfr
  PORT( left, sw : IN std_logic;
        right : OUT std_logic);
 END COMPONENT;

 SIGNAL s1, s2 : std_logic;

BEGIN

 X1 : dep
  PORT MAP( top => s2, bottom => O1 );

 X2 : uxfr
  PORT MAP( left => s1, sw => I1, right => O1 );

 X3 : ground
  PORT MAP( x => s1 );

 X4 : vcc
  PORT MAP( x => s2 );

END structural;
```

This architecture can then be configured with the following configuration:

```
CONFIGURATION inv_transcon OF struc_inv IS
```

```
    FOR X1 : dep USE CONFIGURATION WORK.depcon;
    END FOR;

    FOR X2 : uxfr USE CONFIGURATION WORK.uxfrcon;
    END FOR;

    FOR X3 : ground USE CONFIGURATION WORK.groundcon;
    END FOR;

    FOR X4 : vcc USE CONFIGURATION WORK.vcccon;
    END FOR;
END inv_transcon;
```

Now, configuration **decode_map_con** of entity **decode**, described earlier, can be modified as follows:

```
CONFIGURATION decode_map_con OF decode IS
  FOR structural
    FOR I1 : inv USE ENTITY WORK.inv(behave);
     PORT MAP( a => x, b => y );
    END FOR;

    FOR I2 : struc_inv USE CONFIGURATION WORK.inv_transcon;
     PORT MAP( a => I1, b => O1 );
    END FOR;

    FOR ALL : and3 USE ENTITY WORK.and3(behave);
    END FOR;

  END FOR;
END decode_map_con;
```

This configuration maps the first inverter, **I1**, to entity **inv**, and the second inverter, **I2**, to the structural inverter, **struc_inv**. Also, the **I1** and **O1** ports of **struc_inv** are mapped to ports **a** and **b** of the component declaration for component **inv**.

Generics in Configurations

Generics are parameters that are used to pass information into entities. Typical applications include passing in a generic value for the rise and fall delay of output signals of the entity. Other applications include passing in temperature, voltage, and loading to calculate delay values in the model. (Modeling efficiency delay calculations should be done prior to simulation and the calculated delay values can then be passed back into the model through generics.) A description of generics can be found in Chapter 3, "Sequential Processing." This section concentrates on how configurations can be used to specify the value of generics.

Generics can be declared in entities, but can have a value specified in a number of places, as listed in the following:

- A default value can be specified in the generic declaration.

- A value can be mapped in the architecture, in the component instantiation.

- A default value can be specified in the component declaration.

- A value can be mapped in the configuration for the component.

Default values specified in the generic declaration, or the component declaration, can be overridden by mapped values in the architecture or configuration sections. If no overriding values are present, the default values are used; but if a value is mapped to the generic with a generic map, the default value is overridden.

To see an example of this, let's modify the decoder example, used previously in this chapter, to include two generics. The first specifies a timing mode to run the simulation, and the second is a composite type containing the delay values for the device. These two types are declared in the package **p_time_pack**, as shown in the following:

```
LIBRARY IEEE;
USE IEEE.std_logic_1164.ALL;
PACKAGE p_time_pack IS
  TYPE t_time_mode IS (minimum, typical, maximum);
  TYPE t_rise_fall IS
   RECORD
     rise : TIME;
     fall : TIME;
   END RECORD;

  TYPE t_time_rec IS ARRAY(t_time_mode'LOW TO
           t_time_mode'HIGH) OF t_rise_fall;

  FUNCTION calc_delay(newstate : IN std_logic; mode : IN
       t_time_mode;
             delay_tab : IN t_time_rec ) return time;

END p_time_pack;
```

```
PACKAGE BODY p_time_pack IS
 FUNCTION calc_delay(newstate : IN std_logic; mode : IN
     t_time_mode;
             delay_tab : IN t_time_rec ) return time IS
 BEGIN
  CASE f_state(newstate) IS
    WHEN '0' =>
     RETURN delay_tab(mode).fall;
    WHEN '1' =>
     RETURN delay_tab(mode).rise;
    WHEN 'X' =>
     IF (delay_tab(mode).rise <= delay_tab(mode).fall) THEN
       RETURN delay_tab(mode).rise;
     ELSE
       RETURN delay_tab(mode).fall;
     END IF;
  END CASE;
 END calc_delay;
END p_time_pack;
```

This package declares types **t_time_mode** and **t_time_rec**, which are used for the generics of the inverter and three input AND gates. It also includes a new function, **calc_delay**, which is used to retrieve the proper delay value from the delay table, depending on the type of transition occurring.

The **and3** and **inv** gates of the decoder example have been rewritten to include the generics discussed previously, as well as the delay calculation function. Following are the new models:

```
LIBRARY IEEE;
USE IEEE.std_logic_1164.ALL;
USE WORK.p_time_pack.ALL;
ENTITY inv IS
 GENERIC( mode : t_time_mode;
     delay_tab : t_time_rec :=
        (( 1 ns, 2 ns),      -- min
         ( 2 ns, 3 ns),      -- typ
         ( 3 ns, 4 ns)));    -- max

 PORT( a : IN std_logic;
       b : OUT std_logic);
END inv;

ARCHITECTURE inv_gen OF inv IS
BEGIN
 inv_proc : PROCESS(a)
   VARIABLE state : std_logic;
 BEGIN
```

```
      state := NOT(a);
    b <= state after calc_delay( state, mode, delay_tab);
  END PROCESS inv_proc;
END inv_gen;

LIBRARY IEEE; USE IEEE.std_logic_1164.ALL;
USE WORK.p_time_pack.ALL;
ENTITY and3 IS
  GENERIC( mode : t_time_mode;
    delay_tab : t_time_rec :=
        (( 2 ns, 3 ns),   -- min
          ( 3 ns, 4 ns),   -- typ
          ( 4 ns, 5 ns)));  -- max

  PORT( a1, a2, a3 : IN std_logic;
        o1 : OUT std_logic);
END and3;

ARCHITECTURE and3_gen OF and3 IS
BEGIN
  and3_proc : PROCESS( a1, a2, a3 )
    VARIABLE state : std_logic;
  BEGIN
    state := a1 AND a2 AND a3;
    o1 <= state after calc_delay( state, mode, delay_tab);
  END PROCESS and3_proc;
END and3_gen;
```

After the entities and architectures for the gates have been defined, configurations that provide specific values for the generics are defined.

These models can have their generic values specified by two methods. The first method is to specify the generic values in the architecture where the components are being instantiated. The second method is to specify the generic values in the configuration for the model, where the components are instantiated.

Generic Value Specification in Architecture

Specifying the generic values in the architecture of an entity allows the designer to delay the specification of the generic values until the architecture of the entity is created. Different generic values can be specified for each instance of an entity allowing one entity to represent many different physical devices. Following is an example of an architecture with the generic values specified in it:

```
ARCHITECTURE structural OF decode IS
 COMPONENT inv
  GENERIC( mode : t_time_mode;
           delay_tab : t_time_rec);
  PORT( a : IN std_logic;
        b : OUT std_logic);
 END COMPONENT;

 COMPONENT and3
  GENERIC( mode : t_time_mode;
           delay_tab : t_time_rec);
  PORT( a1, a2, a3 : IN std_logic;
        o1 : OUT std_logic);
 END COMPONENT;

 SIGNAL nota, notb : std_logic;
BEGIN
 I1 : inv
  GENERIC MAP( mode => maximum,
          delay_tab => ((1.3 ns, 1.9 ns),
                        (2.1 ns, 2.9 ns),
                        (3.2 ns, 4.1 ns)))
  PORT MAP( a, nota );

 I2 : inv
  GENERIC MAP( mode => minimum,
          delay_tab => ((1.3 ns, 1.9 ns),
                        (2.1 ns, 2.9 ns),
                        (3.2 ns, 4.1 ns)))
  PORT MAP( b, notb );

 A1 : and3
  GENERIC MAP( mode => typical,
          delay_tab => ((1.3 ns, 1.9 ns),
                        (2.1 ns, 2.9 ns),
                        (3.2 ns, 4.1 ns)))
  PORT MAP( nota, en, notb, q0 );

 A2 : and3
  GENERIC MAP( mode => minimum,
          delay_tab => ((1.3 ns, 1.9 ns),
                        (2.1 ns, 2.9 ns),
                        (3.2 ns, 4.1 ns)))
  PORT MAP( a, en, notb, q1 );
```

```
A3 : and3
  GENERIC MAP( mode => maximum,
         delay_tab => ((1.3 ns, 1.9 ns),
                       (2.1 ns, 2.9 ns),
                       (3.2 ns, 4.1 ns)))
  PORT MAP( nota, en, b, q2 );

A4 : and3
  GENERIC MAP( mode => maximum,
         delay_tab => ((2.3 ns, 2.9 ns),
                       (3.1 ns, 3.9 ns),
                       (4.2 ns, 5.1 ns)))
  PORT MAP( a, en, b, q3 );
END structural;
```

Generics are treated in the same manner as ports with respect to how they are mapped. If a component port in a component declaration has a different name than the actual entity compiled into the library, then a port map clause is needed in the configuration specification, for the containing entity. The same is true for a generic. If a generic declaration in a component declaration has a different name than the actual generic for the component, then a generic map clause is needed to make the appropriate mapping.

In the preceding example, the generic names are the same in the entity declaration and the component declaration; therefore, the default mapping provides the appropriate connection between the two.

The configuration for the preceding example needs only to specify which actual entities will be used for the component instantiations in the architecture. No generic information needs to be provided, because the generics have been mapped in the architecture. The configuration can be specified as shown in the following:

```
CONFIGURATION decode_gen_con2 OF decode IS
 FOR structural
  FOR i1, i2 : inv USE ENTITY WORK.inv(inv_gen);
  END FOR;

  FOR a1, a2, a3, a4 : and3 USE ENTITY
     WORK.and3(and3_gen);
  END FOR;
 END FOR;
END decode_gen_con2;
```

The lower-level configuration cannot specify values for the generics if the architecture has mapped values to the generics in the architecture.

Generic Specifications in Configurations

The method of specifying generic values with the most flexibility is to specify generic values in the configuration for the entity. This method allows the latest binding of all the methods for specifying the values for generics. Usually, the later the values are specified, the better. Late binding allows back annotation of path delay generics to occur in the configuration.

For instance, there are a number of steps involved in the design of an ASIC. Following are the steps required:

1. Create the logic design model of a device.
2. Simulate the model.
3. Add estimated delays to device model.
4. Simulate model.
5. Create physical layout of the model.
6. Calculate physical delays from the layout.
7. Feed back physical delays to the device model.
8. Resimulate using actual delays.

The process of feeding back the physical delays into the model can be accomplished by modifying the architecture or by creating a configuration to map the delays back to the model. Modifying the architecture involves changing the values in all of the generic map clauses used to map the delays in the architecture. This method has a big drawback. Modifying the architecture that contains the component instantiation statements requires recompilation of the architecture and the configuration for the design unit. This can be an expensive proposition in a very large design.

The second method, which creates a configuration that maps all of the delays to the generics of the entity, is much more efficient. A configuration of this type contains a generic map value for each generic to be specified in the configuration. Any generics not specified in the configuration are mapped in the architecture or defaulted.

Let's use the decoder example again but now assume that it represents part of an ASIC that has delays back annotated to it. The **inv** and **and3** devices have an intrinsic propagation delay through the device that is based on the internal characteristics of the device, and these devices have an external delay that is dependent on the driver path and device loading. The intrinsic and external delays are passed into the model as generic values. The intrinsic delay is passed into the model to allow a single model to be used for model processes. The external delay is passed to the model, because it may vary for every instance, as loading may be different for each instance. (A more accurate model of delays is obtained using input delays.)

The entity and architecture for the **inv** and **and3** gates look like this:

```
LIBRARY IEEE;
USE IEEE.std_logic_1164.ALL;
ENTITY inv IS
  GENERIC(int_rise, int_fall, ext_rise,
          ext_fall : time);
  PORT( a: IN std_logic; b: OUT std_logic);
END inv;

ARCHITECTURE inv_gen1 OF inv IS
BEGIN
  inv_proc : PROCESS(a)
   VARIABLE state : std_logic;
  BEGIN
   state := NOT(a);
   IF state = '1' THEN
     b <= state AFTER (int_rise + ext_rise);
   ELSIF state = '0' THEN
     b <=state AFTER (int_fall + ext_fall);
   ELSE
     b <= state AFTER (int_fall + ext_fall);
   END IF;
  END PROCESS inv_proc;
END inv_gen1;
--------------------------------------------------
LIBRARY IEEE; USE IEEE.std_logic_1164.ALL;
ENTITY and3 IS
   GENERIC(int_rise, int_fall, ext_rise, ext_fall : time);
  PORT( a1, a2, a3: IN std_logic;
        o1: OUT std_logic);
END and3;

ARCHITECTURE and3_gen1 OF and3 IS
BEGIN
  and3_proc : PROCESS(a1, a2, a3)
   VARIABLE state : std_logic;
  BEGIN
   state := a1 AND a2 AND a3;

   IF state = '1' THEN
```

```
       o1 <= state AFTER (int_rise + ext_rise);
   ELSIF state = `0' THEN
     o1 <= state AFTER (int_fall + ext_fall);
   ELSE
     o1 <= state AFTER (int_fall + ext_fall);
   END IF;

 END PROCESS and3_proc;
END and3_gen1;
```

There are no local configurations specified at this level in the design because this has nearly the same effect of mapping the generic values in the architecture. Instead, a full configuration for entity decode is specified that maps the generics at all levels of the decoder. The entity and architecture for the decoder, as shown in the following, are very similar to the original example used earlier:

```
LIBRARY IEEE;
USE IEEE.std_logic_1164.ALL;
ENTITY decode IS
 PORT( a, b, en : IN std_logic;
       q0, q1, q2, q3 : OUT std_logic);
END decode;

ARCHITECTURE structural OF decode IS
 COMPONENT inv
  PORT( a : IN std_logic;
        b : OUT std_logic);
 END COMPONENT;

 COMPONENT and3
  PORT( a1, a2, a3 : IN std_logic;
        o1 : OUT std_logic);
 END COMPONENT;

 SIGNAL nota, notb : std_logic;
BEGIN
 I1 : inv
  PORT MAP( a, nota);

 I2 : inv
  PORT MAP( b, notb);

 AN1 : and3
  PORT MAP( nota, en, notb, q0);

 AN2 : and3
  PORT MAP( a, en, notb, q1);

 AN3 : and3
```

```
     PORT MAP( nota, en, b, q2);

 AN4 : and3
   PORT MAP( a, en, b, q3);
END structural;
```

Notice that the component declarations for components **inv** and **and3** in the architecture declaration section do not contain the generics declared in the entity declarations for entities **inv** and **and3**. Because the generics are not being mapped in the architecture, there is no need to declare the generics for the components in the architecture.

Following is the configuration to bind all of these parts together into an executable model:

```
CONFIGURATION decode_gen1_con OF decode IS
 FOR structural
   FOR I1 : inv USE ENTITY WORK.inv(inv_gen1)
     GENERIC MAP( int_rise => 1.2 ns,
                  int_fall => 1.7 ns,
                  ext_rise => 2.6 ns,
                  ext_fall => 2.5 ns);
   END FOR;

   FOR I2 : inv USE ENTITY WORK.inv(inv_gen1)
     GENERIC MAP( int_rise => 1.3 ns,
                  int_fall => 1.4 ns,
                  ext_rise => 2.8 ns,
                  ext_fall => 2.9 ns);
   END FOR;

   FOR AN1 : and3 USE ENTITY WORK.and3(and3_gen1)
     GENERIC MAP( int_rise => 2.2 ns,
                  int_fall => 2.7 ns,
                  ext_rise => 3.6 ns,
                  ext_fall => 3.5 ns);
   END FOR;

   FOR AN2 : and3 USE ENTITY WORK.and3(and3_gen1)
     GENERIC MAP( int_rise => 2.2 ns,
                  int_fall => 2.7 ns,
                  ext_rise => 3.1 ns,
                  ext_fall => 3.2 ns);
   END FOR;

   FOR AN3 : and3 USE ENTITY WORK.and3(and3_gen1)
     GENERIC MAP( int_rise => 2.2 ns,
                  int_fall => 2.7 ns,
                  ext_rise => 3.3 ns,
```

```
                              ext_fall => 3.4 ns);
          END FOR;

          FOR AN4 : and3 USE ENTITY WORK.and3(and3_gen1)
            GENERIC MAP( int_rise => 2.2 ns,
                         int_fall => 2.7 ns,
                         ext_rise => 3.0 ns,
                         ext_fall => 3.1 ns);
          END FOR;
         END FOR;
        END decode_gen1_con;
```

Each component instance is configured to the correct entity and architecture, and the generics of the entity are mapped with a generic map clause. Using this type of configuration allows each instance to have unique delay characteristics. Of course, the generics passed into the device can represent any type of data the designer wants, but typically the generics are used to represent delay information. VITAL uses generics to pass delay information to library components. We examine this more closely in later chapters.

The power of this type of configuration is realized when the delay values are updated. For instance, in the ASIC example, the estimated delays are included in the configuration initially, but after the ASIC device has been through the physical layout process, the actual delay information can be determined. This information can be fed back into the configuration so that the configuration has the actual delay information calculated from the layout tool. Building a new simulatable device, including the new delay information, requires only a recompile of the configuration. The entities and architectures do not need to be recompiled.

If the delay information was included in the architecture for the device, then a lot more of the model would need to be recompiled to build the simulatable entity. All of the architectures that included the generics would need to be recompiled, and so would the configuration for the entity. A lot of extra code would be recompiled unnecessarily.

The information in this section on generics can be summarized by the charts shown in Figures 7-4 and 7-5. (These charts were originally created by Paul Krol.)

These charts shows the effect of the declarations and mapping of generics on the values actually obtained in the model. The first four columns of Figure 7-4 describe where a particular generic, G, can be declared and mapped to a value. The next column describes the error/warning number returned from a particular combination of declaration and mapping. The

Figure 7-4
Configuration
Generic Table.

Declaration		Mapping		Error / Warning	Generic Values	
Entity	Component	Instance	Configuration		Same	Other
D	D	A			I	E
D	N	A			I	E
D	N		A		C	E
N	D	A			I	M
N	D		A		C	M
X	D / N		A	1		
X	D / N			2		
X	D / N	A		2		
D / N	X	A		3		
D / N	X		A		C	E
		A	A	4		
		X	X	5		

next two columns describe the values obtained by the generic, G, and any other generics for the entity for a particular declaration and mapping combination. At the bottom of Figure 7-4 and in Figure 7-5, are the tables of translations used to translate the character values used to the appropriate action taken.

Board-Socket-Chip Analogy

A good analogy for describing how entity declarations, architectures, component declarations, and configuration specifications all interact is the board-socket-chip analogy. (This analogy was originally presented to me by Dr. Alec Stanculescu.) In this analogy, the architecture of the top-level entity represents the board being modeled. The component instance represents a socket on the board, and the lower-level entity being instantiated in the architecture represents the chip.

This analogy helps describe how the ports and generics are mapped at each level. At the board (architecture) level component socket pins are interconnected with signals. The chip pins are then connected to socket

Figure 7-5
Configuration
Generic Table Transla-
tions.

Declarations / Mapping	
D	Declared, with default value
N	Declared, with no default value
X	Not Declared
A	Actual Mapped

Errors / Warnings	
1	Can only map generic in configuration if declared in the entity
2	Generic declared in component but not in entity, hence it is not used
3	Can only map generic in component instance if declared in component declarations
4	Can't map a generic in the component instance and the configuration
5	Must map at least once generic to get the default value for other generics

Generic Values	
E	Default taken from Entity
M	Default taken form configuration
I	Actual taken from component instance
C	Actual taken from configuration

pins when the chip is plugged into the socket. Following is an example of how this works:

```
LIBRARY IEEE;
USE IEEE.std_logic_1164.ALL;
ENTITY board IS
  GENERIC (qdelay, qbdelay : time);
  PORT( clk, reset, data_in : IN std_logic;
        data_out : OUT std_logic);
END board;

ARCHITECTURE structural OF board IS
  COMPONENT dff
    GENERIC( g1, g2 : time);
    PORT( p1, p2, p3, p4 : IN std_logic;
```

```
        p5, p6 : OUT std_logic);
    END COMPONENT;

    SIGNAL ground : std_logic := '1';
    SIGNAL int1, nc : std_logic;
BEGIN
    U1 : dff
      GENERIC MAP( g1 => qdelay,
                   g2 => qbdelay)
      PORT MAP( p1 => clk,
                p2 => data_in,
                p3 => reset,
                p4 => ground,
                p5 => int1,
                p6 => nc);

    U2 : dff
      GENERIC MAP( g1 => qdelay,
                   g2 => qbdelay)
      PORT MAP( p1 => clk,
                p2 => int1,
                p3 => reset,
                p4 => ground,
                p5 => data_out,
                p6 => nc);
    END structural;
```

The entity and architecture shown are a simple 2-bit shift register made from two D flip-flop (**DFF**) component instantiations. This example, though relatively simple, shows how ports and generics are mapped at different levels.

The component instance for component **DFF** in the architecture statement part acts like a socket in the architecture for the board. When a component instance is placed in the architecture, signals are used to connect the component to the board, which is the architecture. The actual chip is not connected to the socket until a configuration is specified for the board entity. If all of the names of the socket ports and generics match the names of the actual entity being used, then no mapping is needed. The default mapping connects the chip to the socket. If the names are different, or the number of ports are not the same, for the component instantiation and the actual entity, then a mapping between the socket (component instantiation) and the chip (actual entity) is needed.

The actual chip to be mapped is described by the entity and architecture shown here:

```
LIBRARY IEEE;
USE IEEE.std_logic_1164.ALL;
```

```
ENTITY dff IS
 GENERIC( q_out, qb_out : time);
 PORT( preset, clear, din,
       clock : IN std_logic;
       q, qb : OUT std_logic);
END dff;

ARCHITECTURE behave OF dff IS
BEGIN
 dff_proc : PROCESS(preset, clear, clock)
  VARIABLE int_q : std_logic;
 BEGIN
  IF preset = '0' and clear = '0' THEN
    IF (clock'EVENT) AND (clock = '1') THEN
     int_q := din;
    END IF;

  ELSIF preset = '1' AND clear = '0' THEN
   int_q := '1';

  ELSIF clear = '1' AND preset = '0' THEN
   int_q := '0';

  ELSE
   int_q := 'X';

  END IF;

  q <= int_q after q_out;

  int_q := not(int_q);
  qb <= int_q after qb_out;

 END PROCESS dff_proc;
END behave;
```

The names of the ports and generics are completely different than the component declaration; therefore, mapping is required. Following is a configuration that places the actual chip in the socket (maps the ports and generics):

```
CONFIGURATION board_con OF board IS
 FOR structural
  FOR U1,U2: dff USE WORK.dff(behave)
   GENERIC MAP( q_out => g1, qb_out => g2)
   PORT MAP( preset => ground, clear => p3,
             din => p2, clock => p1,
             q => p5, qb => p6);
  END FOR;
 END FOR;
END board_con;
```

Block Configurations

When an architecture contains block statements, the configuration must reflect this fact. (Block statements are discussed in Chapter 2, "Behavioral Modeling.") Blocks act like another level of hierarchy between the containing architecture and any components being configured. The configuration must specify which block of a configuration is being configured when the architecture is being configured.

Following shows an architecture fragment that contains three blocks:

```
LIBRARY IEEE;
USE IEEE.std_logic_1164.ALL;
ENTITY cpu IS
  PORT( clock : IN std_logic;
        addr  : OUT std_logic_vector(0 to 3);
        data  : INOUT std_logic_vector(0 to 3);
        interrupt : IN std_logic;
        reset : IN std_logic);
END cpu;

ARCHITECTURE fragment OF cpu IS
  COMPONENT int_reg
   PORT( data : IN std_logic;
         regclock : IN std_logic;
         data_out : OUT std_logic);
  END COMPONENT;

  COMPONENT alu
   PORT( a, b : IN std_logic;
         c, carry : OUT std_logic);
  END COMPONENT;

  SIGNAL a, c, carry : std_logic_vector(0 TO 3);
BEGIN
 reg_array : BLOCK
 BEGIN
  R1 : int_reg
    PORT MAP( data(0), clock, data(0));

  R2 : int_reg
    PORT MAP( data(1), clock, data(1));

  R3 : int_reg
    PORT MAP( data(2), clock, data(2));

  R4 : int_reg
    PORT MAP( data(3), clock, data(3));

  END BLOCK reg_array;
```

```
shifter : BLOCK
BEGIN
 A1 : alu
    PORT MAP( a(0), data(0), c(0), carry(0));

 A2 : alu
    PORT MAP( a(1), data(1), c(1), carry(1));

 A3 : alu
    PORT MAP( a(2), data(2), c(2), carry(2));

 A4 : alu
    PORT MAP( a(3), data(3), c(3), carry(3));

 shift_reg : BLOCK
 BEGIN
  R1 : int_reg
    PORT MAP( data, shft_clk, data_out);

 END BLOCK shift_reg;
END BLOCK shifter;
END fragment;
```

The architecture consists of three blocks, each containing component instantiations. The first block contains four **int_reg** components, and the second contains an **alu** component, plus another **BLOCK** statement. The last block contains a single **int_reg** component.

The configuration for this architecture must take into account the fact that **BLOCK** statements exist in the architecture. Following is a simple configuration for the architecture:

```
CONFIGURATION cpu_con OF cpu IS
 FOR fragment
  FOR reg_array
   FOR ALL: int_reg USE CONFIGURATION WORK.int_reg_con;
  END FOR;
 END FOR;
  FOR shifter
   FOR A1 : alu USE CONFIGURATION WORK.alu_con;
   END FOR;
   FOR shift_reg
    FOR R1 : int_reg USE CONFIGURATION WORK.int_reg_con;
     END FOR;
    END FOR;
  END FOR;
 END FOR;
END cpu_con;
```

In the configuration **cpu_con** of entity **cpu**, architecture **fragment** is used for the entity. Inside of block **reg_array**, all (**R1** through **R4**) of the

Figure 7-6
Block Diagram of
Autopilot Example.

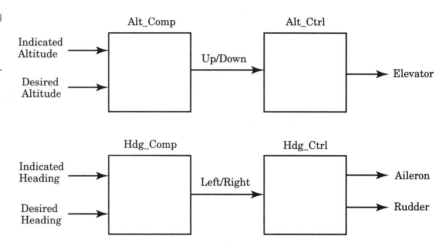

int_reg components use configuration **int_reg_con**. In block **shifter**, the **alu** component (**A1**) uses configuration **alu_con**. For block **shift_reg** inside of block **shifter**, the **int_reg** component use configuration **int_reg_con**.

Architecture Configurations

The last type of configuration we discuss is the architecture configuration. This configuration exists in the architecture declarative region and specifies the configurations of parts used in the architecture. If this type of configuration is used, a separate configuration declaration is not needed to configure the components used in the architecture.

The next example configuration is for a very high-level description of an autopilot. The autopilot block diagram is shown in Figure 7-6. Following is an example of this type of configuration:

```
PACKAGE ap IS
  TYPE alt IS INTEGER RANGE 0 TO 50000;
  TYPE hdg IS INTEGER RANGE 0 TO 359;
  TYPE vdir IS INTEGER RANGE 0 TO 9;
  TYPE hdir IS INTEGER RANGE 0 TO 9;
  TYPE control IS INTEGER RANGE 0 TO 9;
END ap;

USE WORK.ap.ALL;
ENTITY autopilot IS
  PORT( altitude : IN alt;
```

```
            altitude_set : IN alt;
            heading : IN hdg;
            heading_set : IN hdg;
            rudder : OUT control;
            aileron : OUT control;
            elevator : OUT control);
END autopilot;

ARCHITECTURE block_level OF autopilot IS
 COMPONENT alt_compare
  PORT( alt_ref : IN alt;
        alt_ind : IN alt;
        up_down : OUT vdir);
 END COMPONENT;

 COMPONENT hdg_compare
  PORT( hdg_ref : IN hdg;
        hdg_ind : IN hdg;
        left_right : OUT hdir);
 END COMPONENT;

 COMPONENT hdg_ctrl
  PORT( left_right : IN hdir;
        rdr : OUT control;
        alrn : OUT control);
 END COMPONENT;

 COMPONENT alt_ctrl
  PORT( up_down : IN vdir;
        elevator : OUT control);
 END COMPONENT;

 SIGNAL up_down : vdir;
 SIGNAL left_right : hdir;

 FOR M1 : alt_compare USE CONFIGURATION WORK.alt_comp_con;

 FOR M2 : hdg_compare USE CONFIGURATION WORK.hdg_comp_con;

 FOR M3 : hdg_ctrl USE ENTITY WORK.hdg_ctrl(behave);

 FOR M4 : alt_ctrl USE ENTITY WORK.alt_ctrl(behave);

BEGIN
 M1 : alt_compare
  PORT MAP( alt_ref => altitude,
            alt_ind => alt_set,
            up_down => up_down);

 M2 : hdg_compare
  PORT MAP( hdg_ref => heading,
            hdg_ind => hdg_set,
            left_right => left_right);
```

```
M3 : hdg_ctrl
  PORT MAP( left_right => left_right,
            rdr => rudder,
            alrn => aileron);

M4 : alt_ctrl
  PORT MAP( up_down => up_down,
            elevator => elevator);

END block_level;
```

This model is a top-level description of an autopilot. There are four instantiated components that provide the necessary functionality of the autopilot. This model demonstrates how component instantiations can be configured in the architecture declaration section of an architecture. Notice that after the component declarations in the architecture declaration section of architecture **block_level**, there are four statements similar to the following:

```
FOR M1 : alt_compare USE CONFIGURATION WORK.alt_comp_con;
```

These statements allow the designer to specify either the configuration or the entity-architecture pair to use for a particular component type. This type of configuration does not provide the same flexibility to the designer as the separate configuration declaration, but it is useful for small designs.

Configurations are a useful tool for managing large designs. With proper use of configurations, a top-down design approach can be implemented that allows all levels of description of the design to be used for the most efficient model needed at any point in the design process.

SUMMARY

In this chapter, we discussed the following:

- How default configurations can be used to bind architectures to entities.

- How component configurations can be used to specify which entity to use for each component instantiation.

- How port maps within configurations allow mapping entities with different names to component instances.

- How generics can be specified in configurations to allow late binding of generic information.

- How block configurations can be used to configure architectures with block statements in them.

- How architecture configurations allow specification of configurations for component instantiations in the architecture declaration section.

The basic features of VHDL have now been introduced. In the next chapter, we examine some of the more esoteric but useful features that exist in VHDL.

CHAPTER 8

Advanced Topics

In this chapter, some of the more esoteric features of VHDL are discussed. Some of the features may be useful for certain types of designs, and not for others. Typical usage examples are presented to show how these features might be taken advantage of.

Some of the features discussed include overloading, qualified expressions, user-defined attributes, generate statements, aliases, and TextIO. All of these features provide the user with an advanced environment with which to do modeling.

Overloading

Overloading allows the designer to write much more readable code. An object is overloaded when the same object name exists for multiple subprograms or type values. The VHDL compiler selects the appropriate object to use in each instance.

In VHDL, a number of types of overloading are possible. Subprograms can be overloaded, operators can be overloaded, and enumeration types can be overloaded. Overloading subprograms allows subprograms to operate on objects of different types. Overloading an operator allows the operator to perform the same operation on multiple types. Overloading frees the designer from the necessity of generating countless unique names for subprograms that do virtually the same operation. The result of using overloaded subprograms and operators is models that are easier to read and maintain.

Subprogram Overloading

Subprogram overloading allows the designer to write multiple subprograms with the same name, but the number of arguments, the type of arguments, and return value (if any) can be different. The VHDL compiler, at compile time, selects the subprogram that matches the subprogram call. If no subprogram matches the call, an error is generated.

The following example illustrates how a subprogram can be overloaded by the argument type:

```
LIBRARY IEEE;
USE IEEE.std_logic_1164.ALL;
PACKAGE p_shift IS
  TYPE s_int IS RANGE 0 TO 255;
  TYPE s_array IS ARRAY(0 TO 7) OF std_logic;

  FUNCTION shiftr( a : s_array) return s_array;
  FUNCTION shiftr( a : s_int) return s_int;
END p_shift;

PACKAGE BODY p_shift IS
  FUNCTION shiftr( a : s_array) return s_array IS
   VARIABLE result : s_array;
  BEGIN
   FOR i IN a'RANGE LOOP
    IF i = a'HIGH THEN
```

```
        result(i) := '0';
    ELSE
        result(i) := a(i + 1);
    END IF;
  END LOOP;

  RETURN result;
END shiftr;

FUNCTION shiftr( a : s_int) return s_int IS
BEGIN
  RETURN (a/2);
END shiftr;
END p_shift;
```

The package **p_shift** contains two functions both named **shiftr**. Both functions provide a right-shift capability, but each function operates on a specific type. One function works only with type **s_int**, and the other works only with type **s_array**. The compiler picks the appropriate function based on the calling argument(s) and return argument.

In the following example, different types of function calls are shown, and the results obtained with each call:

```
USE WORK.p_shift.ALL;
ENTITY shift_example IS
END shift_example;

ARCHITECTURE test OF shift_example IS
  SIGNAL int_signal : s_int;
  SIGNAL array_signal : s_array;
BEGIN
  -- picks function that works with s_int type
  int_signal <= shiftr(int_signal);

  -- picks function that works with
  -- s_array type
  array_signal <= shiftr(array_signal);

  -- produces error because no function
  -- will match
  array_signal <= shiftr(int_signal);
END test;
```

The architecture test contains three calls to function **shiftr**. The first calls **shiftr** with an argument type of **s_int** and a return type of **s_int**. This call uses the second function described in package body **p_shift**, the function with input arguments, and return type of **s_int**.

The second call to **shiftr** uses the array type **s_array**, and therefore picks the first function defined in package **p_shift**. Both the input argument(s) type(s) and return type must match for the function to match the call.

The third call to function **shiftr** shows an example of a call where the input argument matches the **s_int** type function, but the return type of the function does not match the target signal. With the functions currently described in package **p_shift**, no function matches exactly, and therefore the compilation of the third line produces an error.

To make the third call legal, all that is needed is to define a function that matches the types of the third call. An example of the function declaration is shown in the following code line. The function body for this function is left as an exercise for the reader:

```
FUNCTION shiftr( a : s_int) return s_array;
```

OVERLOADING SUBPROGRAM ARGUMENT TYPES

To overload argument types, the base type of the subprogram parameters or return value must differ. For example, base types do not differ when two subtypes are of the same type. Two functions that try to overload these subtypes produce a compile error. Following is an example:

```
PACKAGE type_error IS
  SUBTYPE log4 IS BIT_VECTOR( 0 TO 3);
  SUBTYPE log8 IS BIT_VECTOR( 0 TO 7);

  -- this function is Ok
  FUNCTION not( a : log4) return integer;

  -- this function declaration will cause an
  -- error
  FUNCTION not( a : log8) return integer;

END type_error;
```

This package declares two subtypes **log4** and **log8** of the unconstrained **BIT_VECTOR** type. Two functions named **not** are then declared using these subtypes. The first function declaration is legal, but the second function declaration causes an error. The error is that two functions have been declared for the same base type. The two types **log4** and **log8** are not distinct, because they both belong to the same base type.

All of the examples shown so far have been overloading of functions. Overloading of procedures works in the same manner.

SUBPROGRAM PARAMETER OVERLOADING Two or more subprograms with the same name can have a different number of parameters. The types of the parameters can be the same, but the number of
parameters can be different. This is shown by the following example:

```
LIBRARY IEEE;
USE IEEE.std_logic_1164.ALL;
PACKAGE p_addr_convert IS
 FUNCTION convert_addr(a0, a1 : std_logic) return integer;

 FUNCTION convert_addr(a0, a1, a2 : std_logic) return
     integer;

 FUNCTION convert_addr(a0, a1, a2, a3 : std_logic) return
     integer;

END p_addr_convert;

PACKAGE BODY p_addr_convert IS
 FUNCTION convert_addr(a0, a1 : std_logic) RETURN
     INTEGER IS
  VARIABLE result : INTEGER := 0;
 BEGIN
  IF (a0 = '1') THEN
    result := result + 1;
  END IF;

  IF (a1 = '1') THEN
    result := result + 2;
  END IF;

  RETURN result;
 END convert_addr;

 FUNCTION convert_addr(a0, a1, a2 : std_logic) RETURN
     INTEGER IS
  VARIABLE result : INTEGER := 0;
 BEGIN

result := convert_addr(a0, a1);

  IF (a2 = '1') THEN
    result := result + 4;
  END IF;
  RETURN result;
 END convert_addr;

 FUNCTION convert_addr(a0, a1, a2, a3 : std_logic) RETURN
     INTEGER IS
  VARIABLE result : INTEGER := 0;
 BEGIN
```

```
    result := convert_addr(a0, a1, a2);

      IF (a3 = '1') THEN
        result := result + 8;
      END IF;
      RETURN result;
    END convert_addr;

  END p_addr_convert;
```

This package declares three functions that convert 2, 3, or 4 input bits into integer representation. Each function is named the same, but the appropriate function is called depending on the number of input arguments that are passed to the function. If 2 bits are passed to the function, then the function with two arguments is called. If 3 bits are passed, the function with three arguments is called, and so on.

Following is an example using these functions:

```
LIBRARY IEEE;
USE IEEE.std_logic_1164.ALL;
USE WORK.p_addr_convert.ALL;
ENTITY test IS
  PORT(i0, i1, i2, i3 : in std_logic);
END test;

ARCHITECTURE test1 OF test IS
  SIGNAL int1, int2, int3 : INTEGER;
BEGIN
  -- uses first function
  int1 <= convert_addr(i0, i1);

  -- uses second function
  int2 <= convert_addr(i0, i1, i2);

  -- uses third function
  int3 <= convert_addr(i0, i1, i2, i3);
END test1;
```

The first call to the **convert_addr** function has only two arguments in the argument list, and therefore the first function in package **p_addr_convert** is used. The second call has three arguments in its argument list and calls the second function. The last call matches the third function from package **p_addr_convert**.

Overloading Operators

One of the most useful applications of overloading is the overloading of operators. The need for overloading operators arises because the operators supplied in VHDL only work with specific types. For instance, the + operator only works with integer, real, and physical types, while the & (concatenation) operator only works with array types. If a designer wants to use a particular operator on a user-defined type, then the operator must be overloaded to handle the user type. A complete listing of the operators and the types supported by them can be found in the LRM.

An example of a typical overloaded operator is the + operator. The + operator is defined for the numeric types, but if the designer wants to add two **BIT_VECTOR** objects, the + operator does not work. The designer must write a function that overloads the operator to accomplish this operation. The following package shows an overloaded function for operator + that allows addition of two objects of **BIT_VECTOR** types:

```
PACKAGE math IS
  FUNCTION "+"( l,r : BIT_VECTOR) RETURN INTEGER;
END math;

PACKAGE BODY math IS
 FUNCTION vector_to_int( S : BIT_VECTOR) RETURN INTEGER IS
  VARIABLE result : INTEGER := 0;
  VARIABLE prod : INTEGER := 1;
 BEGIN
  FOR i IN s'RANGE LOOP
   IF s(i) = '1' THEN
     result := result + prod;
   END IF;
   prod := prod * 2;
  END LOOP;

  RETURN result;
 END vector_to_int;

 FUNCTION "+"(l,r : BIT_VECTOR) RETURN INTEGER IS
 BEGIN
   RETURN ( vector_to_int(l) + vector_to_int(r));
 END;
END math;
```

Whenever the + operator is used in an expression, the compiler calls the + operator function that matches the types of the operands. When the operands are of type **INTEGER**, the built-in + operator function is called. If the operands are of type **BIT_VECTOR**, then the function from package **math** is called. The following example shows uses for both functions:

```
USE WORK.math.ALL;
ENTITY adder IS
  PORT( a, b : IN BIT_VECTOR(0 TO 7);
        c : IN INTEGER;
        dout : OUT INTEGER);
END adder;

ARCHITECTURE test OF adder IS
  SIGNAL internal : INTEGER;
BEGIN
  internal <= a + b;
  dout <= c + internal;
END test;
```

This example illustrates how overloading can be used to make very readable models. The value assigned to signal **internal** is the sum of inputs **a** and **b**. Since **a** and **b** are of type **BIT_VECTOR**, the overloaded operator function that has two **BIT_VECTOR** arguments is called. This function adds the values of **a** and **b** together and returns an integer value to be assigned to signal **internal**.

The second addition uses the standard built-in addition function that is standard in VHDL because both operands are of type **INTEGER**. This model could have been written as shown in the following, but would still function in the same manner:

```
PACKAGE math IS
  FUNCTION addvec( l,r : bit_vector) RETURN INTEGER;
END math;

PACKAGE BODY math IS
  FUNCTION vector_to_int( S : bit_vector) RETURN INTEGER IS
   VARIABLE result : INTEGER := 0;
   VARIABLE prod : INTEGER := 1;
  BEGIN
   FOR i IN s'RANGE LOOP
    IF s(i) = '1' THEN
     result := result + prod;
    END IF;
    prod := prod * 2;
   END LOOP;
   RETURN result;
  END vector_to_int;

  FUNCTION addvec(l,r : bit_vector) RETURN INTEGER IS
  BEGIN
```

```
        RETURN ( vector_to_int(l) + vector_to_int(r));
  END addvec;
END math;

USE WORK.math.ALL;
ENTITY adder IS
  PORT( a, b : IN BIT_VECTOR(0 TO 7);
        c : IN INTEGER;
        dout : OUT INTEGER);
END adder;

ARCHITECTURE test2 OF adder IS
  SIGNAL internal : INTEGER;
BEGIN
  internal <= addvec(a,b);
  dout <= c + internal;
END test2;
```

In this example, a function called **advec** is used to add **a** and **b**. Both coding styles give exactly the same results, but the first example using the overloaded + operator is much more readable and easier to maintain. If another person besides the designer of a model takes over the maintenance of the model, it is much easier for the new person to understand the model if overloading was used.

OPERATOR ARGUMENT TYPE OVERLOADING Arguments to overloaded operator functions do not have to be of the same type, as the previous two examples have shown. The parameters to an overloaded operator function can be of any type. In some cases, it is preferable to write two functions so that the order of the arguments is not important.

Let's examine the functions for an overloaded logical operator that mixes signals of type **BIT** and signals of a nine-state value system:

```
PACKAGE p_logic_pack IS
  TYPE t_nine_val IS (Z0, Z1, ZX,
                      R0, R1, RX,
                      F0, F1, FX);

  FUNCTION "AND"( l, r : t_nine_val) RETURN BIT;

  FUNCTION "AND"( l : BIT; r : t_nine_val) RETURN BIT;

  FUNCTION "AND"( l : t_nine_val; r : BIT) RETURN BIT;

END p_logic_pack;
```

```
PACKAGE BODY p_logic_pack IS
 FUNCTION nine_val_2_bit( t : IN t_nine_val) RETURN BIT IS
 TYPE t_nine_val_conv IS ARRAY(t_nine_val) OF BIT;
 CONSTANT nine_2_bit : t_nine_val_conv :=
       ('0',     -- Z0
        '1',     -- Z1
        '1',     -- ZX
        '0',     -- R0
        '1',     -- R1
        '1',     -- RX
        '0',     -- F0
        '1',     -- F1
        '1');    -- FX
BEGIN
 RETURN nine_2_bit(t);
END nine_val_2_bit;

 FUNCTION "AND"(l,r : t_nine_val) RETURN BIT IS
 BEGIN
    RETURN (nine_val_2_bit(l) AND nine_val_2_bit(r));
 END;

 FUNCTION "AND"(l :BIT; r : t_nine_val) RETURN BIT IS
 BEGIN
    RETURN ( l AND nine_val_2_bit(r));
 END;

 FUNCTION "AND"(l : t_nine_val; r : BIT) RETURN BIT IS
 BEGIN
    RETURN (nine_val_2_bit(l) AND r);
 END;
END p_logic_pack;
```

The package **p_logic_pack** declares three overloaded functions for the
AND operator. In one function, both input types are type **t_nine_val**. In
the other two functions, only one input is type **t_nine_val**, and the other
input is type **BIT**. All functions return a result of type **BIT**. Notice that, to
overload the **AND** operator, the syntax is the same as overloading the +
operator from the previous example.

When the **AND** operator is used in a model, the appropriate function is
called based on the types of the operands. In the following code fragments,
we can see the differences:

```
SIGNAL a, b : t_nine_val;
SIGNAL c,e  : bit;

e <= a AND b;
```

```
-- calls first function

e <= a AND c;
-- calls third function

e <= c AND b;
-- calls second function
```

By having three functions called **AND**, we do not need to worry about which side of the operator an expression resides on. All of the possible combinations of operator order are covered with three functions, because the function for two inputs of type **BIT** are built in.

Aliases

An alias creates a new name for all or part of the range of an array type. It is very useful for naming parts of a range as if they were subfields. For example, in a CPU model, an instruction is fetched from memory. The instruction may be an array of 32 bits that is interpreted as a number of smaller fields to represent the instruction opcode, source register 1, source register 2, and so on. Aliases provide a mechanism to name each of the subfields of the instruction and to reference these fields directly by the alias names. This is illustrated by the following example:

```
SIGNAL instruction : BIT_VECTOR(31 DOWNTO 0);

ALIAS opcode : BIT_VECTOR(3 DOWNTO 0) IS instruction(31
     DOWNTO 28);
ALIAS src_reg : BIT_VECTOR(4 DOWNTO 0) IS instruction(27
     DOWNTO 23);
ALIAS dst_reg : BIT_VECTOR(4 DOWNTO 0) IS instruction(22
     DOWNTO 18);
```

In this example, the aliases have been created for a signal object. Using the alias name in an assignment or referencing operation is the same as using the piece of the instruction object being aliased, but much more convenient.

Remember that the semantics in place for the object being aliased are applied to the alias as well. If an alias is created for a constant object, the alias cannot have an assignment for the same reasons that a constant cannot have an assignment.

Qualified Expressions

One of the side effects of overloading is that multiple functions or procedures may match in a particular instance because the types are ambiguous. For the compiler to figure out which subprogram to use, a qualified expression may be required. A qualified expression states the exact type that the expression should attain. For instance, when evaluating an expression containing a mixture of overloaded subprograms and constant values, the designer may need to qualify an expression to produce correct results. Following is an example of such a situation:

```
PACKAGE p_qual IS
  TYPE int_vector IS ARRAY(NATURAL RANGE <>) OF INTEGER;

  FUNCTION average( a : int_vector) RETURN INTEGER;

  FUNCTION average( a : int_vector) RETURN REAL;

END p_qual;

USE WORK.p_qual.ALL;
ENTITY normalize IS
  PORT( factor : IN REAL;
        points : IN int_vector;
        result : OUT REAL);
END normalize;

ARCHITECTURE qual_exp OF normalize IS
BEGIN
  result <= REAL'(average(points)) * factor;
END qual_exp;
```

Package **p_qual** defines two overloaded functions named **average** and an unconstrained type, **int_vector**. The package body is left as an exercise for the reader.

Architecture **qual_exp** has a single concurrent signal assignment statement that calls function **average**. Because there are two functions named **average**, there are two possible functions that can be used by this call. To clarify which function to use, the expression has been qualified to return a **REAL** type. The keyword **REAL** followed by a ' specifies that the expression inside the parentheses return a type **REAL**.

The expression was qualified to make sure that the **average** function returning a **REAL** number was called instead of the **average** function that returns an **INTEGER**. In this example, the expression required a qualified expression to allow the architecture to compile. The compiler does not make any random guesses about which function to use. The designer must specify exactly which one to use in cases where more than one function can match; otherwise, an error is generated.

Another use for a qualified expression is to build the source value for an assignment statement. Based on the type of the signal assignment target, the source value can be built. Following is an example:

```
PACKAGE p_qual_2 IS
  TYPE vector8 IS ARRAY( 0 TO 7) OF BIT;
END p_qual_2;

USE WORK.p_qual_2.ALL;
ENTITY latch IS
  PORT( reset, clock : IN BIT;
        data_in : IN vector8;
        data_out : OUT vector8);
END latch;

ARCHITECTURE behave OF latch IS
BEGIN
  PROCESS(clock)
  BEGIN
   IF (clock = '1') THEN
    IF (reset = '1') THEN
     data_out <= vector8'(others => '0');
    ELSE
     data_out <= data_in;
    END IF;
   END IF;
  END PROCESS;
END behave;
```

This example is an 8-bit transparent **latch**, with a **reset** line to set the **latch** to zero. When the clock input is a '1' value, the **latch** is transparent, and input values are reflected on the output. When the clock input is '0', the **data_in** value is latched. When **reset** is a '1' value while clock input is a '1', the **latch** is reset. This is accomplished by assigning all '0's to **data_out**. One method to assign all '0's to **data_out** is to use an aggregate assignment. Because **data_out** is 8 bits, the following aggregate assignment sets **data_out** to all '0's:

```
data_out <= ('0', '0', '0', '0', '0', '0', '0', '0');
```

This aggregate works fine unless the type of **data_out** changes. If the type of output **data_out** was suddenly changed to 16 bits instead of 8, the aggregate could no longer be used.

Another method to accomplish the assignment to output **data_out** is to use a qualified expression. The assignment to **data_out** when **reset = '1'** in the preceding example shows how this might be done. The following expression:

```
(others => '0')
```

can be qualified with the type of the target signal (**data_out**). This allows the compiler to determine how large the target signal is and how large to make the source being assigned to the signal. Now, whenever the target signal type is changed, the source changes to match.

User-Defined Attributes

VHDL user-defined attributes are a mechanism for attaching data to VHDL objects. The data attached can be used during simulation or by another tool that reads the VHDL description. Data such as the disk file name of the model, loading information, driving capability, resistance, capacitance, physical location, and so on can be attached to objects. The type and value of the data is completely user-definable. The value, when specified, is constant throughout the simulation.

User-defined attributes can behave similar to entity generic values, with one exception. Generics are only legal on entities, but user-defined attributes can be assigned to the following list of objects:

- Entity
- Architecture
- Configuration
- Procedure
- Function
- Package
- Type and Subtype
- Constant

■ Signal

■ Variable

■ Component

■ Label

To see how user-defined attributes operate, let's examine the following description:

```
PACKAGE p_attr IS
 TYPE t_package_type IS ( leadless,
                          pin_grid,
                          dip);

  ATTRIBUTE package_type : t_package_type;
  ATTRIBUTE location : INTEGER;
END p_attr;

USE WORK.p_attr.ALL;
ENTITY board IS
 PORT(
    .
    .
    .
       );
END board;

ARCHITECTURE cpu_board OF board IS
 COMPONENT mc68040
   GENERIC(   . .  . . .   );
   PORT(
    .
    .
    .
       );
 END COMPONENT;
 SIGNAL a : INTEGER;
 SIGNAL b : t_package_type;

 ATTRIBUTE package_type OF mc68040 : COMPONENT IS pin_grid;

 ATTRIBUTE location OF mc68040 : COMPONENT IS 20;
BEGIN
 a <= mc68040'location;
 -- returns 20

 b <= mc68040'package_type;
 -- returns pin_grid

END cpu_board;
```

This is a very simple example of how attributes can be attached to objects. Much more complicated types and attributes can be created. What this example shows is a code fragment of a CPU board design in which the package type and location information are specified as attributes of the single microprocessor used in the design.

The **package_type** attribute is used to hold the kind of packaging used for the microprocessor. Attributes that have values specified do not have to be used in the simulation. Other tools such as physical layout tools or fault simulation can make use of attributes that a logic simulator cannot.

In this example, a physical layout tool could read the package type information from the **package_type** attribute and, based on the value assigned to the attribute, fill in the value for the location attribute.

The package **p_attr** defines the type used for one of the attributes and contains the attribute declarations for two attributes. The attribute declarations make the name and type of the attribute visible to any object for use if needed.

In the architecture **cpu_board** of entity **board** are the attribute specifications. The attribute specification describes the attribute name to be used, the name of the object to which the attribute is attached, the object kind, and finally the value of the attribute.

To access the value of a user-defined attribute, use the same syntax for a predefined attribute. In the signal assignment statements of architecture **cpu_board**, the attribute value is retrieved by specifying the name of the object, followed by a ' and the attribute name.

Generate Statements

Generate statements give the designer the ability to create replicated structures, or select between multiple representations of a model. Generate statements can contain **IF-THEN** and looping constructs, nested to any level, that create concurrent statements.

Typical applications include memory arrays, registers, and so on. Another application is to emulate a conditional compilation mechanism found in other languages such as C.

Following is a simple example showing the basics of generate statements:

```
LIBRARY IEEE;
USE IEEE.std_logic_1164.ALL;
ENTITY shift IS
```

```
PORT( a, clk : IN std_logic;
      b :  OUT std_logic);
END shift;

ARCHITECTURE gen_shift OF shift IS
  COMPONENT dff
    PORT( d, clk : IN std_logic;
          q : OUT std_logic);
  END COMPONENT;

  SIGNAL z : std_logic_vector( 0 TO 4 );
BEGIN
  z(0) <= a;

  g1 : FOR i IN 0 TO 3 GENERATE
    dffx : dff PORT MAP( z(i), clk, z(i + 1));
  END GENERATE;

  b <= z(4);
END gen_shift;
```

This example represents the behavior for a 4-bit shift register. Port **a** is the input to the shift register, and port **b** is the output. Port **clk** shifts the data from **a** to **b**.

Architecture **gen_shift** of entity **shift** contains two concurrent signal assignment statements and one **GENERATE** statement. The signal assignment statements connect the internal signal **z** to input port **a** and output port **b**. The generate statement in this example uses a **FOR** scheme to generate four DFF components. The resultant schematic for this architecture is shown in Figure 8-1.

The **FOR** in the generate statement acts exactly like the **FOR** loop sequential statement in that variable **i** need not be declared previously, **i** is not visible outside the generate statement, and **i** cannot be assigned inside the generate statement.

The result of the generate statement is functionally equivalent to the following architecture:

Figure 8-1
Schematic Representing Generate Statement.

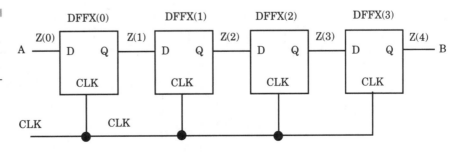

```
ARCHITECTURE long_way_shift OF shift IS
  COMPONENT dff
    PORT( d, clk : IN std_logic;
          q : OUT std_logic);
  END COMPONENT;

  SIGNAL z : std_logic_vector( 0 TO 4 );
  BEGIN
  z(0) <= a;

  dff1: dff PORT MAP( z(0), clk, z(1) );
  dff2: dff PORT MAP( z(1), clk, z(2) );
  dff3: dff PORT MAP( z(2), clk, z(3) );
  dff4: dff PORT MAP( z(3), clk, z(4) );

  b <= z(4);
END long_way_shift;
```

The difference between the two architectures is that architecture **gen_shift** could be specified with generic parameters such that different size shift registers could be generated based on the value of the generic parameters. Architecture **long_way_shift** is fixed in size and cannot be changed.

Irregular Generate Statement

The last example showed how a regular structure could be generated, but in practice most structures are not completely regular. Most regular structures have irregularities at the edges. This is shown by Figure 8-2.

In the last example, the irregularities were handled by the two concurrent signal assignment statements. Following is another way to handle the irregularities:

```
LIBRARY IEEE;
USE IEEE.std_logic_1164.ALL;
```

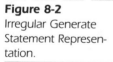

Figure 8-2
Irregular Generate Statement Representation.

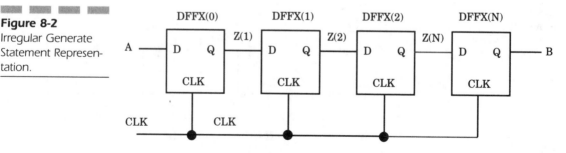

```
ENTITY shift IS
 GENERIC ( len : INTEGER);
 PORT( a, clk : IN std_logic;
       b : OUT std_logic);
END shift;

ARCHITECTURE if_shift OF shift IS
 COMPONENT dff
  PORT( d, clk : IN std_logic;
        q : OUT std_logic);
 END COMPONENT;

 SIGNAL z : std_logic_vector( 1 TO (len -1) );
BEGIN
 g1 : FOR i IN 0 TO (len -1) GENERATE
  IF i = 0 GENERATE
     dffx : dff PORT MAP( a, clk, z(i + 1));
  END GENERATE;

  IF i = (len -1) GENERATE
     dffx : PORT MAP( z(i), clk, b );
  END GENERATE;

  IF (i > 0) AND i < (len -1) GENERATE
     dffx : PORT MAP( z(i), clk, Z(i + 1) );
  END GENERATE;

 END GENERATE;
END if_shift;
```

This example uses a shift register that has a configurable size. Generic **len** passed in specifies the length of the shift register. (Generic **len** must be at least 2 for the shift register to work properly.) Generic **len** is used in the specification of the length of signal array **z**. This type of array is known as a generically constrained array because the size of the array is specified through one or more generics.

The **FOR** clause of the generate also uses generic **len** to specify the maximum number of DFF components to be generated. Notice that this generate statement uses the conditional form of the generate statement. If the condition is true, the concurrent statements inside the generate statement are generated; otherwise, nothing is generated.

The first **IF-THEN** condition checks for the first flip-flop in the shift register. If this is the first flip-flop, notice that the port map clause maps the input signal **a** directly to the flip-flop instead of through an intermediate signal. The same is true of the next **IF-THEN** condition. It checks for the last flip-flop of the shift register and maps the last output to output port **b**. Any other flip-flops in the shift register are generated by the third conditional generate statement.

Following is another interesting example using the conditional generate statement:

```
PACKAGE gen_cond IS
 TYPE t_checks IS ( onn, off);
END gen_cond;

USE WORK.gen_cond.ALL;

LIBRARY IEEE;
USE IEEE.std_logic_1164.ALL;
ENTITY dff IS
 GENERIC( timing_checks : t_checks;
          setup, qrise, qfall, qbrise, qbfall : time);
 PORT( din, clk : IN std_logic;
       q, qb : OUT std_logic);
END dff;

ARCHITECTURE condition OF dff IS
BEGIN
 G1 : IF (timing_checks = onn) GENERATE
   ASSERT ( din'LAST_EVENT >>setup)
     REPORT "setup violation"
     SEVERITY ERROR;
 END GENERATE;

 PROCESS(clk)
  VARIABLE int_qb : std_logic;
 BEGIN
  IF (clk = '1') AND (clk'EVENT) AND (clk'LAST_VALUE =
     '0') THEN
    int_qb := not din;

   q <= din AFTER f_delay( din, qrise, qfall);

   qb <= int_qb AFTER f_delay( int_qb, qbrise, qbfall);
  END IF;
 END PROCESS;
END condition;
```

In this example, a DFF component is modeled using a generate statement to control whether or not a timing check statement is generated for the architecture. The generic, **timing_checks**, can be passed a value of **onn** or **off**. (Note the spelling of **onn**. We cannot use a value of on because it is a reserved word.) If the value is **onn**, then the generate statement generates a concurrent assertion statement. If the value of generic **timing_checks** is **off**, then no assertion statement is generated. This functionality emulates the conditional compilation capability of some programming languages, such as C and Pascal.

TextIO

One of the predefined packages that is supplied with VHDL is the Textual Input and Output (TextIO) package. The TextIO package contains procedures and functions that give the designer the ability to read from and write to formatted text files. These text files are ASCII files of any format that the designer desires. (VHDL does not impose any limits of format, but the host machine might impose limits.) TextIO treats these ASCII files as files of lines, where a line is a string, terminated by a carriage return. There are procedures to read a line and write a line and a function that checks for end of file.

The TextIO package also declares a number of types that are used while processing text files. Type **line** is declared in the TextIO package and is used to hold a line to write to a file or a line that has just been read from the file. The line structure is the basic unit upon which all TextIO operations are performed. For instance, when reading from a file, the first step is to read in a line from the file into a structure of type **line**. Then the line structure is processed field by field.

The opposite is true for writing to a file. First, the line structure is built field by field in a temporary line data structure, then the line is written to the file.

Following is a very simple example of a TextIO behavior:

```
USE WORK.TEXTIO.ALL;
ENTITY square IS
  PORT( go : IN std_logic);
END square;

ARCHITECTURE simple OF square IS
BEGIN
  PROCESS(go)
    FILE infile : TEXT IS IN "/doug/test/example1";

    FILE outfile : TEXT IS OUT "/doug/test/outfile1";

    VARIABLE out_line, my_line : LINE;
    VARIABLE int_val : INTEGER;
  BEGIN
    WHILE NOT( ENDFILE(infile)) LOOP
    -- read a line from the input file
    READLINE( infile, my_line);

    -- read a value from the line
    READ( my_line, int_val);

    -- square the value
```

```
int_val := int_val **2;

    -- write the squared value to the line
     WRITE( out_line, int_val);

    -- write the line to the output file
     WRITELINE( outfile, out_line);
   END LOOP;
  END PROCESS;
 END simple;
```

This example shows how to read a single integer value from a line, square the value, and write the squared value to another file. It illustrates how TextIO can be used to read values from files and write values to files.

The process statement is executed whenever signal **go** has an event occur. The process then loops until an end-of-file condition occurs on the input file **infile**. The **READLINE** statement reads a line from the file and places the line in variable **my_line**. The next executable line contains a **READ** procedure call that reads a single integer value from **my_line** into variable **int_val**. Procedure **READ** is an overloaded procedure that reads different type values from the line, depending on the type of the argument passed to it.

After the value from the file has been read into variable **int_val**, the variable is squared, and the squared value is written to another variable of type **line**, called **out_line**. Procedure **WRITE** is also an overloaded procedure that writes a number of different value types, depending on the type of the argument passed to it.

The last TextIO procedure call made is the **WRITELINE** procedure call. This procedure writes out the line variable **out_line** to the output file **outfile**.

If the following input file is used as input to this architecture, the second file shown reflects the output generated:

```
10
20
50
16#A              <- hex input
1_2_3             <- underscores ignored
87          52 <- second argument ignored
```

The output from the input file would look like this:

```
100
400
2500
100
```

```
15129
7569
```

The first value in the input file is 10. It is squared to result in 100 and written to the output file. The same is true for the values 20 and 50. The next value in the file is specified in hexadecimal notation. A hexadecimal A value is 10 base ten, which squared results in 100.

The next example in the file shows a number with embedded underscore characters. The underscores are used to separate fields of a number and are ignored in the value of the number. The number 1_2_3 is the same as 123.

The last entry in the input file shows a line with two input values on the line. When the line is read into the **my_line** variable, both values exist in the line, but because there is only one **READ** procedure call, only the first value is read from the line.

More than one data item can be read from a single line, as well as data items of any types. For instance, a TextIO file could be a list of instructions for a microprocessor. The input file could contain the type of instruction, a source address, and a destination address. This is shown by the following simple example:

```
USE WORK.TEXTIO.ALL;
PACKAGE p_cpu IS
 TYPE t_instr IS (jump, load,
                  store, addd,
                  subb, test, noop);

 FUNCTION convertstring( s : STRING) RETURN t_instr;

END p_cpu;

PACKAGE BODY p_cpu IS
 FUNCTION convertstring( s : STRING) RETURN t_instr IS
  SUBTYPE twochar IS string(1 to 2);
  VARIABLE val : twochar;
 BEGIN
  val := s(1 to 2);
  CASE val IS

   WHEN "ju" =>
    RETURN jump;
   WHEN "lo" =>
    RETURN load;
   WHEN "st" =>
    RETURN store;
   WHEN "ad" =>
    RETURN addd;
   WHEN "su" =>
```

```
        RETURN subb;
    WHEN "te" =>
      RETURN test;
    WHEN "no" =>
      RETURN noop;
    WHEN others =>
      RETURN noop;
  END CASE;
 END convertstring;
END p_cpu;


USE WORK.p_cpu.ALL;
USE WORK.TEXTIO.ALL;
ENTITY cpu_driver IS
 PORT( next_instr : IN BOOLEAN;
       instr : OUT t_instr;
       src : OUT INTEGER;
       dst : OUT INTEGER);
END cpu_driver;

ARCHITECTURE a_cpu_driver OF cpu_driver IS
 FILE instr_file : TEXT IS IN "instfile";
BEGIN
 read_instr : PROCESS( next_instr)
  VARIABLE aline : LINE;
  VARIABLE a_instr : STRING(1 to 4);
  VARIABLE asrc, adst : INTEGER;
  BEGIN
   IF next_instr THEN
     IF ENDFILE(instr_file) THEN
       ASSERT FALSE
         REPORT "end of instructions"
         SEVERITY WARNING;
     ELSE
        READLINE( instr_file, aline);
        READ( aline, a_instr);
        READ( aline, asrc);
        READ( aline, adst);
      END IF;

      instr <= convertstring(a_instr);
      src <= asrc;
      dst <= adst;

   END IF;
 END PROCESS read_instr;
END a_cpu_driver;
```

Package **p_cpu** defines type **t_instr**, the enumerated type that represents CPU instructions to be executed. The package also defines a function, **convert_string**, that is used to convert the string value read in using TextIO procedures into a **t_instr** type. The conversion is necessary because the TextIO package does not contain any procedures for reading in user-defined types. (However, a designer can write a user-defined overloaded procedure that has the same basic interface as the procedures in the TextIO package.) This process is usually very straightforward, as seen by the **convert_string** procedure.

Entity **cpu_driver** is the entity that reads in the file of instructions. It has a single input port called **next_instr** which is used to signal the entity to read in the next instruction. When a true event occurs on input port **next_instr**, process **read_instr** executes. If the file is at the end already, the assert statement is called, and a warning message is issued. If we are not at the end of the file, the process reads in a line from the file into variable **aline**.

Successive reads on variable **aline** retrieve the appropriate fields from the line. All of the reads return the value into internal variables, but variables **asrc** and **adst** are not really needed because there exists a TextIO procedure for reading integer values. Variable **ainstr** is used to allow the string read in to be converted into the enumerated type **t_instr** before being assigned to the output port **instr**.

SUMMARY

In this chapter we discussed the following:

- overloading functions, arguments, operators to make VHDL models more readable
- how aliases can be used to name sections of an object
- how qualified expressions are used to direct conversion
- how user-defined attributes can be used to add information to objects
- how generate statements can be used to replicate entity instantiations
- how TextIO is used to read and write text files

This chapter showed some of the more esoteric features of VHDL. This chapter concludes the discussion of VHDL features. The next two chapters concentrate on the synthesis process and how to write VHDL for synthesis. The next few chapters then guide the reader through a top-down description of a device.

Synthesis

One of the most best uses of VHDL today is to synthesize ASIC and FPGA devices. This chapter and the next focus on how to write VHDL for synthesis.

Synthesis is an automatic method of converting a higher level of abstraction to a lower level of abstraction. There are several synthesis tools available currently, including commercial as well as university developed tools. In this discussion, the examples use the commercially available Exemplar Logic Leonardo synthesis tool.

The current synthesis tools available today convert Register Transfer Level (RTL) descriptions to gate level netlists. These gate level netlists consist of interconnected gate level macro cells. Models for the gate level cells are contained in technology libraries for each type of technology supported.

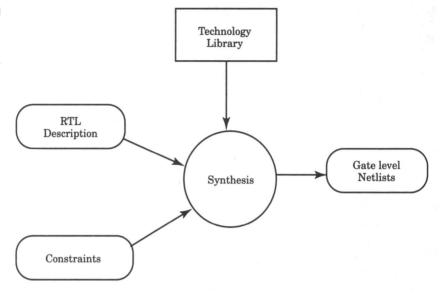

These gate level netlists currently can be optimized for area, speed, testability, and so on. The synthesis process is shown in Figure 9-1.

The inputs to the synthesis process are an RTL (Register Transfer Level) VHDL description, circuit constraints and attributes for the design, and a technology library. The synthesis process produces an optimized gate level netlist from all of these inputs. In the next few sections, each of these inputs is described, and we discuss the synthesis process in more detail.

▬▬▬ ▬▬▬ # Register Transfer Level Description

A register transfer level description is characterized by a style that specifies all of the registers in a design, and the combinational logic between. This is shown by the register and cloud diagram in Figure 9-2. The registers are described either explicitly through component instantiation or implicitly through inference. The registers are shown as the rectangular objects connected to the clock signal. The combinational logic is described by logical equations, sequential control statements (**CASE, IF then ELSE**, and so on), subprograms, or through concurrent statements, which are represented by the cloud objects between registers.

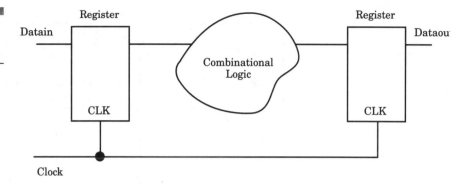

Figure 9-2
Register and Cloud
Diagram.

RTL descriptions are used for synchronous designs and describe the clock by clock behavior of the design. Following is an example of an RTL description that uses component instantiation:

```
ENTITY datadelay IS
  PORT( clk, din, en : IN BIT;
        dout : OUT BIT);
END datadelay;

ARCHITECTURE synthesis OF datadelay IS
  COMPONENT dff
    PORT(clk, din : IN BIT;
         q,qb : OUT BIT);
  END COMPONENT;
  SIGNAL q1, q2, qb1, qb2 : BIT;
BEGIN

  r1 : dff PORT MAP(clk, din, q1, qb1);
  r2 : dff PORT MAP(clk, q1, q2, qb2);

  dout <= q1 WHEN en = '1' ELSE
          q2;

END synthesis;
```

This example is the circuit for a selectable data delay circuit. The circuit delays the input signal **din** by 1 or 2 clocks depending on the value of **en**. If **en** is a 1, then input **din** is delayed by 1 clock. If **en** is a 0, input **din** is delayed by 2 clocks.

Figure 9-3 shows a schematic representation of this circuit. The clock signal connects to the **clk** input of both flip-flops, while the **din** signal connects only to the first flip-flop. The **q** output of the first flip-flop is then

Figure 9-3
Register Transfer
Level with Compo-
nent Instances.

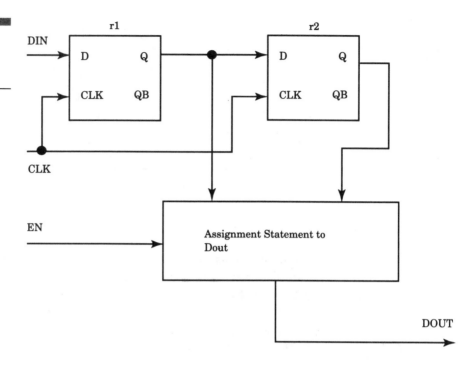

connected to the **d** input of the next flip-flop. The selected signal assignment to signal **dout** forms a mux operation that selects between the two flip-flop outputs.

This example could be rewritten as follows using register inference:

```
ENTITY datadelay IS
  PORT( clk, din, en : IN BIT;
        dout : OUT BIT);
END datadelay;

ARCHITECTURE inference OF datadelay IS
  SIGNAL q1, q2 : BIT;
BEGIN
  reg_proc: PROCESS
  BEGIN

    WAIT UNTIL clk'EVENT and clk = '1';

      q1 <= din;
      q2 <= q1;

  END PROCESS;

  dout <= q1 WHEN en = '1' ELSE
          q2;
```

```
END inference;
```

In the first version, the registers are instantiated using component instantiation statements that instantiate **r1** and **r2**.

In this version, the **dff** components are not instantiated, but are inferred through the synthesis process. Register inference is discussed more in Chapter 10, "VHDL Synthesis." Process **reg_proc** has a **WAIT** statement that is triggered by positive edges on the clock. When the **WAIT** statement is triggered, signal **q1** is assigned the value of **din**, and **q2** is assigned the previous value of **q1**. This, in effect, creates two flip-flops. One flip-flop for signal **q1**, and the other for signal **q2**.

This is a register transfer level description because registers **r1** and **r2** from the first version form the registers, and the conditional signal assignment for port **dout** forms the combinational logic between registers. In the second version, the inferred registers form the register description, while the conditional signal assignment still forms the combinational logic.

The advantage of the second description is that it is technology independent. In the first description, actual flip-flop elements from the technology library were instantiated, thereby making the description technology dependent. If the designer should decide to change technologies, all of the instances of the flip-flops would need to be changed to the flip-flops from the new technology. In the second version of the design, the designer did not specify particular technology library components, and the synthesis tools are free to select flip-flops from whatever technology library the designer is currently using, as long as these flip-flops match the functionality required.

After synthesis, both of these descriptions produce a gate level description, as shown in Figure 9-4.

Notice that the gate level description has two registers (FDSR1) with mux (Mux21S) logic controlling the output signal from each register. Depending on the technology library selected and the constraints, the mux logic varies widely from and-or-invert gates to instantiated 2-input multiplexers.

Following is the netlist generated by the Exemplar Logic Leonardo synthesis tool for the same design:

```
- -
-- Definition of   datadelay
- -
--
- -
- -
- -
```

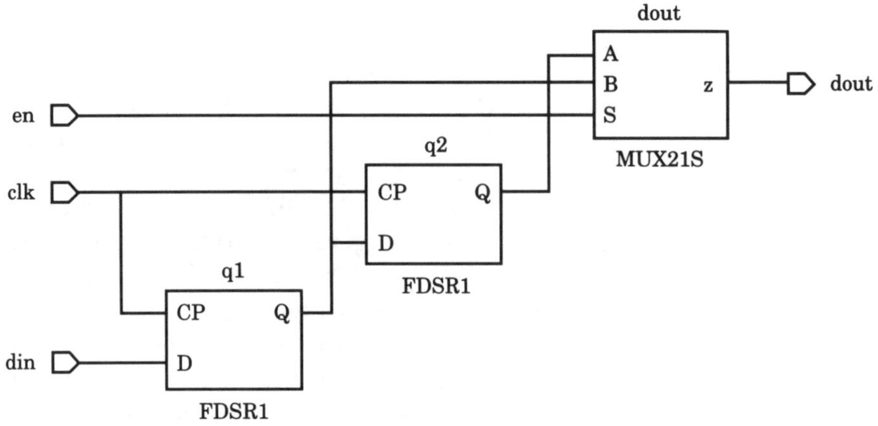

Figure 9-4
A Gate level Description.

```
library IEEE, EXEMPLAR;
use IEEE.STD_LOGIC_1164.all;
use EXEMPLAR.EXEMPLAR_1164.all;

entity datadelay is
 port (
    clk : IN std_logic ;
    din : IN std_logic ;
    en : IN std_logic ;
    dout : OUT std_logic) ;
end datadelay ;

architecture inference of datadelay is
 component FDSR1
    port (
       Q : OUT std_logic ;
       D : IN std_logic ;
       CP : IN std_logic) ;
   end component ;
   component MU«TIM»21S
     port (
       Z : OUT std_logic ;
       A : IN std_logic ;
       B : IN std_logic ;
       S : IN std_logic) ;
   end component ;
   signal q2, q1: std_logic ;

begin
  q2_XMPLR : FDSR1 port map ( Q=>q2, D=>q1, CP=>clk);
  q1_XMPLR : FDSR1 port map ( Q=>q1, D=>din, CP=>clk);
  dout_XMPLR_XMPLR : MU«TIM»21S port map ( Z=>dout, A=>q2,
      B=>q1, S=>en);
end inference ;
```

The netlist matches the gate level generated schematic. The netlist contains two instantiated flip-flops (**FDSR1**) and one instantiated 2-input multiplexer (**Mux21S**).

This very simple example shows how RTL synthesis can be used to create technology-specific implementations from technology-independent VHDL descriptions. In the next few sections, we examine much more complex examples. But first, let's look at some of the ways to control how the synthesized design is created.

Constraints

Constraints are used to control the output of the optimization and mapping process. They provide goals that the optimization and mapping processes try to meet and control the structural implementation of the design. They represent part of the physical environment that the design has to interface with. The constraints available in synthesis tools today include area, timing, power, and testability constraints. In the future, we will probably see packaging constraints, layout constraints, and so on. Today, the most common constraints in use are timing constraints.

A block diagram of a design with some possible constraints is shown in Figure 9-5. Again, the design is shown using the cloud notation. The combinational logic between registers is represented as clouds, with wires going in and out representing the interconnection to the registers.

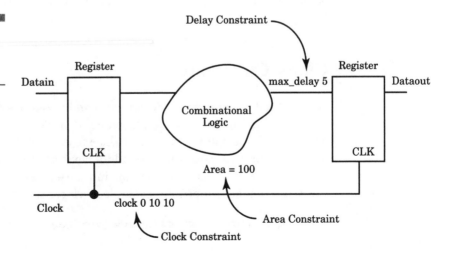

Figure 9-5
Register and Cloud Diagram with Constraints.

There are a number of constraints shown on the diagram including required time constraints, late arrival constraints, and clock cycle constraints.

Required time constraints specify the latest time that a signal can occur. Clock constraints are used to specify the operating frequency of the clock. From the clock constraint, required time constraints of each signal feeding a clocked register can be calculated. Each of these constraints is further described in the next sections.

Timing Constraints

Typical uses for timing constraints are to specify maximum delays for particular paths in a design. For instance, a typical timing constraint is the required time for an output port. The timing constraint guides the optimization and mapping to produce a netlist that meets the timing constraint. Meeting timing is usually one of the most difficult tasks when designing an ASIC or FPGA using synthesis tools. There may be no design that meets the timing constraints specified. A typical delay constraint in Leonardo synthesis format is shown here:

```
set_attribute -port data_out -name required_time -value 25
```

This constraint specifies that the maximum delay for signal `data_out` should be less than or equal to 25 library units. A library unit can be whatever the library designer used when describing the technology from a synthesis point of view. Typically, it is nanoseconds, but can be picoseconds or some other time measurement depending on the technology.

Clock Constraints

One method to constrain a design is to add a `required_time` constraint to every flip-flop input with the value of a clock cycle. The resulting design would be optimized to meet the one clock cycle timing constraint. An easier method, however, is to add a clock constraint to the design. A clock constraint effectively adds an input `required_time` constraint to every flip-flop data input. An example clock constraint is shown here:

```
set_attribute -port clk -name clock_cycle -value 25
```

This example sets a clock cycle constraint on port **clk** with a value of 25 library units.

Some synthesis tools (such as Exemplar Logic Leonardo) do a static timing analysis to calculate the delay for each of the nodes in the design. The static timing analyzer uses a timing model for each element connected in the netlist. The timing analyzer calculates the worst and best case timing for each node by adding the contribution of each cell that it traverses.

The circuit is checked to see if all delay constraints have been met. If so, the optimization and mapping process is done; otherwise, alternate optimization strategies may be applied—such as adding more parallelism or more buffered outputs to the slow paths—and the timing analysis is executed again. More detail about the typical timing analysis is discussed later in the section "Technology Libraries."

Attributes

Attributes are used to specify the design environment. For instance, attributes specify the loading that output devices have to drive, the drive capability of devices driving the design, and timing of input signals. All of this information is taken into account by the static timing analyzer to calculate the timing through the circuit paths. A cloud diagram showing attributes is shown in Figure 9-6.

Figure 9-6

Register and Cloud Diagram with Attributes.

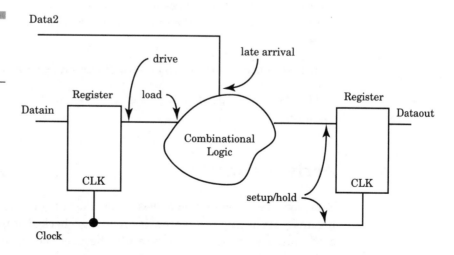

Load

Each output can specify a drive capability that determines how many loads can be driven within a particular time. Each input can have a load value specified that determines how much it will slow a particular driver. Signals that are arriving later than the clock can have an attribute that specifies this fact.

The **Load** attribute specifies how much capacitive load exists on a particular output signal. This load value is specified in the units of the technology library in terms of pico-farads, or standard loads, and so on. For instance, the timing analyzer calculates a long delay for a weak driver and a large capacitive load, and a short delay for a strong driver and a small load. An example of a load specification in Leonardo synthesis format is shown here:

```
set_attribute -port xbus -name input_load -value 5
```

This attribute specifies that signal **xbus** will load the driver of this signal with 5 library units of load.

Drive

The **Drive** attribute specifies the resistance of the driver, which controls how much current it can source. This attribute also is specified in the units of the technology library. The larger a driver is the faster a particular path will be, but a larger driver takes more area, so the designer needs to trade off speed and area for the best possible implementation. An example of a drive specification in Leonardo synthesis format is shown here:

```
set_attribute -port ybus -name output_drive -value 2.7
```

This attribute specifies that signal **ybus** has 2.7 library units of drive capability.

Arrival Time

Some synthesis tools (such as Exemplar Logic Leonardo) use a static timing analyzer during the synthesis process to check that the logic being created matches the timing constraints the user has specified. Setting the arrival time on a particular node specifies to the static timing analyzer when a particular signal will occur at a node. This is especially important for late arriving signals. Late arriving signals drive inputs to the current block at a later time, but the results of the current block still must meet its own timing constraints on its outputs. Therefore, the path to the output of the late arriving input must be faster than any other inputs, or the timing constraints of the current block cannot be met.

Technology Libraries

Technology libraries hold all of the information necessary for a synthesis tool to create a netlist for a design based on the desired logical behavior, and constraints on the design. Technology libraries contain all of the information that allows the synthesis process to make the correct choices to build a design. Technology libraries contain not only the logical function of a ASIC cell, but the area of the cell, the input to output timing of the cell, any constraints on fanout of the cell, and the timing checks that are required for the cell. Other information stored in the technology library may be the graphical symbol of the cell for use in schematics.

Following is an example technology library description of a 2-input AND gate written in Synopsys .lib format:

```
library (xyz) {
cell (and2) {
 area : 5;
 pin (a1, a2) {
   direction : input;
   capacitance : 1;
 }
 pin (o1) {
   direction : output;
   function : "a1 * a2";
   timing () {
     intrinsic_rise : 0.37;
     intrinsic_fall : 0.56;
     rise_resistance : 0.1234;
     fall_resistance : 0.4567;
     related_pin : "a1 a2";
   }
 }
}
}
```

This technology library describes a library named **xyz** with one library cell contained in it. The cell is named **and2** and has two input pins **a1** and

a2 and one output pin **o1**. The cell requires 5 units of area, and the input pins have 1 unit of loading capacitance to the driver driving them. The intrinsic rise and fall delays listed with pin **o1** specify the delay to the output with no loading. The timing analyzer uses the intrinsic delays plus the rise and fall resistance with the output loading to calculate the delay through a particular gate. Notice that the function of pin **o1** is listed as the **AND** of pins **a1** and **a2**. Also, notice that pin **o1** is related to pins **a1** and **a2** in that the timing delay through the device is calculated from pins **a1** and **a2** to pin **o1**.

Most synthesis tools have fairly complicated delay models to calculate timing through an ASIC cell. These models include not only intrinsic rise and fall time, but output loading, input slope delay, and estimated wire delay. A diagram illustrating this is shown in Figure 9-7.

The total delay from gate A1 to gate C1 is:

```
intrinsic_delay  + loading_delay + wire_delay + slope_delay
```

The *intrinsic delay* is the delay of the gate without any loading. The *loading delay* is the delay due to the input capacitance of the gate being driven. The *wire delay* is an estimated delay used to model the delay through a typical wire used to connect cells together. It can be a statistical model of the wire delays usually based on the size of the chip die. Given a particular die size, the wire loading effect can be calculated and added to the overall delay. The final component in the delay equation is the extra delay needed to handle the case of slowly rising input signals due to heavy loading or light drive.

Figure 9-7
Delay Effects Used in
Delay Model.

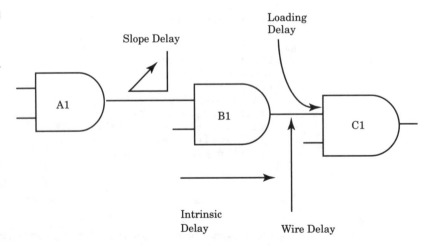

In the preceding technology library, the intrinsic delays are given in the cell description. The loading delay is calculated based on the load applied to the output pin **o1** and the resistance values in the cell description. The value calculated for the wire delay depends on the die size selected by the user. Selecting a wire model scales the delay values. Finally, the input slope delay is calculated by the size of the driver, in this example, A1, and the capacitance of the gate being driven. The capacitance of the gate being driven is in the technology library description.

Technology libraries can also contain data about how to scale delay information with respect to process parameters and operating conditions. Operating conditions are the device operating temperature and power supply voltage applied to the device.

Synthesis

To convert the RTL description to gates, three steps typically occur. First, the RTL description is translated to an unoptimized boolean description usually consisting of primitive gates such as **AND** and **OR** gates, flip-flops, and latches. This is a functionally correct but completely unoptimized description. Next, boolean optimization algorithms are executed on this boolean equivalent description to produce an optimized boolean equivalent description. Finally, this optimized boolean equivalent description is mapped to actual logic gates by making use of a technology library of the target process. This is shown in Figure 9-8.

Translation

The translation from RTL description to boolean equivalent description is usually not user controllable. The intermediate form that is generated is usually a format that is optimized for a particular tool and may not even be viewable by the user.

All **IF**, **CASE**, and **LOOP** statements, conditional signal assignments, and selected signal assignment statements are converted to their boolean equivalent in this intermediate form. Flip-flops and latches can either be instantiated or inferred; both cases produce the same flip-flop or latch entry in the intermediate description.

Figure 9-8
Synthesis Process.

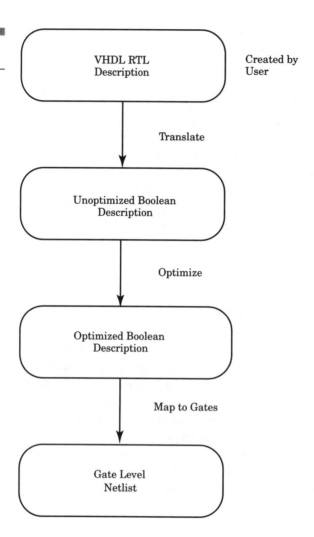

Boolean Optimization

The optimization process takes an unoptimized boolean description and converts it to an optimized boolean description. In many designers' eyes, this is where the real work of synthesis gets done. The optimization process uses a number of algorithms and rules to convert the unoptimized boolean description to an optimized one. One technique is to convert the unoptimized boolean description to a very low-level description (a pla format), optimize that description (using pla optimization techniques), and then try to reduce the logic generated by sharing common terms (introducing intermediate variables).

Flattening

The process of converting the unoptimized boolean description to a pla format is know as *flattening*, because it creates a flat signal representation of only two levels: an **AND** level and an **OR** level. The idea is to get the unoptimized boolean description into a format in which optimization algorithms can be used to optimize the logic. A pla structure is a very easy description in which to perform boolean optimization, because it has a simple structure and the algorithms are well known. An example of a boolean description is shown here:

```
Original equations
a = b and c;
b = x or (y and z);
c = q or w;
```

This description shows an output **a** that has three equations describing its function. These equations use two intermediate variables **b** and **c** to hold temporary values which are then used to calculate the final value for **a**. These equations describe a particular structure of the design that contains two intermediate nodes or signals, **b** and **c**. The flattening process removes these intermediate nodes to produce a completely flat design, with no intermediate nodes. For example, after removing intermediate variables:

```
a = (x and q) or (q and y and z) or (w and x) or (w and y
    and z);
```

This second description is the boolean equivalent of the first, but it has no intermediate nodes. This design contains only two levels of logic gates: an **AND** plane and an **OR** plane. This should result in a very fast design because there are very few logic levels from the input to the output. In fact, the design is usually very fast. There are, however, a number of problems with this type of design.

First, this type of design can actually be slower than one that has more logic levels. The reason is that this type of design can have a tremendous fanout loading on the input signals because inputs fanout to every term. Second, this type of design can be very large, because there is no sharing between terms. Every term has to calculate its own functionality. Also, there are a number of circuits that are difficult to flatten, because the number of terms created is extremely large. An equation that only contains **AND** functions produces one term. A function that contains a large **XOR** function can produce hundreds or even thousands of terms. A 2-input **XOR** has the terms **A and (not B)** or **B and (not A)**. An N-input **XOR** has **2**(N-1)** terms. For instance, a 16-input **XOR** has 32768 terms and a 32-bit XOR has over 2 billion terms. Clearly, designs with these types of functions cannot be flattened.

Flattening gets rid of all of the implied structure of design whether it is good or not. Flattening works best with small pieces of random control logic that the designer wants to minimize. Used in conjunction with structuring, a minimal logic description can be generated.

Usually, the designer wants a design that is nearly as fast as the flattened design, but is much smaller in area. To reduce the fanout of the input pins, terms are shared. Some synthesis vendors call this process *structuring* or factoring.

Factoring

Factoring is the process of adding intermediate terms to add structure to a description. It is the opposite of the flattening process. Factoring is usually desirable because, as was mentioned in the last section, flattened designs are usually very big and may be slower than a factored design because of the amount of fanouts generated. Following is a design before factoring:

```
x = a and b or a and d;
y = z or b or d;
```

After factoring the common term, (**b** or **d**), is factored out to a separate intermediate node. The results are shown here:

```
x = a and q;
y = z or q;
q = b or d;
```

Factoring usually produces a better design but can be very design dependent. Adding structure adds levels of logic between the inputs and outputs. Adding levels of logic adds more delay. The net result is a smaller design, but a slower design. Typically, the designer wants a design that is nearly as fast as the flattened design if it was driven by large drivers, but as small as the completely factored design. The ideal case is one in which the critical path was flattened for speed and the rest of the design was factored for small area and low fanout.

After the design has been optimized at the boolean level, it can be mapped to the gate functions in a technology library.

Mapping to Gates

The mapping process takes the logically optimized boolean description created by the optimization step and uses the logical and timing information from a technology library to build a netlist. This netlist is targeted to the user's needs for area and speed. There are a number of possible netlists that are functionally the same but vary widely in speed and area. Some netlists are very fast but take a lot of library cells to implement, and others take a small number of library cells to implement but are very slow.

To illustrate this point, let's look at a couple of netlists that implement the same functionality. Following is the VHDL description:

```
LIBRARY IEEE;
USE IEEE.std_logic_1164.ALL;
USE IEEE.std_logic_unsigned.ALL;
ENTITY adder IS
  PORT( a,b : IN std_logic_vector(7 DOWNTO 0);
        c : OUT std_logic_vector(7 DOWNTO 0)
        );
END adder;

ARCHITECTURE test OF adder IS
BEGIN
  c <= a + b;
END test;
```

Both of the examples implement an 8-bit adder, but the first implementation is a small but slower design, and the second is a bigger but fast design. The small but slower design is an 8-bit ripple carry adder shown in Figure 9-9. The bigger but faster design is an 8-bit lookahead adder shown in Figure 9-10.

Both of these netlists implement the same function, an 8-bit adder. The ripple carry adder takes less cells to implement but is a slower design because it has more logic levels. The lookahead adder takes more cells to implement but is a faster design because more of the boolean operations are calculated in parallel. The additional logic to calculate the functionality in parallel adds extra logic to the design making the design bigger.

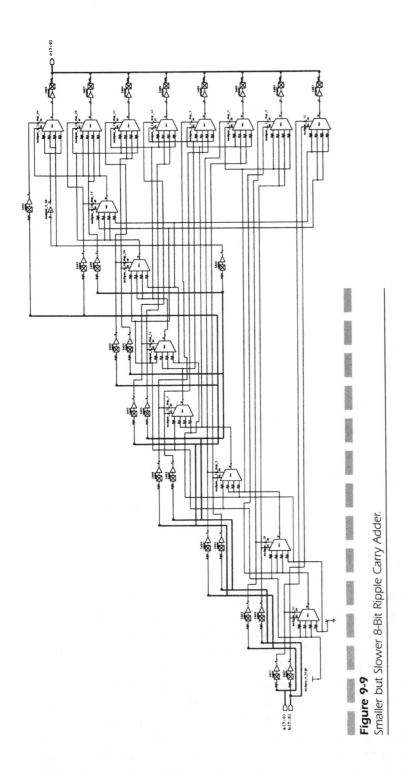

Figure 9-9
Smaller but Slower 8-Bit Ripple Carry Adder.

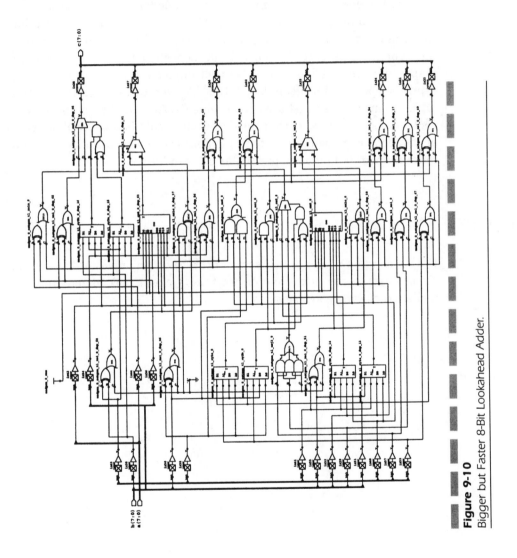

Figure 9-10

Bigger but Faster 8-Bit Lookahead Adder.

In most synthesis tools, the designer has control over which type of adder is selected through the use of constraints. If the designer wants to constrain the design to a very small area and doesn't need the fastest possible speed, then the ripple carry adder probably works. If the designer wants the design to be as fast as possible and doesn't care as much about how big the design gets, then the lookahead adder is the one to select.

The mapping process takes as input the optimized boolean description, the technology library, and the user constraints, and generates an optimized netlist built entirely from cells in the technology library. During the mapping process, cells are inserted that implement the boolean function from the optimized boolean description. These cells are then locally optimized to meet speed and area requirements. As a final step, the synthesis tool has to make sure that the output does not violate any of the rules of the technology being used to implement the design, such as the maximum number of fanouts a particular cell can have.

SUMMARY

In this chapter, we discussed some of the basic principles of the synthesis process. In the next chapter, we take a closer look at how to write models that can be synthesized.

10

VHDL Synthesis

In this chapter, we focus on how to write VHDL that can be read by synthesis tools. We start out with some simple combinational logic examples, move on to some sequential models, and end the chapter with a state machine description.

All of the examples are synthesized with the Exemplar Logic Leonardo synthesis environment. The technology library used is an example library from Exemplar Logic. All of the output data should be treated as purely sample outputs and not representative of how well the Exemplar Logic tools work with real design data and real constraints.

Simple Gate—Concurrent Assignment

The first example is a simple description for a 3-input **OR** gate:

```
LIBRARY IEEE;
USE IEEE.std_logic_1164.ALL;
ENTITY or3 IS
  PORT (a, b, c : IN std_logic;
        d : OUT std_logic);
END or3;

ARCHITECTURE synth OF or3 IS
BEGIN
 d <= a OR b OR c;
END synth;
```

This model uses a simple concurrent assignment statement to describe the functionality of the **OR** gate. The model specifies the functionality required for this entity, but not the implementation. The synthesis tool can choose to implement this functionality in a number of ways, depending on the cells available in the technology library and the constraints on the model. For instance, the most obvious implementation is shown in Figure 10-1.

This implementation uses a 3-input **OR** gate to implement the functionality specified in the concurrent signal assignment statement contained in architecture **synth**.

What if the technology library did not contain a 3-input **OR** device? Two other possible implementations are shown in Figures 10-2 and 10-3.

The first implementation uses a 3-input **NOR** gate followed by an inverter. The synthesis tool may choose this implementation if there are no 3-input **OR** devices in the technology library. Alternatively, if there are no 3-input devices, or if the 3-input devices violate a speed constraint, the 3-

Figure 10-1
Model
Implementation.

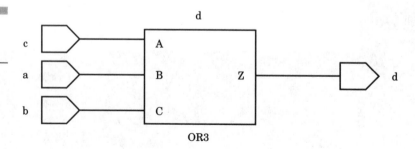

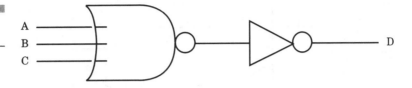

Figure 10-2
3-Input OR.

A
B
C
D

input OR function could be built from four devices, as shown in Figure 10-3. Given a technology library of parts, the functionality desired, and design constraints, the synthesis tool is free to choose among any of the implementations that satisfy all the requirements of a design, if such a design exists. There are lots of cases where the technology or constraints are such that no design can meet all of the design requirements.

IF Control Flow Statements

In the next example, control flow statements such as **IF THEN ELSE** are used to demonstrate how synthesis from a higher level description is accomplished. This example forms the control logic for a household alarm system. It uses sensor input from a number of sensors to determine whether or not to trigger different types of alarms. Following is the input description:

```
LIBRARY IEEE;
USE IEEE.std_logic_1164.ALL;
ENTITY alarm_cntrl IS
   PORT( smoke, front_door, back_door, side_door,
         alarm_disable, main_disable,
         water_detect : IN std_logic;
         fire_alarm, burg_alarm,
         water_alarm : OUT std_logic);
END alarm_cntrl;

ARCHITECTURE synth OF alarm_cntrl IS
BEGIN
   PROCESS(smoke, front_door, back_door, side_door,
         alarm_disable, main_disable,
         water_detect)
   BEGIN
    IF ((smoke = '1') AND (main_disable = '0')) THEN
       fire_alarm <= '1';
    ELSE
```

Figure 10-3
Another 3-Input OR
Implementation.

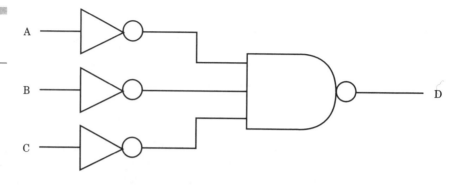

```
      fire_alarm <= `0';
   END IF;

       IF (((front_door = `1') OR (back_door = `1') OR
          (side_door = `1')) AND
            ((alarm_disable = `0') AND (main_disable =
               `0'))) THEN
          burg_alarm <= `1';
       ELSE
          burg_alarm <= `0';
       END IF;

       IF ((water_detect = `1') AND (main_disable = `0'))
           THEN
          water_alarm <= `1';
       ELSE
          water_alarm <= `0';
       END IF;
   END PROCESS;
END synth;
```

The input description contains a number of sensor input ports such as a smoke detector input, a number of door switch inputs, a basement water detector, and two disable signals. The **main_disable** port is used to disable all alarms, while the **alarm_disable** port is used to disable only the burglar alarm system.

The functionality is described by three separate **IF** statements. Each **IF** statement describes the functionality of one or more output ports. Notice that the functionality could also be described very easily with equations, as in the first example. Sometimes, however, the **IF** statement style is more readable. For instance, the first **IF** statement can be described by the following equation:

```
fire_alarm <= smoke and not(main_disable);
```

Because the three **IF** statements are separate and they generate separate outputs, we can expect that the resulting logic would be three separate pieces of logic. However, the **main_disable** signal is shared between the three pieces of logic. Any operations that make use of this signal may be shared by the other logic pieces. How this sharing takes place is determined by the synthesis tool and is based on the logical functionality of the design and the constraints. Speed constraints may force the logical operations to be performed in parallel.

A sample synthesized output is shown in Figure 10-4. Notice that signal **main_disable** connects to all three output gates, while signal **alarm_disable** only connects to the alarm control logic. The logic for the water alarm and smoke detector turn out to be quite simple, but we could have guessed that because our equations were so simple. The next example is not so simple.

Figure 10-4

A sample synthesized output.

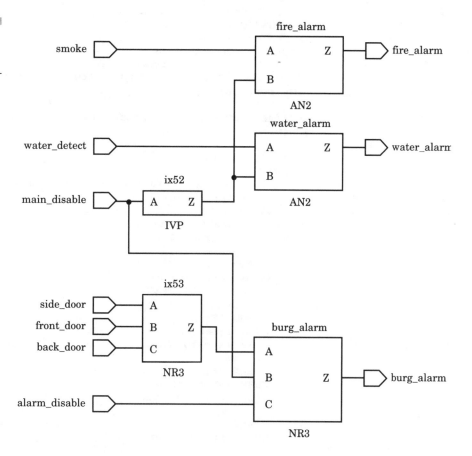

Case Control Flow Statements

The next example is an implementation of a comparator. There are two 8-bit inputs to be compared and a CTRL input that determines the type of comparison made. The possible comparison types are A > B, A < B, A = B, A ≠ B, A ≥ B, and A ≤ B. The design contains one output port for each of the comparison types. If the desired comparison output is true, then the output value on that output port is a '1'. If false, the output port value is a '0'. Following is a synthesizable VHDL description of the comparator:

```
PACKAGE comp_pack IS
  TYPE bit8 is range 0 TO 255;
  TYPE t_comp IS (greater_than, less_than, equal,
                  not_equal, grt_equal, less_equal);
END comp_pack;

LIBRARY IEEE;
USE IEEE.std_logic_1164.ALL;
USE WORK.comp_pack.ALL;
ENTITY compare IS
  PORT( a, b : IN bit8;
        ctrl : IN t_comp;
        gt, lt, eq, neq, gte, lte : OUT std_logic);
 END compare;

ARCHITECTURE synth OF compare IS
BEGIN

  PROCESS(a, b, ctrl)
  BEGIN
   gt <= '0'; lt <= '0'; eq <= '0'; neq <= '0'; gte <=
     '0'; lte <= '0';
   CASE ctrl IS
     WHEN greater_than =>
       IF (a > b) THEN
         gt <= '1';
       END IF;
     WHEN less_than =>
       IF (a < b) THEN
         lt <= '1';
       END IF;
     WHEN equal =>
       IF (a = b) THEN
         eq <= '1';
       END IF;
     WHEN not_equal =>
       IF (a /= b) THEN
         neq <= '1';
       END IF;
     WHEN grt_equal =>
```

```
      IF (a >= b) THEN
         gte <= '1';
      END IF;
   WHEN less_equal =>
      IF (a > b) THEN
         lte <= '1';
      END IF;
   END CASE;
  END PROCESS;
END synth;
```

Notice that, in this example, the equations of the inputs and outputs are harder to write because of the comparison operators. It is still possible to do, but is much less readable than the case statement shown earlier.

When synthesizing a design, the complexity of the design is related to the complexity of the equations that describe the design function. Typically, the more complex the equations, the more complex the design created. There are exceptions to this rule, especially when the equations reduce to nothing.

A sample synthesized output from the preceding description is shown in Figure 10-5. The inputs are shown on the left of the schematic diagram, and the outputs are shown in the lower-right of the schematic. The equations for the comparison operators have all been shared and combined together to produce an optimal design. This design is a very small number of gates for the operation performed.

There are still a number of cases where hand design can create smaller designs, but in most cases today the results of synthesis are very good; and you get the added benefit of using a higher level design language for easier maintainability and a shorter design cycle.

Simple Sequential Statements

Let's take a closer look at an example that we already discussed in the last chapter. This is the inferred D flip-flop. Inferred flip-flops are created by **WAIT** statements or **IF THEN ELSE** statements, which are surrounded by sensitivities to a clock. By detecting clock edges, the synthesis tool can locate where to insert flip-flops so that the design that is ultimately built behaves as the simulation predicts.

Following is an example of a simple sequential design using a **WAIT** statement:

```
LIBRARY IEEE;
```

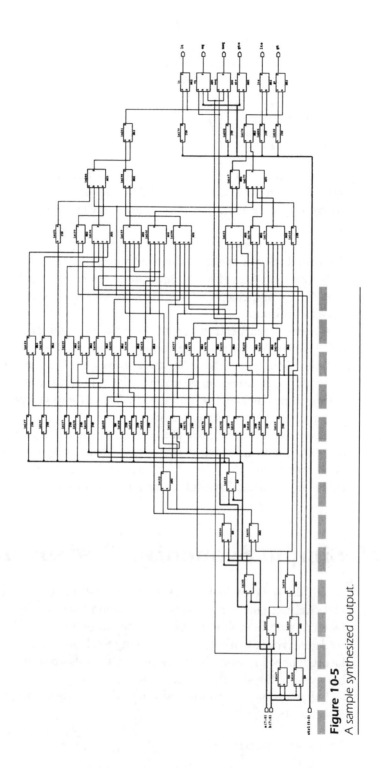

Figure 10-5
A sample synthesized output.

```
USE IEEE.std_logic_1164.ALL;
ENTITY dff IS
  PORT( clock, din : IN std_logic;
        dout : OUT std_logic);
END dff;

ARCHITECTURE synth OF dff IS
BEGIN
  PROCESS
  BEGIN
   WAIT UNTIL ((clock'EVENT) AND (clock = '1'));

   dout <= din;

   END PROCESS;
END synth;
```

The description contains a synthesizable entity and architecture representing a D flip-flop. The entity contains the **clock, din,** and **dout** ports needed for a D flip-flop, while the architecture contains a single process statement with a single **WAIT** statement. When the clock signal has a rising edge occur, the contents of **din** are assigned to **dout.** Effectively, this is how a D flip-flop operates.

The synthesized output of this design matches the functionality of the RTL description. It is very important for the synthesis and simulation results to agree. Otherwise, the resulting synthesized design may not work as planned. Part of the synthesis methodology should require that a final gate level simulation of the design is executed to verify that the gate level functionality is correct. (We perform this step in an example later on.)

The output of the Leonardo synthesis tool is shown in Figure 10-6. As expected, the output of the synthesis tool produced a single D flip-flop. The synthesis tool connected the ports of the entity to the proper ports of actual FPGA library macro so that the device works as expected in the design.

Figure 10-6
The output of the
Leonardo synthesis
tool.

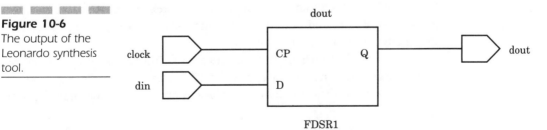

Asynchronous Reset

In a number of instances, D flip-flops are required to have an asynchronous reset capability. The previous D flip-flop did not have this capability. How would we generate a D flip-flop with an asynchronous reset? Remember the simulation and synthesis results must agree. Following is one way to accomplish this:

```
LIBRARY IEEE;
USE IEEE.std_logic_1164.ALL;
ENTITY dff_asynch IS
  PORT( clock, reset, din : IN std_logic;
        dout : OUT std_logic);
END dff_asynch;

ARCHITECTURE synth OF dff_asynch IS
BEGIN
  PROCESS(reset, clock)
  BEGIN
   IF (reset = '1') THEN
     dout <= '0';
   ELSEIF (clock'EVENT) AND (clock = '1') THEN
     dout <= din;
   END IF;
  END PROCESS;
END synth;
```

The **ENTITY** statement now has an extra input, the **reset** port, which is used to asynchronously reset the D flip-flop. Notice that **reset** and **clock** are in the process sensitivity list and cause the process to be evaluated. If an event occurs on signals **clock** or **reset**, the statements inside the process are executed.

First, signal **reset** is tested to see if it has an active value ('1'). If active, the output of the flip-flop is reset to '0'. If **reset** is not active ('0'), then the **clock** signal is tested for a rising edge. If signal **clock** has a rising edge, then input **din** is assigned as the new flip-flop output.

The fact that the **reset** signal is tested first in the **IF** statement gives the **reset** signal a higher priority than the **clock** signal. Also, because the **reset** signal is tested outside of the test for a clock edge, the **reset** signal is asynchronous to the clock.

The Leonardo synthesis tool produces a D flip-flop with an asynchronous **reset** input, as shown in Figure 10-7. The resulting design has an extra inverter (IVP component) in the circuit because the only flip-flop

Figure 10-7
The Leonardo
synthesis tool
produces a
D flip-flop.

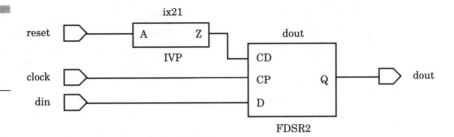

macro that would match the functionality required had a **reset** input that
was active low.

Asynchronous *Preset* and *Clear*

Is it possible to describe a flip-flop with an asynchronous **preset** and
clear? As an attempt, we can use the same technique as in the asynchronous **reset** example. The following example illustrates an attempt to
describe a flip-flop with an asynchronous **preset** and **clear** inputs:

```
LIBRARY IEEE;
USE IEEE.std_logic_1164.ALL;
ENTITY dff_pc IS
   PORT( preset, clear, clock, din : IN std_logic;
         dout : OUT std_logic);
END dff_pc;

ARCHITECTURE synth OF dff_pc IS
BEGIN
PROCESS(preset, clear, clock)
  BEGIN
    IF (preset = '1') THEN
      dout <= '1';

    ELSEIF (clear = '1') THEN
      dout <= '0';

    ELSEIF (clock'EVENT) AND (clock = '1') THEN
      dout <= din;

    END IF;
  END PROCESS;
END synth;
```

The entity contains a **preset** signal that sets the value of the flip-flop to a '1', a clear signal that sets the value of the flip-flop to a '0', and the normal **clock** and **din** ports used for the clocked D flip-flop operation. The architecture contains a single process statement with a single **IF** statement to describe the flip-flop behavior. The **IF** statement assigns a '1' to the output for a '1' value on the **preset** input and a '0' to the output for a '1' on the **clear** input. Otherwise, the **clock** input is checked for a rising edge, and the **din** value is clocked to the output **dout**.

What does the output of the synthesis process produce for this VHDL input? The output is shown in Figure 10-8. We were expecting the output of the synthesis tool in which the design **preset** input was connected to the **preset** input of the flip-flop, and the design **clear** input was connected to the **clear** input of the flip-flop. The output from the synthesis tool is a design in which the design **preset** and **clear** inputs are separated from the flip-flop **preset** and **clear** inputs by some logic.

This logic circuitry performs a prioritization of the **preset** signal with respect to the **clear** signal. Because the **preset** signal occurs before the **clear** signal in the **IF** statement, the **preset** signal is tested before the **clear** signal. If the **preset** signal is active, the flip-flop presets regardless of the state of the **clear** input. Effectively, the **preset** signal has a higher priority than the **clear** signal. There is currently no way to write a VHDL description to generate a design in which the **preset** and **clear** inputs have the same priority.

Figure 10-8
Output of synthesis process.

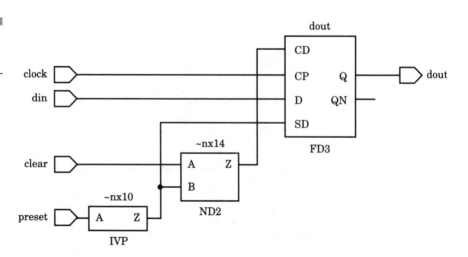

More Complex Sequential Statements

The next example is a more complex sequential design of a 4-bit counter. This example makes use of a two process description style. This style works very well for some synthesis tools, producing very good synthesis results.

Each process has a particular function. One process is clocked and the other is not. The clocked process is used to maintain the present state of the counter, while the unclocked process calculates the next state of the counter.

Following is an example of a counter written in this way:

```
USE IEEE.std_logic_1164.ALL;
USE IEEE.std_logic_unsigned.ALL;
PACKAGE count_types IS
  SUBTYPE bit4 IS std_logic_vector(3 DOWNTO 0);
END count_types;

LIBRARY IEEE;
USE IEEE.std_logic_1164.ALL;
USE IEEE.std_logic_unsigned.ALL;
USE WORK.count_types.ALL;
ENTITY count IS
  PORT(clock, load, clear : IN std_logic;
       din : IN bit4;
       dout : INOUT bit4);
END count;

ARCHITECTURE synth OF count IS
  SIGNAL count_val : bit4;
BEGIN
  PROCESS(load, clear, din, dout)
  BEGIN
   IF (load = '1') THEN
     count_val <= din;
   ELSEIF (clear = '1') THEN
     count_val <= "0000";
   ELSE
     count_val <= dout + "0001";
   END IF;
  END PROCESS;

  PROCESS
  BEGIN
   WAIT UNTIL clock'EVENT and clock = '1';
```

```
        dout <= count_val;
      END PROCESS;
    END synth;
```

The description contains a package that defines a 4-bit range that causes the synthesis tools to generate a 4-bit counter. Changing the size of the range causes the synthesis tools to generate different size counters. By using a constrained universal integer range, the model can take advantage of the built-in arithmetic operators for type universal integer. The other alternative is to define a type that is 4 bits wide and then create a package that overloads the arithmetic operators for the 4-bit type.

The entity contains a **clock** input port to clock the counter, a **load** input port that allows the counter to be synchronously loaded, a **clear** input port that synchronously clears the counter, a **din** input port that allows values to be loaded into the counter, and an output port **dout** that presents the current value of the counter to the outside world.

The architecture for the counter contains two processes. The process labeled **synch** is the process that maintains the current state of the counter. It is the process that is clocked by the clock and transfers the new calculated output **count_val** to the current output **dout**.

The other process contains a single **IF** statement that determines whether the counter is being loaded, cleared, or is counting up.

A sample synthesized output is shown in Figure 10-9. In this example, the generated results are as expected. The left side of the schematic shows the inputs to the counter; the right side of the schematic has the counter output. Notice that the design contains four flip-flops (**FDSR1**), exactly as specified. Also, notice that the logic generated for the counter is very small. This design was optimized for area; thus, the number of levels of logic are probably higher than a design optimized for speed.

Four-Bit Shifter

Another sequential example is a 4-bit shifter. This shifter can be loaded with a value and can be shifted left or right one bit at a time. Following is the model for the shifter:

```
LIBRARY IEEE;
USE IEEE.std_logic_1164.ALL;
PACKAGE shift_types IS
    SUBTYPE bit4 IS std_logic_vector(3 downto 0);
END shift_types;
```

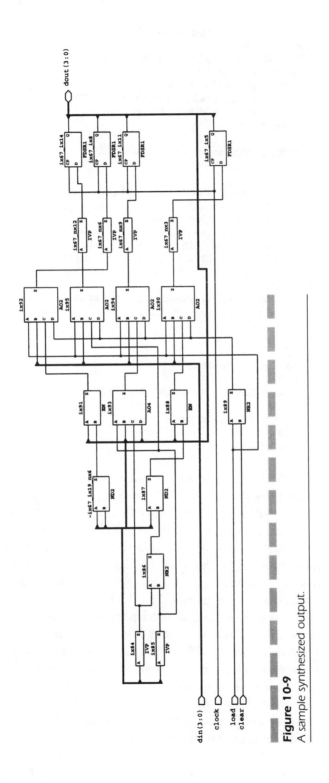

Figure 10-9

A sample synthesized output.

```
USE WORK.shift_types.ALL;
LIBRARY IEEE;
USE IEEE.std_logic_1164.ALL;
ENTITY shifter IS
  PORT( din : IN bit4;
        clk, load, left_right : IN std_logic;
        dout : INOUT bit4);
END shifter;

ARCHITECTURE synth OF shifter IS
  SIGNAL shift_val : bit4;
BEGIN
  nxt: PROCESS(load, left_right, din, dout)
  BEGIN
   IF (load = '1') THEN
     shift_val <= din;
   ELSEIF (left_right = '0') THEN
     shift_val(2 downto 0) <= dout(3 downto 1);
     shift_val(3) <= '0';
   ELSE
     shift_val(3 downto 1) <= dout(2 downto 0);
     shift_val(0) <= '0';
   END IF;
  END PROCESS;

  current: PROCESS
  BEGIN
   WAIT UNTIL clk'EVENT AND clk = '1';

   dout <= shift_val;
  END PROCESS;
END synth;
```

The 4-bit type used for the input and output of the shifter is declared in package **shift_types**. This package is used by entity **shifter** to declare ports **din** and **dout**. Ports **clk**, **load**, and **left_right** are **std_logic** signals used to control the functions of the shifter.

The architecture is organized similarly to the last example, with two processes used to describe the functionality of the architecture. One process keeps track of the current value of the shifter, and the other calculates the next value based on the last value and the control inputs.

Process **current** is used to keep track of the current value of the shifter. It is a process that has a single **WAIT** statement and a single signal assignment statement. When the **clk** signal has a rising edge occur, the signal assignment statement is activated and the next calculated value of the shifter (**shift_val**) is written to the signal that holds the current state of the shifter (**dout**).

Process **nxt** is used to calculate the next value of **shift_val** to be written into **dout**. **Load** is the highest priority input and, if equal to '1', causes shift_val to receive the value of din. Otherwise, signal left_right is tested to see if the shifter is shifting left or right. Because this shifter does not contain a carryin or carryout, '0' values are written into the bits whose value has been shifted over. (A good exercise is to write a shifter that contains a carryin and carryout.)

The synthesis tool produces a schematic for this input description as shown in Figure 10-10. By counting the flip-flops (FDSR1) on the page, it can be seen that this is indeed a 4-bit shifter.

State Machine Example

The next example is a simple state machine used to control a voicemail system. (This example does not represent any real system in use and is necessarily simple to make it easier to fit in the book.) The voicemail controller allows the user to send messages, review messages, save messages, and erase messages. A state diagram showing the possible state transitions is shown in Figure 10-11.

The normal starting state is state main. From main, the user can select whether to review messages or send messages. To get to the Review menu, the user presses the 1 key on the touch-tone phone. To select the Send Message menu, the user presses the 2 key on the touch-tone phone. After the user has selected either of these options, further menu options allow the user to perform other functions such as Save and Erase. For instance, if the user first selects the Review menu by pressing key 1, then pressing key 2 allows the user to save a reviewed message when reviewing is complete.

Following is the VHDL description for the voicemail controller:

```
PACKAGE vm_pack IS
  TYPE t_vm_state IS (main_st, review_st, repeat_st,
                      save_st,
                      erase_st, send_st,
                      address_st, record_st,
                      begin_rec_st, message_st);
    TYPE t_key IS ('0','1','2','3','4','5','6','7','8','9',
                  '*','#');

END vm_pack;

USE WORK.vm_pack.ALL;
LIBRARY IEEE;
USE IEEE.std_logic_1164.ALL;
ENTITY control IS
```

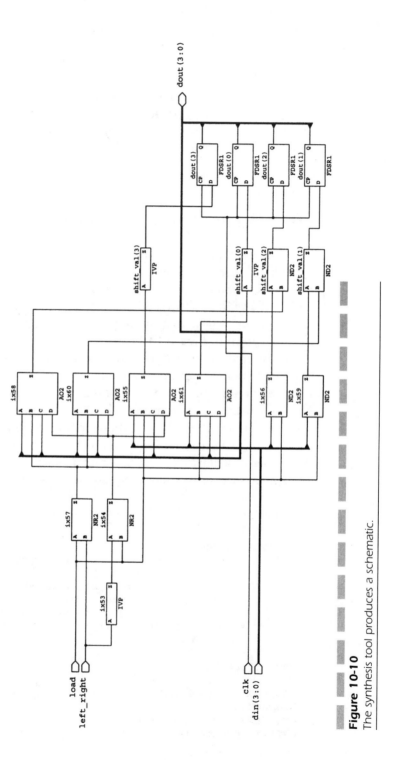

Figure 10-10
The synthesis tool produces a schematic.

274

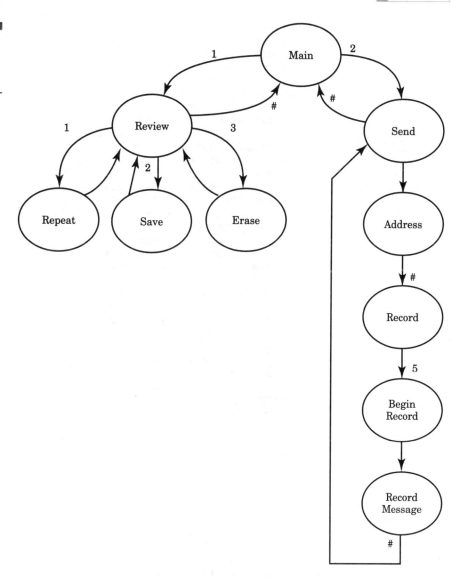

Figure 10-11
State Transition
Diagram for
Voicemail Controller.

```
        PORT( clk : in std_logic;
              key : in t_key;
              play, recrd, erase, save, address : out std_logic);
        END control;

        ARCHITECTURE synth OF control IS
          SIGNAL next_state, current_state : t_vm_state;
        BEGIN
          PROCESS(current_state, key)
          BEGIN
           play <= '0';
```

```
      save <= '0';
      erase <= '0';
      recrd <= '0';
      address <= '0';

  CASE current_state IS
    WHEN main_st =>
      IF (key = '1') THEN
        next_state <= review_st;
      ELSEIF (key = '2') THEN
        next_state <= send_st;
      ELSE
        next_state <= main_st;
      END IF;

    WHEN review_st =>
      IF (key = '1') THEN
        next_state <= repeat_st;
      ELSEIF (key = '2') THEN
        next_state <= save_st;
      ELSEIF (key = '3') THEN
        next_state <= erase_st;
      ELSEIF (key = '#') THEN
        next_state <= main_st;
      ELSE
        next_state <= review_st;
      END IF;

    WHEN repeat_st =>
      play <= '1';
      next_state <= review_st;

    WHEN save_st =>
      save <= '1';
      next_state <= review_st;

    WHEN erase_st =>
      erase <= '1';
      next_state <= review_st;

    WHEN send_st =>
      next_state <= address_st;

    WHEN address_st =>
      address <= '1';
      IF (key = '#') THEN
        next_state <= record_st;
      ELSE
        next_state <= address_st;
      END IF;

    WHEN record_st =>
      IF (key = '5') THEN
        next_state <= begin_rec_st;
      ELSE
```

```
        next_state <= record_st;
      END IF;

    WHEN begin_rec_st =>
      recrd <= '1';
      next_state <= message_st;

    WHEN message_st =>
        recrd <= '1';
        IF (key = '#') THEN
          next_state <= send_st;
        ELSE
          next_state <= message_st;
        END IF;
    END CASE;
  END PROCESS;

  PROCESS
  BEGIN
    WAIT UNTIL clk = '1' AND clk'EVENT;

    current_state <= next_state;
  END PROCESS;
END synth;
```

Package **vm_types** contains the type declarations for the state values and keys allowed by the voicemail controller. Notice that the states are all named something meaningful as opposed to S1, S2, S3, and so on. This makes the model much more readable.

This package is used by the entity to declare local signals and the **key** input port. The entity only has one input, the **key** input, which represents the possible key values from a touch-tone phone keypad. All of the other ports of the entity are output ports (except **clk**) and are used to control the voicemail system operations.

This model uses the two process style to describe the operation of the state machine. This style is very useful for describing state machines as one process represents the current state register, and the other process represents the next state logic.

The next state process starts by initializing all of the output signals to `'0'`. The reason for this is to provide the synthesis tool with a default value to assign the signal if the signal was not assigned in the **CASE** statement.

The rest of the next state process consists of one **CASE** statement. This **CASE** statement describes the action to occur based on the current state of the state machine and any inputs that affect the state machine. The condition that the **CASE** statement keys from is the current state. The state machine can be placed in a different state depending on the inputs

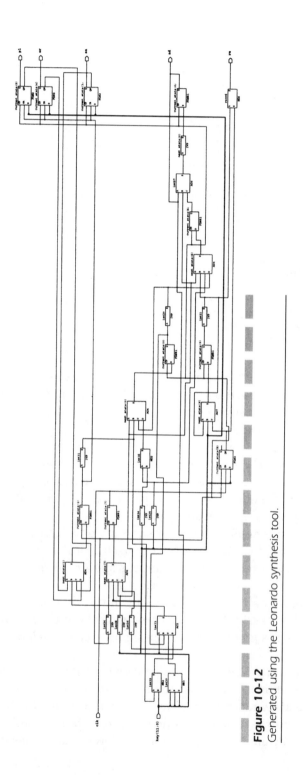

Figure 10-12
Generated using the Leonardo synthesis tool.

that are being tested by the current state. For instance, if the current state is **main_st**, when the **key** input is '1', the next state is **review_st**; when the **key** input is '2', the next state is **send_st**.

When this description is synthesized using the Leonardo synthesis tool, the schematic shown in Figure 10-12 is generated. The **key** and **clk** inputs are shown coming into the left side of the schematic and outputs **save**, **recrd**, **address**, **erase**, and **play** are shown coming out of the right side of the schematic. Intermixed in the design are the state flip-flops that are used to hold the current state of the voicemail controller and the logic used to generate the next state of the controller. This type of output is indicative of state machine descriptions.

SUMMARY

In this chapter, we looked at a number of different VHDL synthesis examples. They ranged from simple gate level descriptions to more complex examples that contained state machines. In the next few chapters, we look at a more complex example that requires a number of state machines, and we follow the process from start to finish.

High Level Design Flow

This chapter describes the design flow used to create complex FPGA and ASIC devices. The designer starts with a design specification, creates an RTL description, verifies that description, synthesizes the description to gates, uses place and route tools to implement the design in the chip, and then verifies that the final result is correct in terms of function and timing. The High Level Design flow is shown in Figure 11-1.

The first step in a high level design flow is the Design Specification process. This process involves specifying the behavior expected of the final design. The designer puts enough detail into the specification so that the design can be built. The specification is usually written in the designer's native language and specifies the expected function and behavior of the design using textual description and graphic elements.

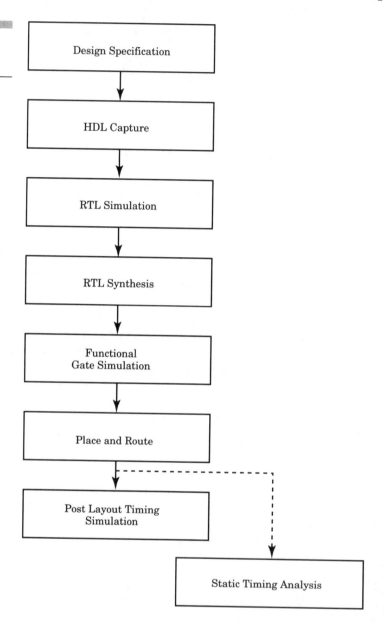

Figure 11-1
High Level Design
Flow.

After the specification has been completed, the designer or designers can begin the process of implementation. Some design teams create a high level behavioral or algorithmic description of the design to verify design intent, then convert that description to RTL (Register Transfer Level) later. However, most design teams skip the behavioral description and implement the RTL directly. The RTL is created during the HDL Capture step. The designer creates the VHDL RTL description that describes the clock-by-clock behavior of the design. The designer most likely uses a common text editor such as Emacs, or vi, whatever is available on the designer's computer. Some designers also use high level entry tools that contain block editors and state machine editors that automatically create the VHDL code.

The designer enters the VHDL code for entities of the design and checks them for correct syntax. After the syntax errors have been removed, the designer can begin the process of verifying the correctness of the VHDL using RTL Simulation.

RTL Simulation

The RTL Simulation step is used to verify the correctness of the RTL VHDL description. The designer has described the clock-by-clock behavior of the design. Now, the designer uses stimulus that represents the design environment to drive the design and check to make sure that the results are correct. A standard VHDL simulator can be used to read the RTL VHDL description and verify the correctness of the design.

The VHDL simulator reads the VHDL description, compiles it into an internal format, and then executes the compiled format using test vectors. The designer can look at the output of the simulation and determine whether or not the design is working properly.

The usual RTL Simulation step looks like Figure 11-2.

The designer creates the VHDL as described earlier and compiles the VHDL RTL description to remove any syntax errors. After the syntax errors have been removed, the design is simulated to verify the correctness of the design. After the simulation has completed, the designer analyzes the results of the simulation to determine if the design is correct or not. If not, the designer must fix the VHDL code and compile and simulate the design again. This process continues until all errors are removed.

The designer loads the compiled VHDL description into the simulator and applies stimulus to the design. This may be a file of input stimulus, a set of commands the designer enters, or an automatic testbench that applies the stimulus and checks the results. (These are discussed in Chapter 14, "RTL Simulation.") After the stimulus has been entered, the designer runs the simulation for as long as needed to generate enough output data to determine if the design is correct. At the beginning of the

Figure 11-2
RTL Simulation Flow.

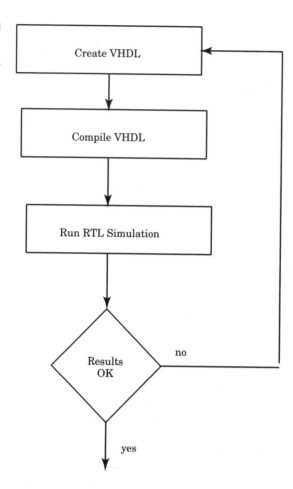

design process, this may be only a few vectors to make sure that the design resets properly. But later, more and more of the vectors are run as the design starts to function properly.

After the simulation has been run, the simulator will have generated output data that can be analyzed. The designer usually has a number of ways to analyze the data. Most common are waveform output and text tabular output. A sample waveform output is shown in Figure 11-3.

A waveform display shows the values of the signals of the design over time. The designer can see the relationships between signal transitions very easily. Using the waveform display, the designer can determine when system clock edges occur and if the proper signal transitions are present.

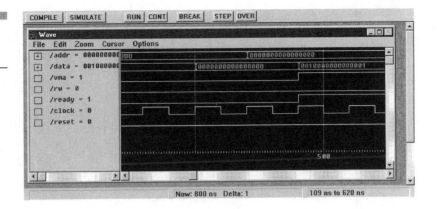

Figure 11-3
Sample waveform output.

The text tabular output is the same data as the waveform display, but in a different format. A sample output is shown in Figure 11-4.

All of the signal transitions are shown from top to bottom instead of left to right. It is also easier to read some of the signal values when the signal has a lot of changes in a short amount of time and the signal values are represented by a number of text characters. Most text table outputs can also filter the output data using a number of different mechanisms such as only on Print on Change or Print on Strobe.

While the output data is being analyzed, the user finds errors in the design description. The user uses the waveform and tabular displays to trace down the source of the errors in the VHDL code, make a change to the VHDL to fix the problem, recompile the design again, and rerun the test. If the problem is fixed, the designer tries to find the next problem, until all problems have been found.

When the designer is happy with the behavior of the design, the designer can start the process of building the real hardware device. To implement the design, the designer uses VHDL synthesis tools. The next step in the process is the VHDL synthesis step.

VHDL Synthesis

The goal of the VHDL synthesis step is to create a design that implements the required functionality and matches the designer's constraints in speed, area, or power.

The VHDL synthesis tools convert the VHDL description into a netlist in the target FPGA or ASIC technology. For the VHDL synthesis tool to

Figure 11-4

Text tabular output.

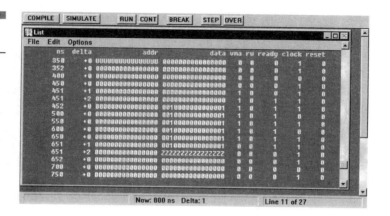

perform this step properly, the VHDL code must be written in a particular style, as discussed in Chapter 10, "VHDL Synthesis."

To synthesize a VHDL description, the designer reads the verified VHDL description into the VHDL synthesis tool in the same way that the designer read the design into the VHDL simulator. The VHDL synthesis tool reports syntax errors and synthesis errors. Synthesis errors usually result from the designer using constructs that are not synthesizable. For instance, ACCESS types in VHDL are not synthesizable, because they could specify hardware that is dynamic in nature. Of course, syntax errors result from improper VHDL syntax being read by the VHDL synthesis tool. Presumably, most all of these errors will already have been taken care of because the VHDL code has already been verified with the VHDL simulator. The VHDL synthesis tool also reports warnings of constructs that have the possibility of generating mismatches between the RTL simulation results and the output netlist simulation results.

The designer reads the VHDL design into the VHDL synthesis tool. If there are no syntax errors, the designer can synthesize the design and map the design to the target technology. If the designer had to make changes to the VHDL description, then the VHDL description needs to be simulated again and the output validated for correctness. First, the designer needs to make sure that the synthesizer is producing an output in the target technology that looks reasonable. The designer looks at the synthesizer output to determine whether or not the synthesizer produced a good result.

The synthesizer produces an output netlist in the target technology and a number of report files. By looking at the netlist, the designer can determine whether or not the design looks reasonable. For most reasonable size designs, however, it can be very difficult to determine how well the synthesizer implemented the function. The designer looks at the report files to determine the quality of the synthesis output. The most common output files are the timing report and the area report. Most synthesis tools produce a number of other reports such as hierarchy reports, instance reports, net reports, power reports, and others. The most useful reports initially are the timing and area reports, because these are usually the most critical factors.

Following is a sample area report:

```
**********************************************************

Cell: adder     View: test     Library: work

**********************************************************

Total accumulated area :
Number of LCs :                              8
Number of CARRYs :                           7

Number of ports :                           24
Number of nets :                           107
Number of instances :                       91
Number of references to this view :          0


            Cell         Library   References       Total Area

            GND          flex10     1 x        1       1 GND
          OUTBUF         flex10     8 x        1       8 OUTBUF
           INBUF         flex10    16 x        1      16 INBUF
           CARRY         flex10     7 x        1       7 CARRYs
            OR2          flex10    14 x        1      14 OR2
           AND2          flex10    21 x        1      21 AND2
          LCELL          flex10     8 x        1       8 LCs
           XOR2          flex10    16 x        1      16 XOR2
```

The area report tells the designer the size of the implemented design. The units of measure are determined by the units used when the synthesis library was implemented. For instance, the typical unit for ASIC designs is equivalent 2-input NAND gates, or gate equivalents. Using this measurement, a 2-input NAND gate would consume one gate equivalent, a 2-input AND gate would also consume one gate equivalent. A 4-input NAND gate would consume two gate equivalents. For FPGA designs, equivalent gate measurements vary from manufacturer to manufacturer because each has a different sized basic cell. In the preceding sample area report, the design produced 8 LC (Logic Cells) and 7 Carry devices. A typical LC is 10 to 12 logic gates; the Carry device is 2 to 3 gates. So, this example would represent about 90 to 120 gates.

The area report shows the designer how much of the resources of the chip the design has consumed. The designer can tell if the design is too big for a particular chip and the designer needs to target a larger chip, if the design should go into a smaller chip, or if the current chip will work fine. The designer can also get a relative size of the design to use in later stages of the design process.

The timing report shows the timing of critical paths or specified paths of the design. The designer examines the timing of the critical paths closely because these paths ultimately determine how fast the design can run. If the longest path is a timing critical part of the design and is not meeting the speed requirements of the designer, then the designer may have to modify the VHDL code or try new timing constraints to make the path meet timing.

The following is a sample timing report:

Critical Path Report

Critical path #1, (unconstrained path)

NAME	GATE	ARRIVAL		LOAD
a(0)/		0.00	up	0.00
ix30/OUT	INBUF	2.40	up	0.00
modgen_0_11_10_10_0_10_c1/Y	AND2	2.40	up	0.00
modgen_0_11_10_10_0_10_c3/Y	OR2	2.40	up	0.00
modgen_0_11_10_10_0_10_c4/Y	OR2	2.40	up	0.00
modgen_0_11_10_10_0_10_c5/Y	CARRY	2.90	up	0.00
modgen_0_11_10_10_1_10_c1/Y	AND2	2.90	up	0.00
modgen_0_11_10_10_1_10_c3/Y	OR2	2.90	up	0.00
modgen_0_11_10_10_1_10_c4/Y	OR2	2.90	up	0.00
modgen_0_11_10_10_1_10_c5/Y	CARRY	3.40	up	0.00
modgen_0_11_10_10_2_10_c2/Y	AND2	3.40	up	0.00
modgen_0_11_10_10_2_10_c4/Y	OR2	3.40	up	0.00
modgen_0_11_10_10_2_10_c5/Y	CARRY	3.90	up	0.00
modgen_0_11_10_10_3_10_c1/Y	AND2	3.90	up	0.00
modgen_0_11_10_10_3_10_c3/Y	OR2	3.90	up	0.00
modgen_0_11_10_10_3_10_c4/Y	OR2	3.90	up	0.00
modgen_0_11_10_10_3_10_c5/Y	CARRY	4.40	up	0.00
modgen_0_11_10_10_4_10_c1/Y	AND2	4.40	up	0.00
modgen_0_11_10_10_4_10_c3/Y	OR2	4.40	up	0.00
modgen_0_11_10_10_4_10_c4/Y	OR2	4.40	up	0.00
modgen_0_11_10_10_4_10_c5/Y	CARRY	4.90	up	0.00
modgen_0_11_10_10_5_10_c1/Y	AND2	4.90	up	0.00
modgen_0_11_10_10_5_10_c3/Y	OR2	4.90	up	0.00
modgen_0_11_10_10_5_10_c4/Y	OR2	4.90	up	0.00

NAME	GATE	ARRIVAL	LOAD
modgen_0_11_10_10_5_10_c5/Y	CARRY	5.40 up	0.00
modgen_0_11_10_10_6_10_c1/Y	AND2	5.40 up	0.00
modgen_0_11_10_10_6_10_c3/Y	OR2	5.40 up	0.00
modgen_0_11_10_10_6_10_c4/Y	OR2	5.40 up	0.00
modgen_0_11_10_10_6_10_c5/Y	CARRY	5.90 up	0.00
modgen_0_11_10_10_7_10_sum0/Y	XOR2	5.90 up	0.00
modgen_0_11_10_10_7_10_sum1/Y	XOR2	5.90 up	0.00
modgen_0_11_10_10_7_10_sum2/Y	LCELL	10.00 up	0.00
ix39/OUT	OUTBUF	13.80 up	0.00
c(7)/		13.80 up	0.00
data arrival time		13.80	

In this report, the worst case path is listed shown with estimated time values for each node traversed in the design. The timing analyzer calculates the time for a path from an input pin to a flip-flop or output, or from a flip-flop output to a flip-flop input, or output pin.

The designer has the ability to ask for the timing for particular paths of interest, or of the paths that have the longest timing value, and how many to display. As mentioned previously, the worst case paths ultimately determine the speed of the design. For instance, in this case, the worst case path is 13.8 nanoseconds; therefore, the fastest this design would be able to run is about 72 MHz.

The last type of output data that the designer can examine is the netlist for the design in the target technology. This output is a gate or macro level output in a format compatible with the place and route tools that are used to implement the design in the target chip. For instance, most place and route tools for FPGA technologies take in an EDIF netlist as an input format. The primitives used in the netlist are those used in the synthesis library to describe the technology. The place and route tools understand what to do with these primitives in terms of how to place a primitive and how to route wires to them. The following example uses a VHDL netlist for ease of understanding. To save space (and boredom), this is not a complete netlist, but gives the reader an idea how a netlist is structured. The complete netlist can be found on the included CD:

```
--
-- Definition of   adder
--
--

library IEEE, EXEMPLAR; use IEEE.STD_LOGIC_1164.all; use
    EXEMPLAR.EXEMPLAR_1164.all;
```

```vhdl
-- Library use clause for technology cells
library altera ;
use altera.all ;

entity adder is
   port (
      a : IN std_logic_vector (7 DOWNTO 0) ;
      b : IN std_logic_vector (7 DOWNTO 0) ;
      c : OUT std_logic_vector (7 DOWNTO 0)) ;
end adder ;

architecture test of adder is
   component XOR2
      port (
         Y : OUT std_logic ;
         IN1 : IN std_logic ;
         IN2 : IN std_logic) ;
   end component ;
   component LCELL
      port (
         Y : OUT std_logic ;
         IN1 : IN std_logic) ;
   end component ;
   component AND2
      port (
         Y : OUT std_logic ;
         IN1 : IN std_logic ;
         IN2 : IN std_logic) ;
   end component;
   .
   .
   .

signal c_dup0_7, c_dup0_6, c_dup0_5, c_dup0_4, c_dup0_3,
   c_dup0_2,
     c_dup0_1, c_dup0_0, modgen_0_11_10_c_int_7,
       modgen_0_11_10_c_int_6,
     modgen_0_11_10_c_int_5, modgen_0_11_10_c_int_4,
       modgen_0_11_10_c_int_3,
     modgen_0_11_10_c_int_2, modgen_0_11_10_c_int_1,
     modgen_0_11_10_10_0_10_s1, modgen_0_11_10_10_0_10_s2,
     modgen_0_11_10_10_0_10_w1, modgen_0_11_10_10_0_10_w2,
     modgen_0_11_10_10_0_10_w3, modgen_0_11_10_10_0_10_w4,
       b_2_int, b_1_int, b_0_int, U_0: std_logic ;
   .
   .
   .
begin
  modgen_0_11_10_10_0_10_sum0 : XOR2 port map ( Y=>
     modgen_0_11_10_10_0_10_s1, IN1=>a_0_int, IN2=>U_0);
  modgen_0_11_10_10_0_10_sum1 : XOR2 port map ( Y=>
     modgen_0_11_10_10_0_10_s2,
```

```
        IN1=>modgen_0_11_10_10_0_10_s1, IN2=>
    b_0_int);
modgen_0_11_10_10_0_10_sum2 : LCELL port map (
    Y=>c_dup0_0, IN1=>
    modgen_0_11_10_10_0_10_s2);
modgen_0_11_10_10_0_10_c0 : AND2 port map (
    Y=>modgen_0_11_10_10_0_10_w1,
    IN1=>a_0_int, IN2=>b_0_int);
modgen_0_11_10_10_0_10_c1 : AND2 port map (
    Y=>modgen_0_11_10_10_0_10_w2,
    IN1=>a_0_int, IN2=>U_0);
modgen_0_11_10_10_0_10_c2 : AND2 port map (
    Y=>modgen_0_11_10_10_0_10_w3,
    IN1=>U_0, IN2=>b_0_int);
    .
    .
    .
ix43 : OUTBUF port map ( \OUT\=>c(3), \IN\=>c_dup0_3);
ix44 : OUTBUF port map ( \OUT\=>c(2), \IN\=>c_dup0_2);
ix45 : OUTBUF port map ( \OUT\=>c(1), \IN\=>c_dup0_1);
ix46 : OUTBUF port map ( \OUT\=>c(0), \IN\=>c_dup0_0);
U_0_XMPLR : GND port map ( Y=>U_0);
end test ;
```

Notice that all of the other interconnect signal names have names such as **modgen_0_11_xx** or **ix123**. There is no corresponding signal name in the source file to specify the signal name; therefore, the synthesis tool generates names for these signals. The netlist can be used to figure out how well the synthesizer implemented a part of the design, or to track down a problem net. It can be very useful to find out why a critical path was implemented too slowly.

When the netlist meets the designer's timing, area, power, and other constraints, the next step is to pass the netlist to the Gate Level Simulator. This simulator checks the functionality of the synthesized design.

Functional Gate Level Verification

Some designers might want to do a quick check on the output of the synthesis tool to make sure that the synthesis tool produced a design that is functionally correct. If proper design rules are followed for the input VHDL description, the synthesis tool should never generate an output that is functionally different from the RTL VHDL input, unless the tool has a bug. However, if some of the warnings or error are ignored or some part of the design is written using a strange VHDL style, the synthesizer can produce an output netlist that does not exactly match the RTL input in terms of functionality. Most designers like to run a quick check on the results of the synthesis tool to make sure the synthesis tool produced a functionally correct output.

To do this, the designer runs a functional gate level verification. The designer reads the output VHDL netlist from the synthesis tool plus a library of the synthesis primitives into the VHDL simulator and runs the simulation using the RTL verification vectors. If the design matches, then the synthesis tool did not produce logic mismatches; if it does not match, the designer needs to debug the VHDL RTL description to see what is wrong.

The most common method for performing this step is to run a VITAL simulation of the netlist from the synthesis tool. For a completely functional simulation, no timing is back annotated. If the synthesis tool supports estimated timing and SDF file generation, the synthesis tool could write the VHDL netlist and an SDF timing file for the design. The designer could use these two files to run a VITAL simulation with estimated timing. After the design has been functionally verified, it is passed to the place and route tools to implement the design.

Place and Route

Place and route tools are used to take the design netlist and implement the design in the target technology device. The place and route tools place each primitive from the netlist into an appropriate location on the target device and then route signals between the primitives to connect the devices according to the netlist. Place and route tools are typically very architecture and device dependent. These tools are tuned to take advantage of each architectural and routing advantage the device contains. FPGA vendors provide these tools because the differences in architectures are large enough that writing a common tool for all architectures would be very difficult. Place and route tools for ASIC devices can be obtained from the ASIC vendor or EDA (Electronic Design Automation) vendors. ASIC architectures do not have as wide of variation between architectures as FPGA architectures and, therefore, place and route tools exist that can handle lots of different ASIC architectures.

Figure 11-5 shows a dataflow diagram of the place and route tools.

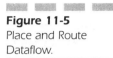

Figure 11-5
Place and Route
Dataflow.

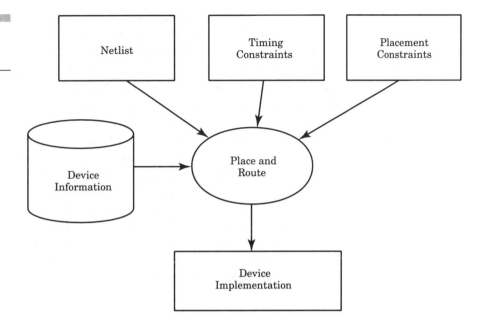

Inputs to the place and route tools are the netlist in EDIF or another netlist format, and possibly timing constraints. The format of the netlist input file varies from manufacturer to manufacturer. Some tools use EDIF; others use proprietary formats such as XNF.

Another input to some place and route tools is the timing constraints, which give the place and route tools an indication about which signals have critical timing associated with them and to route these nets in the most timing efficient manner. These nets are typically identified during the static timing analysis process during synthesis. These constraints tell the place and route tool to place the primitives in close proximity to one another and to use the fastest routing. The closer the cells are, the shorter the routed signals will be and the shorter the time delay.

Some place and route tools allow the designer to specify the placement of large parts of the design. This process is also known as floorplanning. Floorplanning allows the user to pick locations on the chip for large blocks of the design so that routing wires are as short as possible. The designer lays out blocks on the chip as general areas. The floorplanner feeds this information to the place and route tools so that these blocks are placed properly. After the cells are placed, the router makes the appropriate connections.

After all the cells are placed and routed, the output of the place and route tools consists of data files that can be used to implement the chip. In the case of FPGAs, these files describe all of the connections needed to make the FPGA macrocells implement the functionality required. Antifuse FPGAs use this information to burn the appropriate fuses, while reprogrammable devices download this information to the device to turn on the appropriate transistor connections.

The other output from the place and route software is a file used to generate the timing file. This file describes the actual timing of the programmed FPGA device or the final ASIC device. This timing file, as much as possible, describes the timing extracted from the device when it is plugged into the system for testing. The most common format of this file for most simulators is SDF (Standard Delay Format). Sometimes, proprietary formats are generated and later translated to SDF. SDF is used to back annotate the post route timing information from place and route tools into the post layout timing simulation.

Post Layout Timing Simulation

After the place and route process has completed, the designer will want to verify the results of the place and route process. There are a number of methods to accomplish this task but the most common is to use post route gate level simulation. This simulation combines the netlist used for place and route with the timing file from the place and route process into a simulation that checks both functionality and timing of the design. The designer can run the simulation and generate accurate output waveforms that show whether or not the device is operating properly and if the timing is being met.

If the design has been properly structured, the same test vectors used for the RTL simulation can be used for the post route gate level simulation. In this way, the designer is saved the process of generating a new set of vectors to check the gate level design and verifying the new vector output values.

Post route gate level simulation, if done properly, also uses the same simulator as the RTL simulation. For VHDL simulations, this requires a VITAL-compliant (standard way of describing designs with designs that allow SDF timing back annotation) VHDL simulator. VHDL simulators that are not VITAL compliant do not accelerate the execution of the gate level primitives and cannot accept SDF to back annotate the timing.

Static Timing

For designs of 10,000 gates to 100,000 gates, post route timing simulation can be a good method of verifying design functionality and timing. However, as designs get larger, or if the designer does not have test vectors, the designer can use static timing analysis to make sure the design meets the timing requirements. A static timing analyzer traces each path in the design and keeps track of the timing from a clock edge or an input. A timing report is then generated in a number of formats. For instance, the designer can ask for all paths and get an enormous listing of every path in the design. A more intelligent method, however, is to ask for the most timing critical paths in the design and make sure the timing constraints have been met.

Typical static timing analyzers have a number of report types that can be generated so that the designer can make sure the critical paths of the design can be found and verified to be within the required specifications. If paths are not within the specifications, the static timing analyzer shows the entire path so that the designer can try to fix the problem.

SUMMARY

In this chapter, the complete VHDL design process using synthesis was described. This process is very similar no matter which VHDL synthesis or simulation tool is used. The designer must follow a number of steps that add more detail to the design. At each step, the designer has checks to make sure that the correct behavior is being implemented. At the beginning of the process, RTL simulation is used to verify correctness. After synthesis, the netlist, timing report, and area report are all examined to make sure the design fits the designer's constraints. Functional simulation is then run to verify that the synthesis tool produced a functionally correct design. The design is put through the place and route process to implement the design in the target technology. The final check is then to verify using post route gate level simulation that the design is functionally correct and meets timing.

Top-Level
System Design

In the last few chapters, we have discussed VHDL language features and the VHDL synthesis process. In the next few chapters, we tie all of these ideas together by developing a top-down design for a small CPU design, verify its functionality, verify that it can be synthesized, and implement the design in an FPGA device.

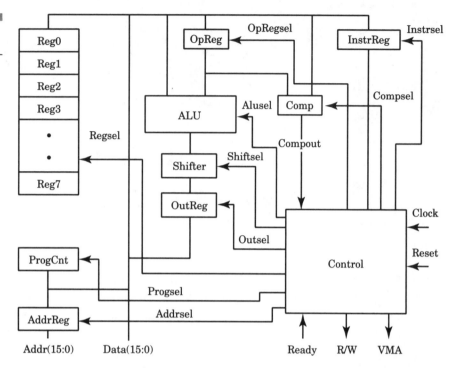

Figure 12-1
CPU Block Diagram.

CPU Design

The example is a small, 16-bit microprocessor. A block diagram is shown in Figure 12-1.

The processor contains a number of basic pieces. There is a register array of eight 16-bit registers, an ALU (Arithmetic Logic Unit), a shifter, a program counter, an instruction register, a comparator, an address register, and a control unit. All of these units communicate through a common, 16-bit tristate data bus.

Top-Level System Operation

The top-level design consists of the processor block and a memory block communicating through a bidirectional databus, an address bus, and a few control lines. The processor fetches instructions from the external memory and executes these instructions to run a program. These instructions are stored in the instruction register and decoded by the control unit. The control unit causes the appropriate signal interactions to make the processor unit execute the instruction.

If the instruction is an add of two registers, the control unit would cause the first register value to be written to register OpReg for temporary storage. The second register value would then be placed on the data bus. The ALU would be placed in add mode and the result would be stored in register OutReg. Register OutReg would store the resulting value until it is copied to the final destination.

When executing an instruction, a number of steps take place. The program counter holds the address in memory of the current instruction. After an instruction has finished execution, the program counter is advanced to where the next instruction is located. If the processor is executing a linear stream of instructions, this is the next instruction. If a branch was taken, the program counter is loaded with the next instruction location directly.

The control unit copies the program counter value to the address register, which outputs the new address on the address bus. At the same time, the control unit sets the R/W (read write signal) to a '0' value for a read operation and sets signal VMA (Valid Memory Address) to a '1', signaling the memory that the address is now valid. The memory decodes the address and places the memory data on the data bus. When the data has been placed on the data bus, the memory has set the READY signal to a '1' value indicating that the memory data is ready for consumption.

The control unit causes the memory data to be written into the instruction register. The control unit now has access to the instruction and decodes the instruction. The decoded instruction executes, and the process starts over again.

Instructions

Instructions can be divided into a number of different types as follows:

- *Load*—These instructions load register values from other registers, memory locations, or with immediate values given in the instruction.

- *Store*—These instructions store register values to memory locations.

- *Branch*—These instructions cause the processor to go to another location in the instruction stream. Some branch instructions test values before branching; others branch without testing.

■ *ALU*—These instructions perform arithmetic and logical operations such as ADD, SUBTRACT, OR, AND, and NOT.

■ *Shift*—These instructions use the shift unit to perform shift operations on the data passed to it.

Sample Instruction Representation

Instructions share common attributes, but come in a number of flavors. Sample instructions are shown in Figure 12-2.

All instructions contain the opcode in the five most significant bits of the instruction. Single word instructions also contain two 3-bit register fields in the lowest 6 bits of the instruction. Some instructions, such as INC (Increment), only use one of the fields, but other instructions, such as MOV (Move), use both register fields to specify the From register and the To register. In double word instructions, the first word contains the opcode and destination register address, and the second word contains the immediate instruction location or data value to be loaded. For instance, a LoadI (Load Immediate) instruction would look like this:

Figure 12-2
Instruction Words.

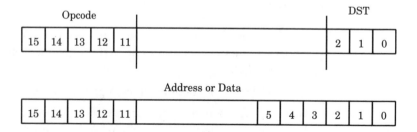

```
LoadI 1, 16#15
```

This instruction loads the hex value 15 into register 1. The instruction words look like those shown in Figure 12-3.

When the control unit decodes the opcode of the first word, it determines that the instruction is two words long and loads the second word to complete the instruction.

The instructions implemented in the processor and their opcodes are listed in Figure 12-4.

Not all of the possible instructions have been implemented in this processor example to limit the complexity for ease of publication. Typical commercial processors are much more complicated and have pipelined instructions streams for faster execution. To reduce complexity, this example is not pipelined.

CPU Top-Level Design

The next few sections contain the VHDL description for each of the CPU components. First of all, a top-level package cpulib.vhd is needed to describe the signal types that are used to communicate between the CPU components. Following is this package:

```
library IEEE;

use IEEE.std_logic_1164.all;

use IEEE.std_logic_arith.all;
package cpu_lib is
   type t_shift is (shftpass, shl, shr, rotl, rotr);
   subtype t_alu is unsigned(3 downto 0);
   constant alupass : unsigned(3 downto 0) := "0000";
   constant andOp : unsigned(3 downto 0) := "0001";
```

Figure 12-3
Instruction Data.

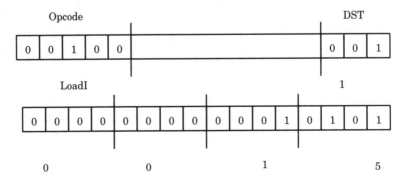

■■■ ■■■ ■■■ ■■■
Figure 12-4
Opcode Table.

OPCODE	INSTRUCTION	NOTE
00000	NOP	No operation
00001	LOAD	Load register
00010	STORE	Store register
00011	MOVE	Move value to register
00100	LOADI	Load register with immediate value
00101	BRANCHI	Branch to immediate address
00110	BRANCHGTI	Branch greater than to immediate address
00111	INC	Increment
01000	DEC	Decrement
01001	AND	And two registers
01010	OR	Or two registers
01011	XOR	Xor two registers
01100	NOT	Not a register value
01101	ADD	Add two register
01110	SUB	Subtract two registers
01111	ZERO	Zero a register
10000	BRANCHLTI	Branch less than to immediate address
10001	BRANCHLT	Branch less than
10010	BRANCHNEQ	Branch not equal
10011	BRANCHNEQI	Branch not equal to immediate address
10100	BRANCHGT	Branch greater than
10101	BRANCH	Branch all the time
10110	BRANCHEQ	Branch if equal
10111	BRANCHEQI	Branch if equal to immediate address
11000	BRANCHLTEI	Branch if less or equal to immediate
11001	BRANCHLTE	Branch if less or equal
11010	SHL	Shift Left
11011	SHR	Shift Right
11100	ROTR	Rotate Right
11101	ROTL	Rotate Left

```vhdl
constant orOp : unsigned(3 downto 0) := "0010";
constant notOp : unsigned(3 downto 0) := "0011";
constant xorOp : unsigned(3 downto 0) := "0100";
constant plus : unsigned(3 downto 0) := "0101";
constant alusub : unsigned(3 downto 0) := "0110";
constant inc : unsigned(3 downto 0) := "0111";
constant dec : unsigned(3 downto 0) := "1000";
constant zero : unsigned(3 downto 0) := "1001";

type t_comp is (eq, neq, gt, gte, lt, lte);
subtype t_reg is std_logic_vector(2 downto 0);
type state is (reset1, reset2, reset3, reset4, reset5,
               reset6, execute, nop, load, store, move,
```

```
            load2, load3, load4, store2, store3,
            store4, move2, move3, move4,incPc, incPc2,
            incPc3, incPc4, incPc5, incPc6, loadPc,
            loadPc2,loadPc3, loadPc4, bgtI2, bgtI3,
            bgtI4, bgtI5, bgtI6, bgtI7,bgtI8, bgtI9,
            bgtI10, braI2, braI3, braI4, braI5, braI6,
            loadI2,loadI3, loadI4, loadI5, loadI6,
            inc2, inc3, inc4);

    subtype bit16 is std_logic_vector(15 downto 0);

end cpu_lib;
```

This package describes a number of types that are used to specify the alu functionality, the shifter operation, and the states needed for the control of the cpu.

The highest level of the design is described by the file top.vhd as shown in the following:

```
library IEEE;
use IEEE.std_logic_1164.all;
use work.cpu_lib.all;

entity top is
end top;

architecture behave of top is
  component mem
    port (addr : in bit16;
            sel, rw : in std_logic;
            ready : out std_logic;
            data : inout bit16);
  end component;
  component cpu
    port(clock, reset, ready : in std_logic;
            addr : out bit16;
            rw, vma : out std_logic;
            data : inout bit16);
  end component;
  signal addr, data : bit16;
  signal vma, rw, ready : std_logic;
  signal clock, reset : std_logic := '0';
begin

  clock <= not clock after 50 ns;
  reset <= '1', '0' after 100 ns;

  m1 : mem port map (addr, vma, rw, ready, data);
  u1 : cpu port map(clock, reset, ready, addr, rw, vma,
                    data);
end behave;
```

This model instantiates components **cpu** and **mem** and specifies the necessary signals to connect the components, as shown in Figure 12-5.

Component **mem** is a memory device and contains the instructions and data for the CPU to execute. Component **cpu** is an RTL implementation of the CPU device that is simulated for correctness and synthesized to implement the design.

Let's now take a look at the description for the memory component to see how it works. The memory is described in file mem.vhd shown in the following:

```
library IEEE;
use IEEE.std_logic_1164.all;
use IEEE.std_logic_arith.all;
use IEEE.std_logic_unsigned.all;
```

Figure 12-5
Top Level of CPU
Design.

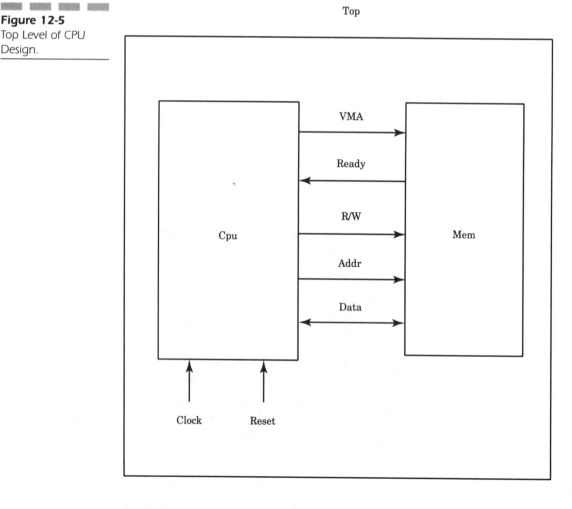

```
use work.cpu_lib.all;

entity mem is
port (addr : in bit16;
      sel, rw : in std_logic;
      ready : out std_logic;
      data : inout bit16);
end mem;

architecture behave of mem is
begin
memproc: process(addr, sel, rw)
 type t_mem is array(0 to 63) of bit16;
 variable mem_data : t_mem :=
  ("0010000000000001",   --- 0  loadI 1, #   -- load source
      address
   "0000000000010000",   --- 1  10
   "0010000000000010",   --- 2  loadI 2, #   -- load
      destination address
   "0000000000110000",   --- 3  30
   "0010000000000110",   --- 4  loadI 6, #   -- load data
      end address
   "0000000000101111",   --- 5  2F
   "0000100000001011",   --- 6  load 1, 3    -- load reg3
      with source element
   "0001000000011010",   --- 7  store 3, 2   -- store reg3
      at destination
   "0011000000001110",   --- 8  bgtI 1, 6, # -- compare to
      see if at end of data
   "0000000000000000",   --- 9  00           -- if so just
      start over
   "0011100000000001",   --- A  inc 1        -- move source
      address to next
   "0011100000000010",   --- B  inc 2        -- move
      destination address to next
   "0010100000001111",   --- C  braI #       -- go to the
      next element to copy
   "0000000000000110",   --- D  06
   "0000000000000000",   --- E
   "0000000000000000",   --- F
   "0000000000000001",   --- 10             --- Start of source
      array
   "0000000000000010",   --- 11
   "0000000000000011",   --- 12
   "0000000000000100",   --- 13
   "0000000000000101",   --- 14
   "0000000000000110",   --- 15
   "0000000000000111",   --- 16
   "0000000000001000",   --- 17
   "0000000000001001",   --- 18
   "0000000000001010",   --- 19
   "0000000000001011",   --- 1A
   "0000000000001100",   --- 1B
   "0000000000001101",   --- 1C
   "0000000000001110",   --- 1D
```

```
          "0000000000001111",   --- 1E
          "0000000000010000",   --- 1F
          "0000000000000000",   --- 20
          "0000000000000000",   --- 21
          "0000000000000000",   --- 22
          "0000000000000000",   --- 23
          "0000000000000000",   --- 24
          "0000000000000000",   --- 25
          "0000000000000000",   --- 26
          "0000000000000000",   --- 27
          "0000000000000000",   --- 28
          "0000000000000000",   --- 29
          "0000000000000000",   --- 2A
          "0000000000000000",   --- 2B
          "0000000000000000",   --- 2C
          "0000000000000000",   --- 2D
          "0000000000000000",   --- 2E
          "0000000000000000",   --- 2F
          "0000000000000000",   --- 30    --- start of destination
             array
          "0000000000000000",   --- 31
          "0000000000000000",   --- 32
          "0000000000000000",   --- 33
          "0000000000000000",   --- 34
          "0000000000000000",   --- 35
          "0000000000000000",   --- 36
          "0000000000000000",   --- 37
          "0000000000000000",   --- 38
          "0000000000000000",   --- 39
          "0000000000000000",   --- 3A
          "0000000000000000",   --- 3B
          "0000000000000000",   --- 3C
          "0000000000000000",   --- 3D
          "0000000000000000",   --- 3E
          "0000000000000000"); --- 3F

begin
  data <= "ZZZZZZZZZZZZZZZZ";
  ready <= '0';

  if sel = '1' then
    if rw = '0' then
      data <= mem_data(CONV_INTEGER(addr(15 downto 0)))
                      after 1 ns;
      ready <= '1';
    elsif rw = '1' then
      mem_data(CONV_INTEGER(addr(15 downto 0))) := data;
    end if;
  else
    data <= "ZZZZZZZZZZZZZZZZ" after 1 ns;
  end if;
end process;

end behave;
```

Entity **mem** is a large array with a simple bus interface to allow reading and writing to the memory. A memory location is selected for read by placing the appropriate address of the location on signal **addr**, setting input **rw** (read write) to a '0' and putting the value '1' on signal **sel** (select). The value of the memory location appears on signal **data**, and signal **ready** is set to a '1' value signaling that the memory information is available.

To write a location in the memory, the address is placed on signal **addr**, set signal **rw** to a '1' value, set signal **sel** to a '1' value, and put the data to be written on input data.

The memory is divided into two separate sections. The first section is the instruction area, and the second is the data area. The instruction area contains the instructions to be executed, and the second section contains the data area for the instructions to manipulate. The CPU instructions start at location 00 and end at location 0D. The data area starts at location 10 and ends at location 3F, the end of the array. The instructions stored in the memory device are a simple algorithm for moving a block of data from one location to another. This type of program could also be considered a block copy operation.

Block Copy Operation

A diagram showing how a block copy operation looks is shown in Figure 12-6.

The copy operation starts when the CPU gets a **reset** signal. A **reset** signal causes the CPU to reset its internal state and start processing instructions at location 00 of the memory. The first few instructions set up the appropriate CPU registers so that the block copy operation can proceed. Register 1 contains the starting address, or the address of the first element of the memory block to be copied. Register 2 contains the starting address for the destination of the memory block. Register 6 contains the ending address of the memory block to be copied.

The first instruction at location 00 loads register 1 with the starting address of the memory block to be copied. The actual address is contained in **mem** location 01. The value is hexadecimal 10 or 16 decimal. The block copy program starts the copy operation from location 10. The first instruction is a double word instruction. The first word specifies the instruction opcode and the registers to be used in the instruction. The second word contains the absolute address to be used in the operation.

The next instruction is at memory location 02. The first instruction advanced the program counter past location 01, which contained the start-

Figure 12-6
Block Copy
Operation.

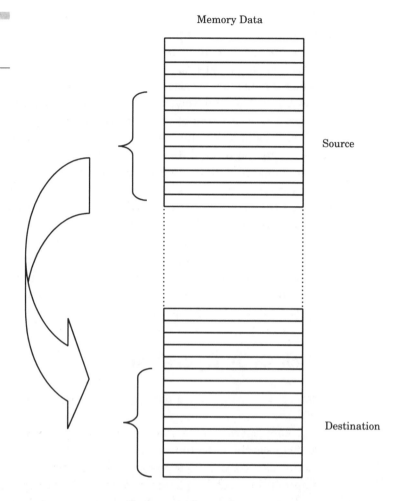

Memory Data

Source

Destination

ing address. This instruction loads register 2 with the destination address for the block copy. The destination is at 30 hex or 48 decimal. These load instructions are load immediate instructions, which load the next memory location into the register specified in the instruction.

The final setup instruction is at **mem** location 04. This instruction loads the last address of the memory block to be copied. This register signals the block copy routine when to stop the copy operation. After this instruction has been executed, all of the registers have been set up for the block copy operation, and the copy loop can start.

Instruction 6 is the start of a loop of instructions that perform the copy operation. Instruction 6 copies the contents of the memory location whose address is contained in register 1 to register 3. Instruction 7 copies the data in register 3 to the memory location specified in register 2. After these two instructions have executed the first time, the first memory element of the block will have been copied.

After the copy operation, the program needs to check if it is done. Instruction 8 compares the address in register 1 versus the end address in register 6 to see if the copy operation has completed the last element. If so, the program should exit because the copy operation has completed. However, in this simple example, there is no other program to execute, so this program branches to instruction 00 and starts the process over again.

If the copy operation is not completed, the CPU executes the instruction at 0A, which increments register 1. This instruction increments register 1 so that it points to the next element to be copied. Instruction 0B increments register 2, which moves the destination address to the next location.

Finally, instruction 0C branches back to instruction 06 and continues the next copy operation. Figures 12-7 and 12-8 show the memory array before the copy and after.

Figure 12-7
The memory array before the copy.

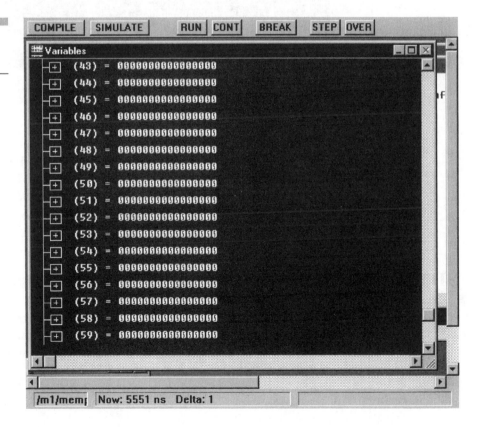

Figure 12-8
The memory array
after the copy.

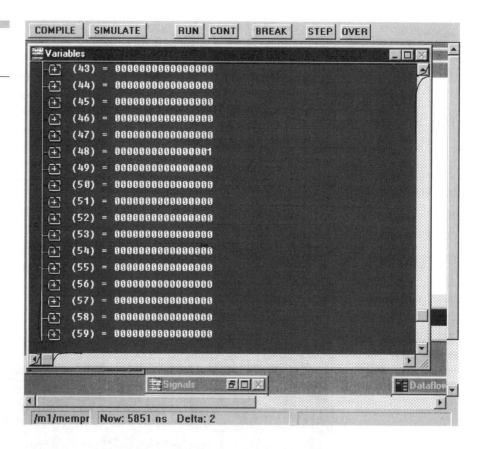

In Figure 12-7, location 48 is all zeroes. In Figure 12-8, location 48 is no longer all zeroes, but has the first value from the block copy operation. If the simulation is run completely, one by one the data from the source array (location16) is copied to the destination (location 48). After the last location is copied, the program repeats the same steps.

SUMMARY

In this chapter, we examined the top level of a design that consisted of a CPU, a memory array, and the top-level instantiation of those components. In the next chapter, we examine the CPU in more detail.

13

CPU: Synthesis Description

In this chapter, we further refine the CPU description and examine the RTL (Register Transfer Level) description of the CPU. The CPU is described by a number of lower-level components that are instantiated to form the CPU design. At the top of the CPU design is an architecture that instantiates all of the lower-level components to form the CPU. The CPU block diagram is shown in Figure 13-1.

Figure 13-1
CPU Block Diagram.

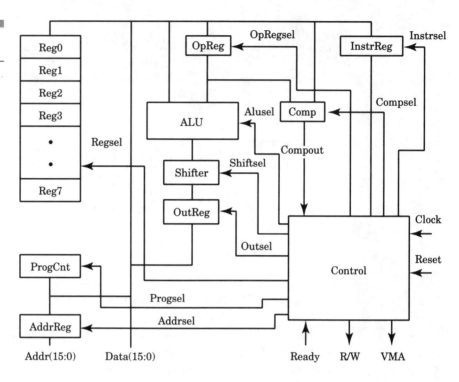

Following is an implementation of this block diagram, shown by file cpu.vhd:

```
library IEEE;
use IEEE.std_logic_1164.all;
use work.cpu_lib.all;

entity cpu is
 port(clock, reset, ready : in std_logic;
      addr : out bit16;
      rw, vma : out std_logic;
      data : inout bit16);
end cpu;

architecture rtl of cpu is
 component regarray
  port( data : in bit16;
        sel : in t_reg;
        en : in std_logic;
        clk : in std_logic;
        q : out bit16);
  end component;
```

```
component reg
 port( a : in bit16;
       clk : in std_logic;
       q : out bit16);
end component;

component trireg
 port( a : in bit16;
       en : in std_logic;
       clk : in std_logic;
       q : out bit16);
end component;

component control
 port( clock : in std_logic;
       reset : in std_logic;
       instrReg : in bit16;
       compout : in std_logic;
       ready : in std_logic;
       progCntrWr : out std_logic;
       progCntrRd : out std_logic;
       addrRegWr : out std_logic;
       outRegWr : out std_logic;
       outRegRd : out std_logic;
       shiftSel : out t_shift;
       aluSel : out t_alu;
       compSel : out t_comp;
       opRegRd : out std_logic;
       opRegWr : out std_logic;
       instrWr : out std_logic;
       regSel : out t_reg;
       regRd : out std_logic;
       regWr : out std_logic;
       rw : out std_logic;
       vma : out std_logic
       );
end component;

component alu
 port( a, b : in bit16;
       sel : in t_alu;
       c : out bit16);
end component;

component shift
 port ( a : in bit16;
        sel : in t_shift;
        y : out bit16);
end component;

component comp
 port( a, b : in bit16;
```

```
        sel : in t_comp;
        compout : out std_logic);
end component;

signal  opdata, aluout, shiftout, instrregOut : bit16;
signal regsel : t_reg;
signal regRd, regWr, opregRd, opregWr, outregRd, outregWr,
       addrregWr, instrregWr, progcntrRd, progcntrWr,
       compout : std_logic;
signal alusel : t_alu;
signal shiftsel : t_shift;
signal compsel : t_comp;
begin

ra1 : regarray port map(data, regsel, regRd, regWr, data);
opreg: trireg port map (data, opregRd, opregWr, opdata);
alu1: alu port map (data, opdata, alusel, aluout);
shift1: shift port map (aluout, shiftsel, shiftout);
outreg: trireg port map (shiftout, outregRd, outregWr,
    data);
addrreg: reg port map (data, addrregWr, addr);
progcntr: trireg port map (data, progcntrRd, progcntrWr,
    data);
comp1: comp port map (opdata, data, compsel, compout);
instr1: reg port map (data, instrregWr, instrregOut);
con1: control port map (clock, reset, instrregOut, com
        pout, ready, progcntrWr, progcntrRd, addrregWr, out
        regWr, outregRd, shiftsel, alusel, compsel, opre
        gRd, opregWr, instrregWr, regsel, regRd, regWr, rw,
        vma);

end rtl;
```

Architecture **rtl** of entity **cpu** is a structural implementation of the block diagram. Architecture **rtl** contains the component declarations of all of the components used to build the design, the signals used to connect the components, and the component instantiations to create the functionality.

After the component and signal declarations are the component instantiation statements that instance the components and connect the appropriate signals. In the next few sections, each of the VHDL component descriptions is described in more detail.

ALU

The first entity described is the ALU. This entity performs a number of arithmetic or logical operations on one or more input busses. A symbol for the ALU is shown in Figure 13-2.

Figure 13-2
ALU Interface.

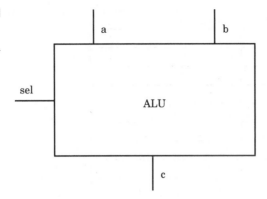

Inputs **a** and **b** are the two input busses upon which the ALU operations are performed. Output bus **c** returns the result of the ALU operation. Input **sel** determines which operation is performed as specified by Figure 13-3.

As we can see, the ALU can perform a number of arithmetic operations, such as add and subtract, and some logical operations, such as AND, OR, and XOR. Following is a VHDL description of the ALU entity:

```
library IEEE;
use IEEE.std_logic_1164.all;
use IEEE.std_logic_unsigned.all;
use work.cpu_lib.all;

entity alu is
  port( a, b : in bit16;
        sel : in t_alu;
        c : out bit16);
end alu;

architecture rtl of alu is
begin
  aluproc: process(a, b, sel)
  begin
  case sel is
    when alupass =>
      c <= a after 1 ns;

    when andOp =>
      c <= a and b after 1 ns;

    when orOp =>
      c <= a or b after 1 ns;

    when xorOp =>
      c <= a xor b after 1 ns;
```

Figure 13-3
ALU Function Table.

Sel Input	Operation
0000	C = A
0001	C = A AND B
0010	C = A OR B
0011	C = NOT A
0100	C = A XOR B
0101	C = A + B
0110	C = A – B
0111	C = A + 1
1000	C = A –1
1001	C = 0

```
      when notOp =>
        c <= not a after 1 ns;

      when plus =>
        c <= a + b after 1 ns;

      when alusub =>
        c <= a - b after 1 ns;

      when inc =>
        c <= a +  "0000000000000001" after 1 ns;

      when dec =>
        c <= a - "0000000000000001" after 1 ns;

      when zero =>
        c <= "0000000000000000" after 1 ns;

      when others =>
        c <= "0000000000000000" after 1 ns;

    end case;
    end process;

  end rtl;
```

The architecture uses a large **case** statement on input **sel** to determine which of the arithmetic or logical operations to perform. The possible values of signal **sel** are determined by type **t_alu** described in package **cpu_lib** in file cpulib.vhd. After the new value for output **c** is calculated, all of the resulting values are assigned with a 1 nanosecond time delay to eliminate delta delay problems during RTL simulation.

Comp

The next component described is the comparator entity **comp**. This entity compares two values and returns either a '1' or '0' depending on the type of comparison requested and the values being compared. A symbol showing the ports of the comparator is shown in Figure 13-4.

The comparison type is determined by the value on input port **sel**. For instance, to compare if inputs **a** and **b** are equal, apply the value **eq** to port **sel**. If ports **a** and **b** have the same value, port **compout** returns '1'. If the values are not equal, '0' is returned. The types of comparisons allowed are described by type **t_comp** in package **cpu_lib** in file cpulib.vhd described earlier. The full table of comparison types and values is shown in Figure 13-5.

All operations work on two input values and return a single bit result. This bit is used to control the flow of operation within the processor while executing instructions. Following is a VHDL description of the **comp** entity:

```
library IEEE;
use IEEE.std_logic_1164.all;
use IEEE.std_logic_arith.all;
```

Figure 13-4
Comp Interface.

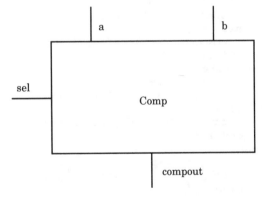

Figure 13-5
Comp Operation
Table.

Sel input value	Comparison
EQ	Compout = 1 when a equals b
NEQ	Compout = 1 when a is not equal b
GT	Compout = 1 when a is greater than b
GTE	Compout = 1 when a is greater than or equal to b
LT	Compout = 1 when a is less than b
LTE	Compout = 1 when a is less than or equal to b

```vhdl
use work.cpu_lib.all;

entity comp is
  port( a, b : in bit16;
          sel : in t_comp;
          compout : out std_logic);
end comp;

architecture rtl of comp is
begin
  compproc: process(a, b, sel)
  begin
    case sel is
      when eq =>
        if a = b then
          compout <= '1' after 1 ns;
        else
          compout <= '0' after 1 ns;
        end if;
      when neq =>
        if a /= b then
          compout <= '1' after 1 ns;
        else
          compout <= '0' after 1 ns;
        end if;
      when gt =>
        if a > b then
          compout <= '1' after 1 ns;
        else
          compout <= '0' after 1 ns;
        end if;
      when gte =>
        if a >= b then
          compout <= '1' after 1 ns;
        else
          compout <= '0' after 1 ns;
        end if;
      when lt =>
        if a < b then
          compout <= '1' after 1 ns;
        else
          compout <= '0' after 1 ns;
        end if;
      when lte =>
        if a <= b then
          compout <= '1' after 1 ns;
        else
          compout <= '0' after 1 ns;
        end if;
    end case;
  end process;

end rtl;
```

The comparator consists of a large **case** statement where each branch of the **case** statement contains an **IF**. If the condition tested is true, a `1` value is assigned; otherwise, a `0` is assigned. Again, each assignment occurs after 1 nanosecond to remove delta delay problems.

Control

The **control** entity provides the necessary signal interactions to make the data flow properly through the CPU and perform the expected functions. Architecture **rt1** contains a state machine that causes all appropriate signal values to update based on the current state and input signals and produce a next state for the state machine. A symbol for the control block is shown in Figure 13-6.

The control symbol has only a few inputs, but a lot of outputs. The control block provides all of the control signals to regulate data traffic for the CPU. Following is the VHDL description for the CPU:

```
library IEEE;
use IEEE.std_logic_1164.all;
use work.cpu_lib.all;

entity control is
  port( clock : in std_logic;
        reset : in std_logic;
        instrReg : in bit16;
        compout : in std_logic;
        ready : in std_logic;
        progCntrWr : out std_logic;
        progCntrRd : out std_logic;
        addrRegWr : out std_logic;
        addrRegRd : out std_logic;
        outRegWr : out std_logic;
        outRegRd : out std_logic;
        shiftSel : out t_shift;
        aluSel : out t_alu;
        compSel : out t_comp;
        opRegRd : out std_logic;
        opRegWr : out std_logic;
        instrWr : out std_logic;
        regSel : out t_reg;
        regRd : out std_logic;
        regWr : out std_logic;
        rw : out std_logic;
        vma : out std_logic
      );
```

Figure 13-6
Control Symbol.

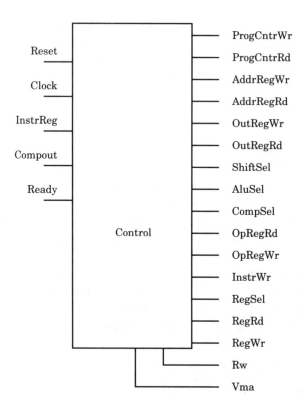

```
end control;

architecture rtl of control is
  signal current_state, next_state : state;
begin
  nxtstateproc: process( current_state, instrReg, compout,
                         ready)
  begin
    progCntrWr <= '0';
    progCntrRd <= '0';
    addrRegWr <= '0';
    outRegWr <= '0';
    outRegRd <= '0';
    shiftSel <= shftpass;
    aluSel <= alupass;
    compSel <= eq;
    opRegRd <= '0';
    opRegWr <= '0';
    instrWr <= '0';
    regSel <= "000";
    regRd <= '0';
    regWr <= '0';
    rw <= '0';
```

```vhdl
              vma <= '0';

              case current_state is
                when reset1 =>
                   aluSel <= zero after 1 ns;
                   shiftSel <= shftpass;
                   next_state <= reset2;

                when reset2 =>
                   aluSel <= zero;
                   shiftSel <= shftpass;
                   outRegWr <= '1';
                   next_state <= reset3;

                when reset3 =>
                   outRegRd <= '1';
                   next_state <= reset4;

                when reset4 =>
                   outRegRd <= '1';
                   progCntrWr <= '1';
                   addrRegWr <= '1';
                   next_state <= reset5;

                when reset5 =>
                  vma <= '1';
                  rw <= '0';
                  next_state <= reset6;

                when reset6 =>
                  vma <= '1';
                  rw <= '0';
                  if ready = '1' then
                    instrWr <= '1';
                    next_state <= execute;
                  else
                    next_state <= reset6;
                  end if;

                when execute =>
                  case instrReg(15 downto 11) is
                    when "00000" =>                    --- nop
                      next_state <= incPc;

                    when "00001" =>              --- load
                      regSel <= instrReg(5 downto 3);
                      regRd <= '1';
                      next_state <= load2;

                    when "00010" =>              --- store
                      regSel <= instrReg(2 downto 0);
                      regRd <= '1';
                      next_state <= store2;
```

```
   when "00011" =>                    ----- move
      regSel <= instrReg(5 downto 3);
      regRd <= '1';
      aluSel <= alupass;
      shiftSel <= shftpass;
      next_state <= move2;

   when "00100" =>            ---- loadI
      progcntrRd <= '1';
      alusel <= inc;
      shiftsel <= shftpass;
      next_state <= loadI2;

   when "00101" =>            ---- BranchImm
      progcntrRd <= '1';
      alusel <= inc;
      shiftsel <= shftpass;
      next_state <= braI2;

   when "00110" =>            ---- BranchGTImm
      regSel <= instrReg(5 downto 3);
      regRd <= '1';
      next_state <= bgtI2;

   when "00111" =>            ------- inc
      regSel <= instrReg(2 downto 0);
      regRd <= '1';
      alusel <= inc;
      shiftsel <= shftpass;
      next_state <= inc2;

   when others =>
      next_state <= incPc;

end case;

when load2 =>
   regSel <= instrReg(5 downto 3);
   regRd <= '1';
   addrregWr <= '1';
   next_state <= load3;

when load3 =>
   vma <= '1';
   rw <= '0';
   next_state <= load4;

when load4 =>
   vma <= '1';
   rw <= '0';
   regSel <= instrReg(2 downto 0);
   regWr <= '1';
   next_state <= incPc;
```

```
      when store2 =>
        regSel <= instrReg(2 downto 0);
        regRd <= '1';
        addrregWr <= '1';
        next_state <= store3;

      when store3 =>
        regSel <= instrReg(5 downto 3);
        regRd <= '1';
        next_state <= store4;

      when store4 =>
        regSel <= instrReg(5 downto 3);
        regRd <= '1';
        vma <= '1';
        rw <= '1';
        next_state <= incPc;

      when move2 =>
        regSel <= instrReg(5 downto 3);
        regRd <= '1';
        aluSel <= alupass;
        shiftsel <= shftpass;
        outRegWr <= '1';
        next_state <= move3;

      when move3 =>
        outRegRd <= '1';
        next_state <= move4;

      when move4 =>
        outRegRd <= '1';
        regSel <= instrReg(2 downto 0);
        regWr <= '1';
        next_state <= incPc;

      when loadI2 =>
        progcntrRd <= '1';
        alusel <= inc;
        shiftsel <= shftpass;
        outregWr <= '1';
        next_state <= loadI3;

      when loadI3 =>
        outregRd <= '1';
        next_state <= loadI4;

      when loadI4 =>
        outregRd <= '1';
        progcntrWr <= '1';
        addrregWr <= '1';
        next_state <= loadI5;
```

```vhdl
    when loadI5 =>
      vma <= '1';
      rw <= '0';
      next_state <= loadI6;

    when loadI6 =>
      vma <= '1';
      rw <= '0';
      if ready = '1' then
        regSel <= instrReg(2 downto 0);
        regWr <= '1';
        next_state <= incPc;
      else
        next_state <= loadI6;
      end if;

    when braI2 =>
      progcntrRd <= '1';
      alusel <= inc;
      shiftsel <= shftpass;
      outregWr <= '1';
      next_state <= braI3;

    when braI3 =>
      outregRd <= '1';
      next_state <= braI4;

    when braI4 =>
      outregRd <= '1';
      progcntrWr <= '1';
      addrregWr <= '1';
      next_state <= braI5;

    when braI5 =>
      vma <= '1';
      rw <= '0';
      next_state <= braI6;

    when braI6 =>
      vma <= '1';
      rw <= '0';
      if ready = '1' then
        progcntrWr <= '1';
        next_state <= loadPc;
      else
        next_state <= braI6;
      end if;

    when bgtI2 =>
      regSel <= instrReg(5 downto 3);
      regRd <= '1';
      opRegWr <= '1';
      next_state <= bgtI3;
```

```
        when bgtI3 =>
          opRegRd <= '1';
          regSel <= instrReg(2 downto 0);
          regRd <= '1';
          compsel <= gt;
          next_state <= bgtI4;

        when bgtI4 =>
          opRegRd <= '1' after 1 ns;
          regSel <= instrReg(2 downto 0);
          regRd <= '1';
          compsel <= gt;
          if compout = '1' then
            next_state <= bgtI5;
          else
            next_state <= incPc;
          end if;

        when bgtI5 =>
          progcntrRd <= '1';
          alusel <= inc;
          shiftSel <= shftpass;
          next_state <= bgtI6;

        when bgtI6 =>
          progcntrRd <= '1';
          alusel <= inc;
          shiftsel <= shftpass;
          outregWr <= '1';
          next_state <= bgtI7;

        when bgtI7 =>
          outregRd <= '1';
          next_state <= bgtI8;

        when bgtI8 =>
          outregRd <= '1';
          progcntrWr <= '1';
          addrregWr <= '1';
          next_state <= bgtI9;

        when bgtI9 =>
          vma <= '1';
          rw <= '0';
          next_state <= bgtI10;

        when bgtI10 =>
          vma <= '1';
          rw <= '0';
          if ready = '1' then
            progcntrWr <= '1';
            next_state <= loadPc;
          else
```

```
      next_state <= bgtI10;
    end if;

when inc2 =>
  regSel <= instrReg(2 downto 0);
  regRd <= '1';
  alusel <= inc;
  shiftsel <= shftpass;
  outregWr <= '1';
  next_state <= inc3;

when inc3 =>
  outregRd <= '1';
  next_state <= inc4;

when inc4 =>
  outregRd <= '1';
  regsel <= instrReg(2 downto 0);
  regWr <= '1';
  next_state <= incPc;

when loadPc =>
  progcntrRd <= '1';
  next_state <= loadPc2;

when loadPc2 =>
  progcntrRd <= '1';
  addrRegWr <= '1';
  next_state <= loadPc3;

when loadPc3 =>
  vma <= '1';
  rw <= '0';
  next_state <= loadPc4;

when loadPc4 =>
  vma <= '1';
  rw <= '0';
  if ready = '1' then
    instrWr <= '1';
    next_state <= execute;
  else
    next_state <= loadPc4;
  end if;

when incPc =>
  progcntrRd <= '1';
  alusel <= inc;
  shiftsel <= shftpass;
  next_state <= incPc2;

when incPc2 =>
  progcntrRd <= '1';
```

```
                      alusel <= inc;
                      shiftsel <= shftpass;
                      outregWr <= '1';
                      next_state <= incPc3;

                  when incPc3 =>
                      outregRd <= '1';
                      next_state <= incPc4;

                  when incPc4 =>
                      outregRd <= '1';
                      progcntrWr <= '1';
                      addrregWr <= '1';
                      next_state <= incPc5;

                  when incPc5 =>
                      vma <= '1';
                      rw <= '0';
                      next_state <= incPc6;

                  when incPc6 =>
                      vma <= '1';
                      rw <= '0';
                      if ready = '1' then
                        instrWr <= '1';
                        next_state <= execute;
                      else
                        next_state <= incPc6;
                      end if;

                  when others =>
                      next_state <= incPc;

              end case;

          end process;

            controlffProc: process(clock, reset)
            begin
              if reset = '1' then
                current_state <= reset1 after 1 ns;
              elsif clock'event and clock = '1' then
                current_state <= next_state after 1 ns;
              end if;
            end process;
          end rtl;
```

Architecture **rtl** contains two processes. The first is a combinational process (not clocked) that examines the current state and all inputs and produces output control values and next state output. The second is a sequential process (has a clock) that is used to store the current state and copy the next state to the current state. The next state transitions occur on rising edges of the clock input. The control block is a very large state machine that contains a number of states for each instruction. Executing all of the states for an instruction perform sthe necessary steps to complete the instruction.

If the **reset** is high, the sequential process labeled **controlffproc** sets signal **current_state** to state value **reset1**. This is the first state of the reset sequence for the CPU. This state starts the process of getting the CPU ready to execute instructions.

If the **reset** signal is not '1' and there is a rising edge on the clock signal, then the **controlffproc** process copies the **next_state** signal generated by the combinational process to signal **current_state**. This is the method for the state machine to advance from one state to another.

After the **reset** signal is set to a value other than '1', the state machine is in state **reset1**. This state causes the **alu** entity to output the value 0, the shift entity to pass the value with no modification, and the next state signal to be updated with the value **reset2**. This can be seen in the VHDL description for entity **control** in the **case** statement starting at the **when** clause for state **reset1**. At the next clock edge, the state machine advances to state **reset2**. State **reset2** leaves the control signals for the **alu** and **shift** entities as before, but also sets the **OutRegWr** signal to a '1', causing the 0 value on the data bus to be written to register **OutReg**. The goal of the reset sequence is to set up the program counter to start reading instructions from memory.

After state **reset2**, the state machine next goes to state **reset3** on the next clock edge. This state sets signal **OutRegRd** to a '1', causing entity **OutReg** to output its value to the data bus. The state machine then advances to state **reset4**. During **reset4**, the value from **OutReg** is copied into register **ProgCntr** and also to register **AddrReg**. The state machine advances to state **reset5**, sets output signal **RW** (read write) to '0' (read mode), and signals VMA (Valid Memory Address) to a '1'. This causes memory entity **mem** to output the data at location 0 to the data bus. The state machine advances to state **reset6** and, depending on the value of the ready signal from the memory, either stays in **reset6** or writes the memory data value to register **InstrReg** and goes to state **execute**.

At this point, the state machine has reset the state of the CPU to a known state and loaded the first instruction into register **InstrReg**. From this point forward, the state machine changes state depending on the instructions encountered.

The reset of the description for the state machine contains the state transitions for the rest of the instructions that have been implemented.

As mentioned previously, not all of the instructions have been implemented and are left as an exercise for the reader.

Reg

The **reg** entity is used for the address register and the instruction register. These registers need to be able to capture the input data on a rising edge of the **clk** input and drive output **q** with the captured data. The value of input **a** is assigned to output **q** when a rising edge occurs on input **clk**. The assignment is delayed by 1 nanosecond to remove delta delay problems during simulation. A symbol for the **reg** entity is shown in Figure 13-7.

The **reg** symbol contains three ports. Port **a** is the data input port, port **q** is the data output port, and port **clk** controls when the data is stored in the **reg** entity. Following is the VHDL description for entity **reg**:

```
library IEEE;
use IEEE.std_logic_1164.all;
use work.cpu_lib.all;

entity reg is
  port( a : in bit16;
        clk : in std_logic;
        q : out bit16);
end reg;

architecture rtl of reg is
begin
  regproc: process
  begin
    wait until clk'event and clk = '1';
    q <= a after 1 ns;
  end process;
end rtl;
```

Figure 13-7
Reg Symbol.

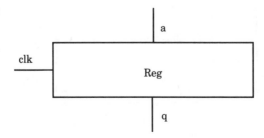

Process **regproc** is triggered when a rising edge occurs on input **clk**. When the process is triggered, input **a** is copied to output **q**.

Regarray

The **regarray** entity is used to model the set of registers within the CPU that are used to store intermediate values during instruction processing. These registers are read from and written to during the execution of instructions. The set of registers is modeled as a RAM of eight 16-bit words. The symbol for the **regarray** entity is shown in Figure 13-8.

To write a location in the **regarray**, set input **sel** to the location to be written, input data with the data to be written, and put a rising edge on input **clk**. To read a location from **regarray**, set input **sel** to the location to be read and set input **en** to a '1'; the data is output on port **q**.

The register array is modeled as two separate processes as shown in the following:

```
library IEEE;
use IEEE.std_logic_1164.all;
use IEEE.std_logic_unsigned.all;
use work.cpu_lib.all;

entity regarray is
  port( data : in bit16;
        sel : in t_reg;
        en : in std_logic;
        clk : in std_logic;
        q : out bit16);
end regarray;
```

Figure 13-8
RegArray Symbol.

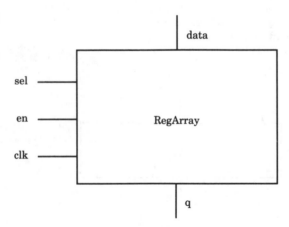

```
architecture rtl of regarray is
  type t_ram is array (0 to 7) of bit16;
  signal temp_data : bit16;
begin
  process(clk,sel)
    variable ramdata : t_ram;
  begin
    if clk'event and clk = '1' then
      ramdata(conv_integer(sel)) := data;
    end if;
    temp_data <= ramdata(conv_integer(sel)) after 1 ns;
  end process;

  process(en, temp_data)
  begin
    if en = '1' then
      q <= temp_data after 1 ns;
    else
      q <= "ZZZZZZZZZZZZZZZZ" after 1 ns;
    end if;
  end process;

end rtl;
```

The first process models the part of the RAM that stores the data. This process contains a local variable **ramdata** that is used to store the data written to the **regarray** entity. When the **clk** signal has a rising edge, the location selected by input **sel** is updated with the new value. This process also writes the location to a signal called **temp_data** to pass the value to the second process. The reason for this is that this model was written using VHDL 87, and variables cannot be shared between processes. In VHDL 93, sharing variables between processes is legal but has other synthesis ramifications.

The second process is used to read data from the **regarray**. Whenever input **sel** changes, the first process updates the value of **temp_data**. Signal **temp_data** is passed to the second process to pass the memory data. The second process outputs the value of **temp_data** if the **en** signal is '1'; otherwise, it puts out Z values. The Z values signify that the **regarray** entity is not driving the output when the **en** input is unasserted.

A smart synthesis tool reading this design can realize that the **regarray** entity can be implemented by a RAM device in the target technology and provide the proper mapping. For instance, if the design were to be mapped to an FPGA technology that included RAM in the architecture, the synthesis tool could map the **regarray** entity to an onboard RAM device. Using such an implementation instead of a set of flip-flops and gates creates a smaller and faster implementation.

Shift

The next device to be described is the **shift** entity. The **shift** entity is used to perform shifting and rotation operations within the CPU. The **shift** entity has a 16-bit input bus, a 16-bit output bus, and a **sel** input that determines which shift operation to perform. This is shown by the symbol in Figure 13-9.

The types of shift operations that can be performed by the **shift** entity are shown in Figures 13-10 and 13-11.

As can be seen by the figures, the **shift** entity can perform a shift left, shift right, rotate left, and rotate right operation. One operation that is not shown by the figures is a pass through operation in which all inputs bits are passed through to the output unchanged. Following is an entity that performs these operations:

```
library IEEE;
use IEEE.std_logic_1164.all;
use work.cpu_lib.all;

entity shift is
  port ( a : in bit16;
         sel : in t_shift;
         y : out bit16);
end shift;

architecture rtl of shift is
begin
  shftproc: process(a, sel)
  begin
    case sel is
      when shftpass =>
        y <= a after 1 ns;

      when shl =>
        y <= a(14 downto 0) & '0' after 1 ns;

      when shr =>
```

Figure 13-9
Shift Symbol.

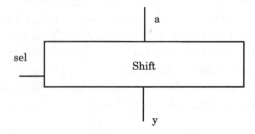

Figure 13-10
Shift Operations.

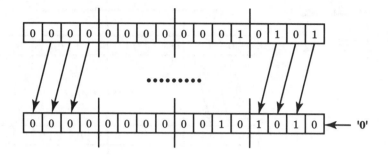

Shift Left

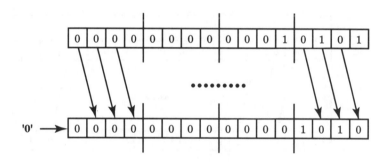

Shift Right

```
            y <= '0' & a(15 downto 1) after 1 ns;

         when rotl =>
            y <= a(14 downto 0) & a(15) after 1 ns;

         when rotr =>
            y <= a(0) & a(15 downto 1) after 1 ns;

      end case;
   end process;
end rtl;
```

The **shftpass** mode allows the shifter to pass the input data to the output without any shift operations. This mode is quite common because all of the ALU operations flow through the **shift** entity, and very few instructions are actually performing a shift operation.

Figure 13-11
Rotate Operations.

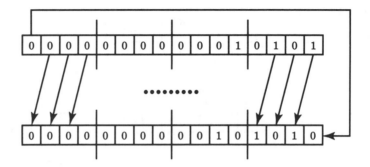

Rotate Left

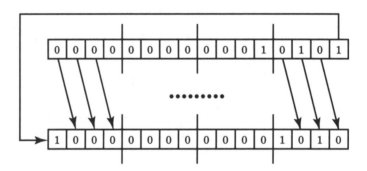

Rotate Right

The **shl** and **shr** selections perform shift left and shift right operations, respectively. The **rotl** and **rotr** selections perform rotate left and rotate right operations, respectively.

Trireg

The last component of the CPU is the tristate register component, **trireg**. The tristate register is connected to the main data bus and can store information from the data bus as well as drive information to the data bus. The **trireg** entity has four ports as shown in Figure 13-12.

Input **a** is the data input to the register, and port **q** is the data output from the register. Input **clk** is used to store a new value into the register.

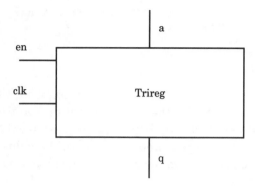

Figure 13-12
Trireg Symbol.

When a rising edge is applied to input **clk**, the data on input **a** is stored in the register.

Input **en** is used to control output **q**. When **en** is a '1' value, the register state is driven to output **q**. When **en** is a '0', output **q** is a high impedance value and not driving. This functionality is implemented by entity **trireg** shown in the following:

```
library IEEE;
use IEEE.std_logic_1164.all;
use work.cpu_lib.all;

entity trireg is
  port( a : in bit16;
        en : in std_logic;
        clk : in std_logic;
        q : out bit16);
end trireg;

architecture rtl of trireg is
  signal val : bit16;
begin
  triregdata: process
  begin
    wait until clk'event and clk = '1';
    val <= a;
  end process;

  trireg3st: process(en, val)
  begin
    if en = '1' then
      q <= val after 1 ns;
    elsif en = '0' then
      q <= "ZZZZZZZZZZZZZZZZ" after 1 ns;
-- exemplar_translate_off
    else
```

```
        q <= "XXXXXXXXXXXXXXXXX" after 1 ns;
-- exemplar_translate_on
      end if;
   end process;
end rtl;
```

The functionality is described by two processes that use a signal to communicate much like the **regarray** entity. The first process controls when signal **val** is written. Signal **val** is written only on the rising edge of input **clk**. The second process transfers the value of signal **val** only when input **en** is a '**1**' value; otherwise, a value of '**z**' is output.

SUMMARY

When all of these entities are connected together correctly, the functionality of the CPU results. The next two chapters focus on simulating the design for proper operation and synthesizing the design to a target device.

CPU:
RTL Simulation

In this chapter, a VHDL simulator is used to verify the functionality of the CPU VHDL RTL description. The VHDL RTL description of the CPU is simulated with a standard VHDL simulator to verify that the description is correct.

A simulator needs two inputs: the description of the design and stimulus to drive the design. Sometimes designs are self stimulating and do not need any external stimulus, but in most cases, VHDL designers use a VHDL testbench of one kind or another to drive the design being tested. The structure of the design looks like Figure 14-1.

The top-level design description instantiates two components: the first being the design under test (DUT) and the second the stimulus driver. These components are connected with signals that represent the external environment of the DUT. The top level of the design does not contain any external ports, just internal signals that connect the two instantiated components.

Figure 14-1
Top-Level Design
Structure.

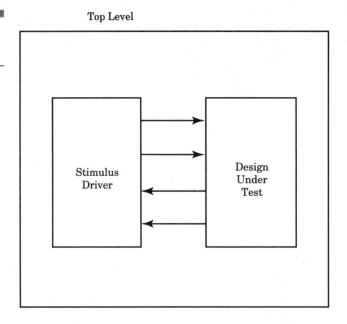

Top Level

Testbenches

A testbench is used to verify the functionality of a design. The testbench allows the design to verify the functionality of the design at each step in the HDL synthesis-based methodology. When the designer makes a small change to fix an error, the change can be tested to make sure that it did not affect other parts of the design. New versions of the design can be verified against known good results to verify compatibility.

A testbench is at the highest level in the hierarchy of the design. The testbench instantiates the design under test (DUT). The testbench provides the necessary input stimulus to the DUT and examines the output from the DUT. Figure 14-2 shows a block diagram of how this process appears.

The testbench encapsulates the stimulus driver, known good results, and DUT, and contains internal signals to make the proper connections. The stimulus driver drives inputs into the DUT. The DUT responds to the input signals and produces output results. Finally, a compare function within the testbench compares the results from the DUT against those known good results and reports any discrepancies. That is the basic function of a testbench, but there are a number of methods of writing a testbench and each method has advantages and disadvantages.

Testbench

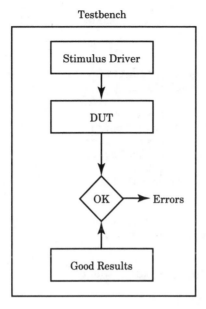

Figure 14-2
Testbench Block
Diagram.

Kinds of Testbenches

There is a myriad of ways to write a testbench, but some of the most common are described in this section. The following are the most common testbench types:

- *Stimulus Only*—Contains only the stimulus driver and DUT; does not contain any results verification.
- *Full Testbench*—Contains stimulus driver, known good results, and results comparison.
- *Simulator Specific*—Testbench is written in a simulator specific format.
- *Hybrid testbench*—Combines techniques from more than one testbench style.
- *Fast testbench*—Testbench written to get ultimate speed from simulation.

To show the different types of testbenches, a common example is used. To make it simple to understand the stimulus and response, a counter example is used. The following description is the package, entity, and architecture for an 8-bit counter:

```
PACKAGE count_types IS
   SUBTYPE bit8 is INTEGER RANGE 0 to 255;
END count_types;

LIBRARY IEEE;
USE IEEE.std_logic_1164.all;
USE work.count_types.all;
ENTITY count IS
   PORT (clk    : IN std_logic;
         ld     : IN std_logic;
         up_dwn : IN std_logic;
         clk_en : IN std_logic;
         din    : IN bit8;
         qout   : INOUT bit8);
END count;

ARCHITECTURE synthesis OF count IS
   SIGNAL count_val : bit8;
BEGIN

   PROCESS(ld, up_dwn, din, qout)
   BEGIN
     IF ld = '1' THEN
       count_val <= din;
     ELSIF up_dwn = '1' THEN
       IF (qout >= 255) THEN
         count_val <= 0;
       ELSE
         count_val <= count_val + 1;
       END IF;
     ELSE
       IF (qout <= 0) THEN
         count_val <= 255;
       ELSE
         count_val <= count_val - 1;
       END IF;
     END IF;
   END PROCESS;

   PROCESS
   BEGIN
     WAIT UNTIL clk'EVENT AND clk = '1';

     IF clk_en = '1' THEN
       qout <= count_val;
     END IF;
   END PROCESS;

END synthesis;
```

Package **count_types** contains the type declaration for the 8-bit signal type used in the counter. The counter is loadable, counts up and down, and contains a clock enable. The counter is implemented as two processes: a combinational process and a sequential process. The combinational process calculates the next state of the counter, and the sequential process keeps track of the current state of the counter and updates the next state of the counter on a rising edge of the `clk` input. We use the counter to discuss a number of different types of testbenches.

Stimulus Only

The stimulus only testbench contains the stimulus driver and DUT blocks of a testbench. The verification process is left to the designer. This type of testbench is useful at the beginning of a design project when no known good vectors exist, or for a quick check of an entity.

Following is an example stimulus only testbench:

```
ENTITY testbench IS END;

----------------------------------------------------------
-- STIMULUS ONLY
-- testbench for 8-bit loadable counter
-- reads from file "counter.txt"
----------------------------------------------------------
LIBRARY ieee;
USE ieee.std_logic_1164.ALL;
USE std.textio.ALL;
USE ieee.std_logic_textio.all;
USE WORK.count_types.all;
ARCHITECTURE stimonly OF testbench IS

        -------------------------------------
        -- component declaration for counter
        -------------------------------------
        COMPONENT count
        PORT (clk    : IN std_logic;
              ld     : IN std_logic;
              up_dwn : IN std_logic;
              clk_en : IN std_logic;
              din    : IN bit8;
              qout : INOUT bit8);
        END COMPONENT;

        SIGNAL clk, ld, up_dwn, clk_en : std_logic;
        SIGNAL qout, din : bit8;

BEGIN
    -- instantiate the component
    uut: count PORT MAP(clk => clk,
            ld => ld,
            up_dwn => up_dwn,
            clk_en => clk_en,
```

```
          din => din,
          qout => qout);


-- provide stimulus and check the result
test: PROCESS

VARIABLE tmpclk, tmpld, tmpup_dwn, tmpclk_en :
  std_logic;
VARIABLE tmpdin : integer;

FILE vector_file : text IS IN "counter.txt";
VARIABLE l : line;
VARIABLE vector_time : time;
VARIABLE r : real;
VARIABLE good_number, good_val : boolean;
VARIABLE space : character;

BEGIN
  WHILE NOT endfile(vector_file) LOOP
    readline(vector_file, l);

    -- read the time from the beginning of the line
    -- skip the line if it doesn't start with a number
    read(l, r, good => good_number);
    NEXT WHEN NOT good_number;

    vector_time := r * 1 ns;            -- convert real
                                           number to time
    IF (now < vector_time) THEN         -- wait until the
                                           vector time
    WAIT FOR vector_time - now;
    END IF;

    read(l, space);   --- skip a space

    -- read clk value
    read(l, tmpclk, good_val);
    assert good_val REPORT "bad clk value ";

    -- read ld value
    read(l,tmpld, good_val);
    assert good_val REPORT "bad ld value ";

    -- read up_dwn value
    read(l,tmpup_dwn, good_val);
    assert good_val REPORT "bad up_dwn value ";

    -- read clk_en value
    read(l,tmpclk_en, good_val);
    assert good_val REPORT "bad clk_en value ";

    read(l, space);   --- skip a space

    -- read din value
```

```
                read(l, tmpdin, good_val);
                assert good_val REPORT "bad din value ";

                  clk <= tmpclk;
                  ld <= tmpld;
                  up_dwn <= tmpup_dwn;
                  clk_en <= tmpclk_en;
                  din <= tmpdin;

          END LOOP;
          ASSERT false REPORT "Test complete";
          WAIT;
          END PROCESS;
      END;
```

The beginning of the testbench declares entity **testbench** as an entity with no ports. This is completely legal as the testbench is the topmost entity and does not interract with any other entities.

Next is the architecture declaration. The architecture uses a number of packages including IEEE standard packages and counter. The next section in the model declares the component for the DUT (Device Under Test), the counter. The ports and types on this component should match the DUT. Next, the local interconnect signals are declared. After the architecture declaration section, the DUT component is instantiated and connected to the local interconnect signals.

A process called **test** is declared which contains the stimulus generation capability. First, a number of local variables are declared that receive the data from the TextIO procedures used to read the stimulus information from a file. TextIO can only assign to variable objects not signals; therefore, local variables are assigned by the TextIO procedures, and these variables are assigned to the internal interconnect signals.

Inside the process is a single **while** loop that reads data from the stimulus file until an end-of-file condition is reached. Each pass through the loop reads another line from the file and reads the appropriate data from that line.

The first data read from the line is the time that this vector is to be applied. The process checks to make sure that the value read is a valid number. If not, the line is discarded because it does not represent a valid stimulus line. This allows comment lines to be inserted in the vector files. If a valid number was not read, the process skips this iteration through the loop and goes to the next iteration using the **next** clause.

If the value read was a good number, then the vector is assumed to be valid. The process reads each data value from the vector and applies the values to the locally declared variables.

In the counter example, the first value read is the **clk** signal. The **TextIO** statement reads a **STD_LOGIC** value from line l and assigns the value **read** to variable **tmpclk**. Later, the **tmpclk** variable is assigned to the signal **clk**.

The process continues to read a line, read a time value, wait until that time value occurs, read all vector values, and apply vector values until the end of the file is reached. When the end of file is reached, the loop terminates, an assertion message is written to standard output, and the process waits forever. The **WAIT** statement after the assertion at the end of the loop doesn't have a termination condition and, therefore, waits forever, effectively stopping execution of this process.

The **TEXTIO** readline statement inside the **while** loop reads a vector line from a vector file. Following is an example vector file:

```
--- vector file for counter
-- time clk ld up_dwn clk_en din
10 0001 0
20 1101 50
30 0001 0
40 1001 0
50 0001 0
60 1001 0
70 0001 0
80 1001 0
90 0001 0
100 1101 10
110 0001 0
120 1001 0
130 0001 0
140 1001 0
150 0001 0
160 1001 0
```

The first two lines of the vector file do not start with valid numbers and are treated as comment lines. Comment lines can be embedded anywhere in the file. Comments can also be placed at the end of a vector because any data after the last field of the vector are ignored.

Each vector line starts with a time value and then contains a string of values to be assigned to the DUT at that time. Spaces can be embedded between vector values if a corresponding read function exists in the **while** loop to skip the space.

For the stimulus only testbench, the test process reads a vector from the file and applies the stimulus to the DUT. The stimulus only testbench does not check the output results of the DUT in reaction to the applied stimulus. The stimulus only testbench is most useful for a quick check of a piece of a design that is easy for the designer to verify manually or for early in the design process when no known good results exist to verify against. When the results are verified, these results become the known good results to verify future versions or minor changes to the design.

Full Testbench

A full testbench is very similar to a stimulus only testbench except that the full testbench also includes the capability to check the output of the DUT. The full testbench applies the stimulus to the design and then examines the outputs of the design to see if the output results of the DUT match known good results.

Following is a full testbench for the counter:

```
ENTITY testbench IS END;

--------------------------------------------------------------
-- FULL TESTBENCH
-- testbench for counter
-- reads from file "counter.txt"
--------------------------------------------------------------
LIBRARY ieee;
USE ieee.std_logic_1164.ALL;
USE std.textio.ALL;
USE ieee.std_logic_textio.all;
USE WORK.count_types.all;
ARCHITECTURE full OF testbench IS

        ------------------------------------
        -- component declaration for counter
        ------------------------------------

        COMPONENT count
        PORT (clk    : IN std_logic;
              ld     : IN std_logic;
              up_dwn : IN std_logic;
              clk_en : IN std_logic;
              din    : IN bit8;
              qout   : INOUT bit8);
        END COMPONENT;

        SIGNAL clk, ld, up_dwn, clk_en : std_logic;
        SIGNAL qout, din : bit8;

    BEGIN
        -- instantiate the component
        uut: count
            PORT MAP(clk => clk,
                     ld => ld,
                     up_dwn => up_dwn,
                     clk_en => clk_en,
```

```
                        din => din,
                        qout => qout);


    -- provide stimulus and check the result
    test: PROCESS

       VARIABLE tmpclk, tmpld, tmpup_dwn, tmpclk_en :
         std_logic;
       VARIABLE tmpqout, tmpdin : bit8;

       FILE vector_file : text IS IN "counter.txt";
       VARIABLE l : line;
       VARIABLE vector_time : time;
       VARIABLE r : real;
       VARIABLE good_number, good_val : boolean;
       VARIABLE space : character;

    BEGIN
    WHILE NOT endfile(vector_file) LOOP
         readline(vector_file, l);

         -- read the time from the beginning of the line
         -- skip the line if it doesn't start with a number
         read(l, r, good => good_number);
         NEXT WHEN NOT good_number;

         vector_time := r * 1 ns;        -- convert real
                                            number to time
         IF (now < vector_time) THEN     -- wait until the
                                            vector time
         WAIT FOR vector_time - now;
         END IF;

         read(l, space);   --- skip a space

         -- read clk value
         read(l, tmpclk, good_val);
         assert good_val REPORT "bad clk value";

         -- read ld value
         read(l, tmpld, good_val);
         assert good_val REPORT "bad ld value";

         -- read up_dwn value
         read(l, tmpup_dwn, good_val);
         assert good_val REPORT "bad up_dwn value";

         -- read clk_en value
         read(l, tmpclk_en, good_val);
         assert good_val REPORT "bad clk_en value";

         read(l, space);   --- skip a space
```

```
                    -- read din value
                    read(1, tmpdin, good_val);
                    assert good_val REPORT "bad din value";

                    read(1, space);   --- skip a space

---- the difference in the file is below

                    -- read good output value
                    read(1, tmpqout, good_val);
                    assert good_val REPORT "bad qout value";

                    -- Compare outputs
                    assert tmpqout = qout REPORT "vector mismatch";

                    clk <= tmpclk;
                    ld <= tmpld;
                    up_dwn <= tmpup_dwn;
                    clk_en <= tmpclk_en;
                    din <= tmpdin;

              END LOOP;
              ASSERT false REPORT "Test complete";
              WAIT;
              END PROCESS;
         END full;
```

The full testbench looks exactly the same as the stimulus only test-bench for most of the file. The full testbench has a top level entity with no ports, an architecture that instantiates the DUT, and a **while** loop that reads a vector file. The differences are in the **while** loop itself. The first part of the **while** loop is exactly the same. The process reads a time value and waits for that time value to occur. The full testbench is different in that, not only does the full testbench read the input values, but it also reads the output values and then performs a compare operation between the output values from the DUT versus the values read from the file. If a mismatch is found, an assertion message is generated to let the designer know that the output results did not match the known good results.

The full testbench also reads from a vector file to get the stimulus for the design and the expected results. The vector file contains a time value, the input values, and the expected output values. Following is the full testbench vector file:

```
--- vector file for counter
-- time clk ld up_dwn clk_en din dout
0 0001   0 0
10 1001 0 255
```

```
20  0101  10  255
30  1001  0   10
40  0001  0   10
50  1001  0   8
60  0001  0   8
70  1001  0   7
80  0001  0   7
90  1001  0   6
100 0101  100 100
110 1001  0   100
120 0001  0   100
130 1001  0   98
140 0001  0   98
150 1001  0   97
160 0001  0   97
```

Notice that the vector file looks nearly the same as the stimulus only vector file except for the extra columns for the expected results.

The full testbench can be used to verify that a DUT matches a specification. To do so, the specification must include a set of known good results that the testbench can match against.

The full testbench can also be used to verify that a small change or optimization still matches the known good results. A designer may find a small error during verification that only requires a small localized change to the design. The designer can make the change and rerun the testbench to make sure that the change did not affect the rest of the design, and that the design still functions properly.

Testbenches can also be used to sign off designs. After the design matches the testbench results, the design is ready to be put into production, or be signed off.

The stimulus only and full testbench are only a couple examples of the many ways that a testbench can be written. Another example is the simulator specific testbench.

Simulator Specific

The simulator specific testbench is written specifically for one brand of simulator. Most simulators include a command language that allows the designer to control the simulator. The designer can compile designs, load designs, create libraries, set breakpoints, run the simulation, and lots of other tasks using the simulator command language. Most of these simulators also allow the designer to set signals to new values. Using command languages, the designer can write a testbench. Following is an example of a simulator specific testbench:

```
-- setup the clock
force -repeat 20 clk 0 0, 1 10

-- log the results to a file
list *

-- setup initial signal conditions
force ld 0
force up_dwn 0
force clk_en 1
force din 16#00

-- run the simulation
run 100

--- set next signal conditions
force ld 1
force up_dwn 0
force clk_en 1
force din 16#AA

--- run the simulation
run 200

--- set next signal conditions
force ld 1
force up_dwn 0
force clk_en 1
force din 16#55

--- run the simulation
run 200

write list data.out

quit -f
```

The command language used for this testbench is the Model Technology ModelSim command language. This simulator has a very rich command language that allows the designer to perform all of the necessary operations to compile designs, load designs, debug designs, save designs, and so on. The ModelSim simulator also has the capability to generate repeating clock signals to drive the design. The first command in the testbench file creates a repeating clock for signal clk. The clock repeats every 20 time units and is set to a '0' value at time 0 and a '1' value at time 10.

The next command (list /*) allows the designer to write all the signal values to an output file. The /* specifies that all signals be written to the file.

The next few commands in the file set up stimulus values on the counter input signals. The **force** command sets the signal to a value until it is changed by another **force** command. The input signals are all set to an initial value and the **run** command advances simulation time and runs the simulation. All of the input values are propagated appropriately through the design.

After the **run** command has finished, the new input stimulus values are set up with more **force** commands, and the simulation is run again. This process continues until all stimulus has been run through the design. The **write** command near the end of the file writes the results of the simulation to a file. The designer can analyze the output file to determine if the design is correct or use a file compare facility to automatically compare the DUT results to known good results.

The advantages of a simulator specific testbench are that it is fairly quick and easy to generate, and it can be loaded and reloaded into the simulator without shutting the simulator down and starting over every time. A simulation can be run, the results analyzed, simulation time reset to 0, a stimulus file loaded, and the simulator run again.

The disadvantage of the simulator specific testbench is that the testbench is specific to one simulator and cannot be easily migrated. If the design is to be passed to another design group using another simulator, the testbenches need to be rewritten in the new command language.

Hybrid Testbenches

Hybrid testbenches do not utilize only one technique, but a combination of a number of techniques. Hybrid testbenches can use a full testbench approach but have some of the stimulus data generated in the testbench rather than read from a file. Hybrid testbenches can also mix simulator specific commands with stimulus read from a file.

Following is a sample hybrid testbench:

```
ENTITY testbench IS END;

---------------------------------------------------------------
-- HYBRID Testbench
-- testbench for 8-bit loadable updown counter
-- reads from file "counter.txt"
---------------------------------------------------------------
LIBRARY ieee;
USE ieee.std_logic_1164.ALL;
```

```vhdl
USE std.textio.ALL;
USE ieee.std_logic_textio.all;
USE WORK.count_types.all;
ARCHITECTURE hybrid OF testbench IS

    -----------------------------------
    -- component declaration for counter
    -----------------------------------
    COMPONENT count
    PORT (clk    : IN std_logic;
           ld    : IN std_logic;
           up_dwn : IN std_logic;
           clk_en : IN std_logic;
           din   : IN bit8;
           qout  : INOUT bit8);
    END COMPONENT;

    SIGNAL ld, up_dwn, clk_en : std_logic;
    SIGNAL clk : std_logic := '0';
    SIGNAL qout, din : bit8;

BEGIN
    -- instantiate the component
    uut: count
        PORT MAP(clk => clk,
                 ld => ld,
                 up_dwn => up_dwn,
                 clk_en => clk_en,
                 din => din,
                 qout => qout);

    -- Generate the system clock
    clk <= not clk after 10 ns;

    -- provide stimulus and check the result
    test: PROCESS

      VARIABLE tmpclk, tmpld, tmpup_dwn, tmpclk_en :
        std_logic;
      VARIABLE tmpqout, tmpdin : bit8;

      FILE vector_file : text IS IN "counter.txt";
      VARIABLE l : line;
      VARIABLE vector_time : time;
      VARIABLE r : real;
      VARIABLE good_number, good_val : boolean;
      VARIABLE space : character;

    BEGIN
    WHILE NOT endfile(vector_file) LOOP
        readline(vector_file, l);

            -- read the time from the beginning of the line
            -- skip the line if it doesn't start with a number
```

```
read(1, r, good => good_number);
NEXT WHEN NOT good_number;

vector_time := r * 1 ns;          -- convert real
                                     number to time
IF (now < vector_time) THEN       -- wait until the
                                     vector time
WAIT FOR vector_time - now;
END IF;

read(1, space);   --- skip a space

-- read ld value
read(1,tmpld, good_val);
assert good_val REPORT "bad ld value";

-- read up_dwn value
read(1,tmpup_dwn, good_val);
assert good_val REPORT "bad up_dwn value";

-- read clk_en value
read(1,tmpclk_en, good_val);
assert good_val REPORT "bad clk_en value";

read(1, space);   --- skip a space

-- read din value
read(1, tmpdin, good_val);
assert good_val REPORT "bad din value";

ld <= tmpld;
up_dwn <= tmpup_dwn;
clk_en <= tmpclk_en;
din <= tmpdin;

END LOOP;
ASSERT false REPORT "Test complete";
WAIT;
END PROCESS;
END;
```

The hybrid testbench example looks very similar to the stimulus only testbench example except that, right after the Counter component instantiation, the system clock is generated by a signal assignment statement. Signal **clk** is assigned the value of **not clk** after 10 nanoseconds. This statement creates a periodic waveform with a period of 20 nanoseconds.

The testbench does not read signal **clock** from the vector file. The vector file contains changes only on signals other than **clock**. This results in a much smaller file that can be read much faster. Following is the hybrid vector file:

```
--- vector file for counter
-- time ld up_dwn clk_en din
10 001 0
20 101 50
30 001 0
100 101 0
110 001 0
250 101 35
260 001 0
```

If this example were a full testbench, the vector file would not be shorter because a vector would be needed on each clock transition to specify the output results for comparison.

The advantage of the hybrid testbench is that less data needs to be read from a vector file. Stimulus data is instead provided by either simulator command language commands or generated in the testbench.

The disadvantage of the hybrid testbench is that it is more difficult to change data from run to run when the hybrid testbench generates the stimulus in the testbench. In the case where simulator command language commands are used to generate stimulus, the testbench is less portable.

Fast Testbench

All of the testbench styles discussed so far have one common trait: They can become the limiting factor in how fast a simulation can run. This is especially true of the testbenches that read data from vector files. These files can become very large, and the time it takes to read a vector and process the vector can be the limiting factor in how fast the simulator executes. The same can be true of the simulator specific testbench if the simulator does not read the entire command file in at the start of simulation. If the file is read in chunks, the file read operation can significantly slow the simulation.

To get around these problems, a designer can elect to use a fast testbench. The fast testbench is optimized for speed and typically does not limit the speed of the simulation, unless the design is very small.

Following is an example fast testbench:

```
ENTITY testbench IS END;

------------------------------------------------------------------
-- FAST Testbench
-- testbench for 8-bit loadable updown counter
------------------------------------------------------------------
LIBRARY ieee;
USE ieee.std_logic_1164.ALL;
```

```vhdl
USE WORK.count_types.all;
ARCHITECTURE fast OF testbench IS

    -----------------------------------
    -- component declaration for counter
    -----------------------------------
    COMPONENT count
    PORT (clk     : IN std_logic;
          ld      : IN std_logic;
          up_dwn  : IN std_logic;
          clk_en  : IN std_logic;
          din     : IN bit8;
          qout    : INOUT bit8);
    END COMPONENT;

    SIGNAL clk, ld, up_dwn, clk_en : std_logic := '0';
    SIGNAL qout, din : bit8;

BEGIN
    -- instantiate the component
    uut: count
        PORT MAP(clk => clk,
                 ld => ld,
                 up_dwn => up_dwn,
                 clk_en => clk_en,
                 din => din,
                 qout => qout);

    -- generate the clock in the testbench
    clk <= not clk after 10 ns;


    -- provide stimulus and check the result
    test: PROCESS
      TYPE stim_vec is
        RECORD
          event_time : time;
          ld : std_logic;
          up_dwn : std_logic;
          clk_en : std_logic;
          din : bit8;
          qout : bit8;
        END RECORD;
      TYPE vec_array is array(0 to 8) of stim_vec;
      VARIABLE stim_array : vec_array := (
          (0 ns,    '0', '0', '1', 10,   10),
          (20 ns,   '1', '0', '1', 100, 2),
          (30 ns,   '0', '0', '1', 0,    0),
          (100 ns,  '1', '0', '1', 55,  8),
          (110 ns,  '0', '0', '1', 0,    0),
          (150 ns,  '1', '0', '1', 150, 58),
          (160 ns,  '0', '0', '1', 0,    151),
```

```
                      (250 ns, '1', '0', '1', 201, 160),
                      (260 ns, '0', '0', '1', 0,   161));

          VARIABLE ev_time : time;

      BEGIN
      FOR i in stim_array'RANGE LOOP

          ev_time := stim_array(i).event_time;

          IF (now < ev_time) THEN        -- wait until the
                                                vector time
            WAIT FOR ev_time - now;
          END IF;

          -- assign ld value
          ld <= stim_array(i).ld;

          -- assign up_dwn value
          up_dwn <= stim_array(i).up_dwn;

          -- assign clk_en value
          clk_en <= stim_array(i).clk_en;

          -- assign din value
          din <= stim_array(i).din;

          -- check qout value
          assert qout = stim_array(i).qout REPORT "vector
            mismatch";

      END LOOP;
      ASSERT false REPORT "Test complete";
      WAIT;
      END PROCESS;
   END;
```

The fast testbench looks similar to the other testbench styles in that it has a top-level entity that instantiates a DUT and a process that generates the stimulus. What's different is that, instead of reading the stimulus vectors from a file, the vectors are compiled into the testbench model.

The testbench declares a record type that contains a field for each input signal (and output signal, if a full testbench is being modeled). Next, the model declares an array of the record type that contains the vector values. A variable of the array type is declared and then initialized with the vector values. A **while** loop reads each record of the array, waits until the vector time is active, and applies the vector values to the design inputs, similar to the way the file was read using **TextIO**. Notice that array and record indexing is used to select each signal value.

The advantages of the fast testbench are that it executes extremely fast and doesn't suffer from the operating system file overhead of reading a file.

A disadvantage is that the compiled model can get very large if the number of vectors is large, making compile time long and simulator memory usage excessive. Another disadvantage of the fast testbench is that the model is not easily changed between simulation runs. Changing the testbench requires a recompilation step. Therefore, the fast testbench is most useful for models that need fast vector application and the vectors can be run in a small- or medium-sized loop where the vectors are applied again and again.

The advantages and disadvantages of each kind of testbench type is shown in Figure 14-3.

Notice that the stimulus only and full testbenches use TextIO. This can limit their speed if the DUT requires a lot of vector input. However, the advantages of using TextIO is the ease of changing the input data. No recompilation step is required to change the stimulus data. All that is required to make a change to the input stimulus is to change the input file and restart the simulation.

The simulator specific testbench is also very easy to change because it is typically an interpreted command language. Interpreted command languages don't need a separate compile step. Updating the command language file and reloading it in the simulator is all that is required to make a change. The price of this flexibility, however, may be slow execution speed. An interpreted command language doesn't need to be compiled, but may not execute fast depending on how many vectors are needed how quickly. A design that needs a lot of vectors very quickly may be limited by the speed of the interpreter.

The fast testbench really excels at going fast, but is much more difficult to change quickly than some of the other testbench types. To make a change, the vectors must be updated and the testbench recompiled. If the vector file is large, this process can take an excessive amount of time. Now that we have discussed testbenches, let's use one to simulate the CPU for correctness.

Figure 14-3
Testbench
Advantages and
Disadvantages.

	Speed	Flexibility	Portability
Stimulus Only	Slow	High	High
Full	Slow	High	High
Simulator Specific	Medium	High	Low
Hybrid	Medium	Medium	High
Fast	Extremely Fast	Low	High

CPU Simulation

Simulating the CPU design is different from most other entities because the CPU design doesn't need much outside stimulus. The memory device provides the input data for the CPU much as a stimulus file would for other entities. The CPU reads its program from the memory device. The CPU need only have the **clk** signal and **reset** signal stimulated properly, and the CPU reads and executes instructions from that point forward.

The only stimulus needed to start the operation of the CPU is a uniform signal applied to the **clk** input and a pulse applied to the **reset** input for at least 2 clock cycles. This starts the CPU into the reset sequence. After the reset sequence has been started, the CPU is initialized and starts executing the CPU instructions from the **mem** entity.

The CPU is simulated as stimulus only initially to verify that the device seems to be functioning. More complex testbenches need to be created that include comparison against a known good result to verify correctness. The simplest method for doing this is to manually verify the results the first time, capture the output results, and then use them for comparison later.

The first step in simulating the CPU is to compile all the files that make up the design into a format that the simulator can use. The compiled format is loaded into the simulator, and the simulation is executed. The Modelsim simulator from Model Technology is used for the simulation process.

The first step in compiling all of the files in the design is to create one or more libraries to store the compiled data. The default library to store the compiled data is a library called **WORK**. The name **work** is the logical name of the library; the physical location of the library can be anywhere. To create a library, the **VLIB** command is used as shown here:

```
vlib work
```

This creates the **work** library in the current working directory of the current disk. After the library has been created, the VHDL source files for the design can be compiled into the target library. To compile each of the files, the **VCOM** command must be run either from the GUI (Graphical User Interface) or from the command line. Most of the operations of the simulator have a GUI method of performing the command line command. This allows casual users as well as expert users to effectively use the simulator. Normally, casual users use the GUI and experts use the command line and script interface.

To compile a file from the GUI, the file is selected in the compile dialog box as shown in Figure 14-4.

The GUI includes a file browser that allows the designer to select the files to compile and then click the Compile button to compile the file.

To compile a file from the command line interface, the following command is issued:

```
vcom cpu_lib.vhd
```

This checks that the VHDL syntax is correct and converts the VHDL syntax to the binary format needed to simulate the design. Following is a complete script that compiles all of the files in the proper order:

```
vcom cpu_lib.vhd
vcom alu.vhd
vcom comp.vhd
vcom reg.vhd
vcom shift.vhd
vcom control.vhd
vcom regarray.vhd
vcom trireg.vhd
vcom cpu.vhd
vcom mem.vhd
vcom top.vhd
```

Figure 14-4
Compile VHDL
Source Dialog Box.

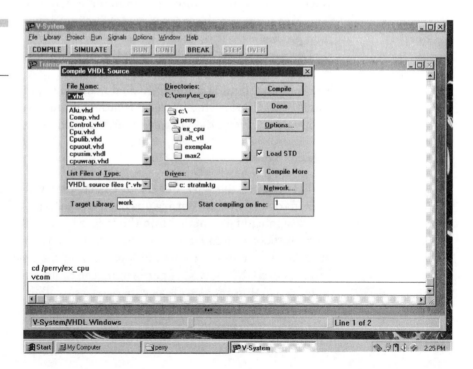

After all of the files have been compiled, the design can be loaded into the simulator for verification. This can be initiated from the GUI or from the command line with the following command:

```
vsim -lib work top behave
```

This command specifies the library (**work**), entity (**top**), and architecture (**behave**) or configuration to simulate. After the design has been loaded, the simulator needs stimulus for the design and specification of what data to monitor. For this simulation, the **current_state**, the memory interface, program counter, and other signals are monitored. Figure 14-5 shows a waveform display of the reset sequence of the CPU.

From this display, we can verify that the CPU is functioning properly. At time 0, the **reset** signal is set to a '1' value, which puts the CPU into state **reset1**, the first state of the reset sequence. After the reset signal is set to '0', the CPU can begin performing the reset sequence. The two most interesting signals to examine are **current_state** and **next_state**. Notice that, while the **reset** input is a '1', the CPU remains in state **reset1**. After signal **reset** is set to a '0', on the next rising edge of signal **clock**, **current_state** advances to state **reset2**.

Each clock rising edge after that causes the CPU to advance to the next state. At state **reset3**, the data bus receives the value 0000 to be used as the starting address for the first instruction. At state **reset4**, register **addreg** is loaded with the data bus value so that the 0000 value can be used to drive the **addr** bus. At state **reset5**, the data bus is driven with

Figure 14-5
Waveform display of the reset sequence.

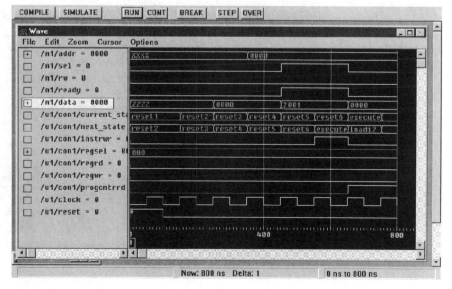

the instruction data from component **mem** at address 0. This data is then loaded into register **instrreg** in **reset6** so that the control entity can use the instruction contents.

The next state after **reset6** is the first execution step of the instruction that was just fetched from the memory. Looking back at the description of the **mem** entity, we can see that the first instruction loads register 1 with the source address of the copy operation. Figure 14-6 shows the waveform display after the reset sequence has completed and the first instruction has started to execute.

This instruction is a LoadI (Load Immediate) instruction that uses two words of the memory. The instruction is shown here:

```
LoadI 1, #
10
```

The first word of the instruction specifies the behavior of the instruction, and the second specifies the data to be loaded into the register specified by the instruction. This instruction first puts the program counter value to the data bus so that the value can be incremented. The program counter is then able to read the second word of the instruction that contains the data to be loaded into reg 1.

During state **execute**, the program counter is incremented and the incremented value can be found as the output of the ALU **aluout**. During

Figure 14-6

Waveform display after the reset sequence has completed.

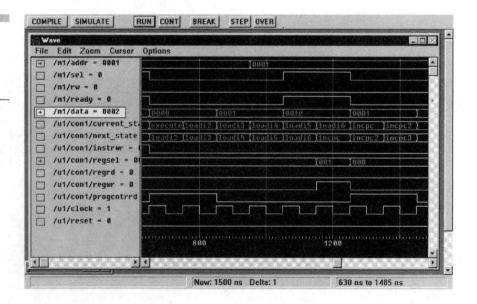

states **loadi2**, **loadi3**, and **loadi4**, this value is transferred to register **addreg** and data is read from **mem** entity in state **loadi5**. During state **loadi6**, data from memory is loaded into register 1.

After the load instruction has executed all of the states, to complete the load instruction, the CPU advances to a set of states that increments the program counter register to point to the next instruction.

The CPU performs three load instructions to load the proper CPU registers before the block copy can proceed. A final load instruction is performed which loads the value to be copied into register 3. At this point, the CPU program counter is pointing to address 7, a store instruction. This instruction uses the address in reg 2 to store the value in reg 3 to the new location. A waveform display showing the store instruction is shown in Figure 14-7.

During state **execute**, the value of reg2 is read to the data bus where it is copied to the address register in state **store2**. During **store3**, register array(3) drives the data bus with the data to be stored. During state **store4**, the value is written to the **mem** address.

After the store instruction is completed, the CPU checks to see if the block copy operation has completed. This is accomplished by the instruction at location 8, which branches back to instruction 00 if reg 1 is greater than reg 6. This instruction execution is shown Figure 14-8.

The first step is to read the value of register 1. This value is stored to register **opreg** during state **bgti2**. Next, the value of **reg6** is read and a

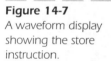

Figure 14-7
A waveform display showing the store instruction.

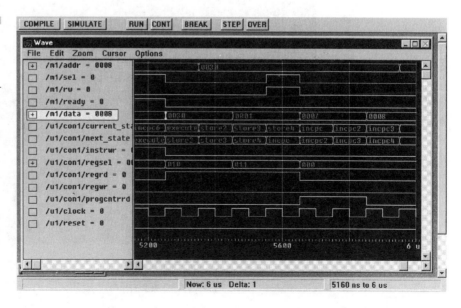

Figure 14-8
Store instruction
execution.

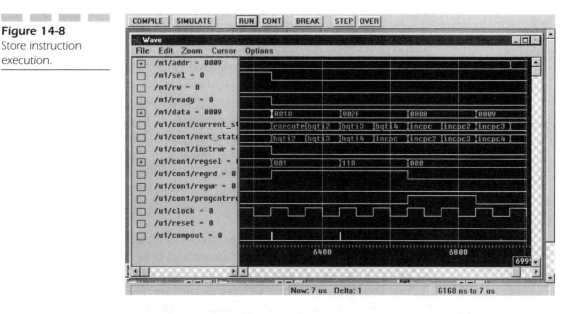

Figure 14-8
Store instruction
execution.

Figure 14-9
The source array.

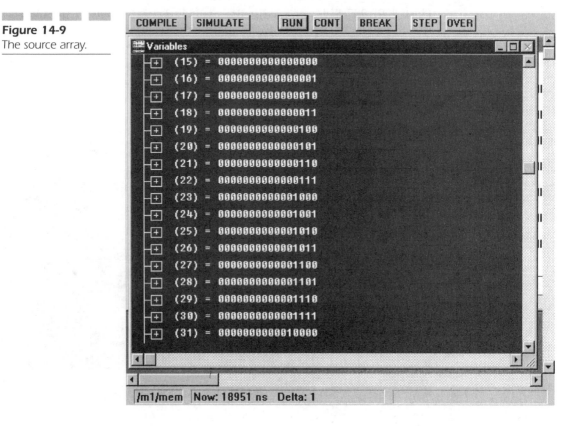

Figure 14-10
The destination array
before the copy
operation has
completed.

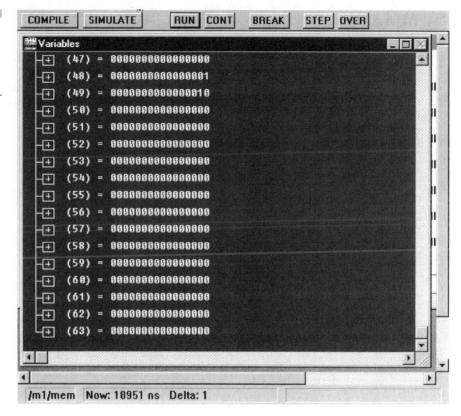

Figure 14-10
The destination array before the copy operation has completed.

comparison is performed. Notice that signal **compout** stays a '0' value because the greater than operation failed; therefore, the branch operation is be performed.

This set of instructions is performed a number of times until the source array is copied to the destination array. The source array is shown in Figure 14-9.

The array starts at location 16 and continues to location 31. The pattern stored in the source array is a very simple one that starts at 1 and ends at 16. Figure 14-10 shows the destination array before the copy operation has completed.

The destination array starts at location 48 and ends at location 63. The destination array is shown after two copy operations have been performed. Notice that location 48 has the first value, and location 49 has the second value. A complete simulation run completely copies one array to another. All of the examples that allow the reader to duplicate the simulation of the CPU are found on the CD that comes with this book.

SUMMARY

In this chapter, we examined what was necessary to perform a functional verification of the CPU design and walked through one loop of the block copy operation CPU simulation. In the next chapter, we synthesize the CPU description to a target FPGA device for implementation.

CPU Design: Synthesis Results

After the CPU has been functionally verified, the design can be implemented in actual hardware. This chapter describes the synthesis process and synthesis results of the CPU RTL description. The VHDL design description is optimized and mapped to a programmable logic device. As opposed to an ASIC device, these devices can be programmed by the designer at his/her desk, and most can be reprogrammed to fix errors later.

A synthesis tool is used to read in the VHDL description and map the description to the target programmable logic device. The synthesis tool reads all of the VHDL source files, links them together (elaborate), optimizes the design, and then maps the optimized description to the target technology. The synthesis tool used is the Leonardo synthesis tool from Exemplar Logic. This is one of the most popular synthesis tools in the FPGA (*Field Programmable Gate Array*) market and produces very good results quickly.

Figure 15-1

The Leonardo GUI.

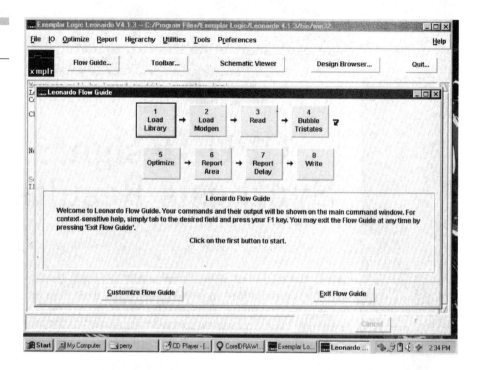

The first step in the synthesis process is to read all of the files of the design into the synthesis tool. This can be accomplished by either using the synthesis tool GUI (*Graphical User Interface*) or by issuing command language commands. First time or casual users will probably use the GUI because no command language syntax knowledge is required, and all operations can be accomplished through menu clicks. Everyday users of the tool quickly learn the command language of the synthesis tool, create scripts that build up the design, and run those scripts to create the design. This provides a repeatable method of creating the design.

The Leonardo GUI is shown in Figure 15-1.

Leonardo contains a flow guide that directs the user step by step in the operations that are required to create a design. The designer clicks each step of the flow guide and executes that step. These steps include reading the design, reading in a library for the target device, setting up the optimize options, and writing out the design in the proper format.

Alternatively, the designer can bring up the menu bar and execute the commands from the menu. The menu bar is shown in Figure 15-2.

The designer can select an operation to perform by clicking the appropriate button on the menu bar.

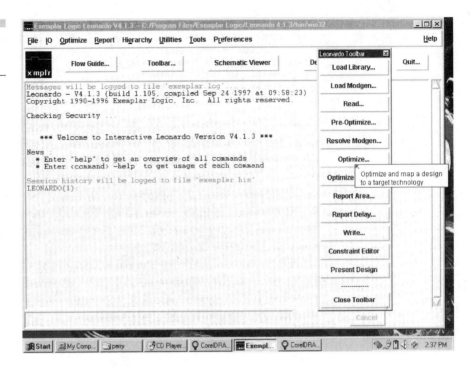

The last way to create the design is via user commands. These can be typed into the command prompt area or read from a script. Following is a typical script that reads a design, optimizes a design, and writes out a design:

```
----- leonardo script

cd /perry/ex_cpu
load_library flex10
analyze -work work cpulib.vhd
analyze -work work alu.vhd
analyze -work work comp.vhd
analyze -work work reg.vhd
analyze -work work shift.vhd
analyze -work work control.vhd
analyze -work work regarray.vhd
analyze -work work trireg.vhd
analyze -work work cpu.vhd
elaborate cpu -work work
pre_optimize .work.cpu.rtl -common_logic -unused_logic -
    extract
load_modgen flex10
resolve_modgen .work.cpu.rtl -default_resolving
```

```
set tristate_map TRUE
optimize .work.cpu.rtl -target flex10 -effort Standard -
    chip -area -pass 1
report_area cpu.area -cell_usage
report_delay -num_paths 2
write -format EDIF cpu.edif
```

This script analyzes the design source files one by one, then links the design together with the elaborate command. After the entire design has been linked together, the design can be boolean optimized. This removes redundant logic and helps balance the design logic. After the design has been optimized, it can be mapped to the target technology. In this example, the design is mapped to the Altera Flex10K technology. This is a LUT-based (*Look Up Table*) architecture that uses SRAM (*Static Ram*) technology to configure the functionality.

To get an idea of what each of the blocks of the CPU generates, each of the entities is synthesized separately. Area reports and timing reports are generated and analyzed to get a feel for what the design will look like. The first entity to be synthesized is `control`.

Control

The `control` entity provides the necessary signal values to direct the CPU operation as described in the last chapter. This entity is written as a state machine and is implemented in the best possible method for a state machine in the target technology. The CPU has six flip-flops to hold the state of the state machine and a number of inputs and outputs to control the operation.

The state machine can be read into the synthesis tool and optimized with the following commands:

```
load_library flex10
analyze cpu_lib.vhd
read control.vhd
set_attribute -port { .work.control.rtl.clock } -name
    CLOCK_CYCLE -value 25
optimize
```

Notice that the clock cycle has been set at 25 nanoseconds. The optimization process produces an area report and a timing report as follows:

```
**********************************************************
Cell: control     View: rtl     Library: work
**********************************************************
              Cell          Library   References   Total Area
            F1_LUT          flex10        4 x
            F4_CAS          flex10        5 x
               GND          flex10        1 x         1      1 GND
               VCC          flex10        1 x         1      1 VCC
          CASCADE2          flex10        9 x
            F2_LUT          flex10       14 x
            F4_LUT          flex10       42 x
            F3_LUT          flex10       12 x
               DFF          flex10       48 x         1     48 DFFs
                                                      1     48 LCs


   Number of ports :                      51
   Number of nets :                      151
   Number of instances :                 136
   Number of references to this view :     0

Total accumulated area :
   Number of GND :                        1
   Number of VCC :                        1
   Number of CASCADEs :                  14
   Number of LCs :                      104
   Number of DFFs :                      48
```

The preceding report is the area report. The following is the timing report:

```
                    Critical Path Report

Critical path #1, (unconstrained path)
NAME                    GATE        ARRIVAL         LOAD
------------------------------------------------------------
instrreg(14)/                       0.00 up         0.00
nx4933/O                F3_LUT      4.10 up         0.00
nx4924/O                F4_LUT      8.20 up         0.00
nx4900/O                F4_LUT     12.30 up         0.00
nx4930/O                F2_LUT     16.40 up         0.00
regsel(0)/O             F4_LUT     20.50 up         0.00
regsel(0)/                         20.50 up         0.00
data arrival time                  20.50

data required time                 not specified
------------------------------------------------------------
data required time                 not specified
data arrival time                  20.50
                                   ----------
                                   unconstrained path
------------------------------------------------------------
```

```
Critical path #2, (unconstrained path)

NAME                    GATE        ARRIVAL         LOAD
--------------------------------------------------------------
instrreg(15)/                       0.00  up        0.00
nx4933/O                F3_LUT      4.10  up        0.00
nx4924/O                F4_LUT      8.20  up        0.00
nx4900/O                F4_LUT      12.30 up        0.00
nx4930/O                F2_LUT      16.40 up        0.00
regsel(0)/O             F4_LUT      20.50 up        0.00
regsel(0)/                          20.50 up        0.00
data arrival time                   20.50

data required time                  not specified
--------------------------------------------------------------
data required time                  not specified
data arrival time                   20.50
                                    ----------
                                    unconstrained path
--------------------------------------------------------------
```

The design takes 104 LCs (*Logic Cells*) to implement, and the critical path through the design is 20.5 nanoseconds. The designer can analyze the output report and determine whether or not the design meets the design requirements. If so, the designer is done; if not, the designer must change the design or add some design constraints to drive the synthesis tool.

If the design did not meet the designer's constraints, it's usually because the design does not meet the speed requirements for the design. The designer analyzes the critical path and determines how to constrain the design so that its speed can be increased or how to modify the VHDL description to increase the speed. Usually, this is accomplished by adding parallel logic to the design to remove high fanout nets, or remove levels of logic by computing subexpressions where needed. The synthesis tool does this automatically when the designer specifies timing constraints.

The designer uses the timing analysis capability from the synthesis tool—that is, if provided; some of the cheaper tools do not have this capability—to find the critical path and add constraints to try and modify the design. In this example, the constraint is 25 nanoseconds, so the design will work as implemented.

Alu

The `alu` implements a number of arithmetic and logical operations. To synthesize the design, the following commands are issued:

```
analyze cpu_lib.vhd
read alu.vhd
optimize
```

Following are the area and timing reports:

```
**********************************************************
Cell: alu     View: rtl     Library: work
**********************************************************
```

Cell	Library	References		Total Area	
F1_LUT	flex10	16	x		
F4_CAS	flex10	16	x		
F4_LUT	flex10	71	x		
F3_LUT	flex10	2	x		
F2_LUT	flex10	1	x		
CASCADE2	flex10	64	x		
VCC	flex10	1	x	1	1 VCC
GND	flex10	1	x	1	1 GND
NOT	flex10	47	x	1	47 NOT
CARRY	flex10	60	x	1	60 CARRYs
OR2	flex10	75	x	1	75 OR2
AND2	flex10	105	x	1	105 AND2
LCELL	flex10	64	x	1	64 LCs
XOR2	flex10	96	x	1	96 XOR2

```
Number of ports :                    52
Number of nets :                    655
Number of instances :               619
Number of references to this view :   0

Total accumulated area :
 Number of CASCADEs :                80
 Number of VCC :                      1
 Number of GND :                      1
 Number of NOT :                     47
 Number of CARRYs :                  60
 Number of OR2 :                     75
 Number of AND2 :                   105
 Number of LCs :                    154
 Number of XOR2 :                    96
```

The preceding is the area report. The following is the delay report:

```
                    Critical Path Report

Critical path #1, (unconstrained path)
NAME                            GATE      ARRIVAL    LOAD
-----------------------------------------------------------
b(0)/                                     0.00 up    0.00
modgen_6_11_10_10_0_10_c0/Y     AND2      0.00 up    0.00
modgen_6_11_10_10_0_10_c3/Y     OR2       0.00 up    0.00
```

```
modgen_6_11_10_10_0_10_c4/Y    OR2        0.00 up    0.00
modgen_6_11_10_10_0_10_c5/Y    CARRY      0.50 up    0.00
modgen_6_11_10_10_1_10_c2/Y    AND2       0.50 up    0.00
modgen_6_11_10_10_1_10_c4/Y    OR2        0.50 up    0.00
modgen_6_11_10_10_1_10_c5/Y    CARRY      1.00 up    0.00
modgen_6_11_10_10_2_10_c1/Y    AND2       1.00 up    0.00
modgen_6_11_10_10_2_10_c3/Y    OR2        1.00 up    0.00
modgen_6_11_10_10_2_10_c4/Y    OR2        1.00 up    0.00
modgen_6_11_10_10_2_10_c5/Y    CARRY      1.50 up    0.00
modgen_6_11_10_10_3_10_c1/Y    AND2       1.50 up    0.00
modgen_6_11_10_10_3_10_c3/Y    OR2        1.50 up    0.00
modgen_6_11_10_10_3_10_c4/Y    OR2        1.50 up    0.00
           .
           .
           .
modgen_6_11_10_10_14_10_c4/Y   OR2        7.00 up    0.00
modgen_6_11_10_10_14_10_c5/Y   CARRY      7.50 up    0.00
modgen_6_11_10_10_15_10_sum0/Y XOR2       7.50 up    0.00
modgen_6_11_10_10_15_10_sum1/Y XOR2       7.50 up    0.00
modgen_6_11_10_10_15_10_sum2/Y LCELL     11.60 up    0.00
nx1031/O                       F4_LUT    15.70 up    0.00
nx1030/O                       CASCADE2  16.80 up    0.00
nx1032/O                       CASCADE2  17.90 up    0.00
nx909/O                        CASCADE2  19.00 up    0.00
c(15)/O                        F1_LUT    19.00 up    0.00
c(15)/                         19.00 up   0.00
data arrival time              19.00

data required time             not specified
-----------------------------------------------------------
data required time             not specified
data arrival time              19.00
                               ----------
                               unconstrained path
-----------------------------------------------------------
```

The **alu** is not triggered by any clock edges and therefore produces only combinational logic. Notice that there are no flip-flops in the design, only gate level logic. There are a lot of logic levels in the design because the design is implementing large add, subtract, and logic functions. The design takes 154 LCs to implement, and the critical path through the design is 19 nanoseconds.

Comp

The **comp** entity implements a number of compare operations. To synthesize the design, the following commands are issued:

```
analyze cpu_lib.vhd
read comp.vhd
optimize
```

Synthesizing the comp entity produces the following reports:

```
**********************************************************

Cell: comp    View: rtl    Library: work

**********************************************************
```

Cell	Library	References		Total Area	
F4_LUT	flex10	3 x			
VCC	flex10	1 x	1	1	VCC
SOFT	flex10	2 x	1	2	LCs
CARRY	flex10	30 x	1	30	CARRYs
BUF	flex10	30 x	1	30	LCs
DELAY	flex10	8 x			
AND2	flex10	102 x	1	102	AND2
NOT	flex10	40 x	1	40	NOT
OR2	flex10	72 x	1	72	OR2
XOR2	flex10	16 x	1	16	XOR2
CASCADE	flex10	6 x	1	6	CASCADEs

```
Number of ports :                        39
Number of nets :                        348
Number of instances :                   310
Number of references to this view : 0

Total accumulated area :
Number of VCC :                           1
Number of CARRYs :                       30
Number of LCs :                          43
Number of AND2 :                        102
Number of NOT :                          40
Number of OR2 :                          72
Number of XOR2 :                         16
Number of CASCADEs :                      6
```

The preceding is the area report. The following is the delay report:

```
                    Critical Path Report

Critical path #1, (unconstrained path)
NAME                               GATE     ARRIVAL     LOAD
-----------------------------------------------------------
a(0)/                              0.00 up  0.00
modgen_5_11_10_10_0_10_11/Y        NOT      0.00 up     0.00
```

```
modgen_5_11_10_10_0_10_c0/Y       AND2      0.00  up    0.00
modgen_5_11_10_10_0_10_c3/Y       OR2       0.00  up    0.00
modgen_5_11_10_10_0_10_c4/Y       OR2       0.00  up    0.00
modgen_5_11_10_10_0_10_13_10/Y    BUF       0.00  up    0.00
modgen_5_11_10_10_0_10_13_c5/Y    CARRY     0.50  up    0.00
modgen_5_11_10_10_1_10_c1/Y       AND2      0.50  up    0.00
modgen_5_11_10_10_1_10_c3/Y       OR2       0.50  up    0.00
modgen_5_11_10_10_1_10_c4/Y       OR2       0.50  up    0.00
modgen_5_11_10_10_1_10_13_10/Y    BUF       0.50  up    0.00
modgen_5_11_10_10_1_10_13_c5/Y    CARRY     1.00  up    0.00
modgen_5_11_10_10_2_10_c1/Y       AND2      1.00  up    0.00
modgen_5_11_10_10_2_10_c3/Y       OR2       1.00  up    0.00
modgen_5_11_10_10_2_10_c4/Y       OR2       1.00  up    0.00
          .
          .
          .
modgen_5_11_10_10_14_10_c3/Y      OR2       7.00  up    0.00
modgen_5_11_10_10_14_10_c4/Y      OR2       7.00  up    0.00
modgen_5_11_10_10_14_10_13_10/Y   BUF       7.00  up    0.00
modgen_5_11_10_10_14_10_13_c5/Y   CARRY     7.50  up    0.00
modgen_5_11_10_10_15_10_c1/Y      AND2      7.50  up    0.00
modgen_5_11_10_10_15_10_c3/Y      OR2       7.50  up    0.00
modgen_5_11_10_10_15_10_c4/Y      OR2       7.50  up    0.00
modgen_5_11_10_11/Y               SOFT     11.60  up    0.00
nx82/O                            F4_LUT   15.70  up    0.00
compout/O                         F4_LUT   19.80  up    0.00
compout/                                   19.80  up    0.00
data arrival time                          19.80

data required time                         not specified
-----------------------------------------------------------------
data required time                         not specified
data arrival time                          19.80
                                           ----------
                                           unconstrained path
-----------------------------------------------------------------
```

The comp entity is also not triggered by any clock edges and produces only combinational logic. Notice that the report file shows no flip-flops in the design, only gate level logic. The design takes 43 LCs to implement, and the critical path through the design is 19.8 nanoseconds. The constraint for the design is 25 nanoseconds, so again the timing constraint has been met.

Reg

The reg entity implements a clocked register function. To synthesize the design, the following commands are issued:

```
analyze cpu_lib.vhd
read reg.vhd
set_attribute -port { .work.reg.rtl.clock } -name
    CLOCK_CYCLE -value 25
optimize
```

Synthesizing the **reg** entity produces the following reports:

```
**********************************************************
Cell: reg    View: rtl    Library: work
**********************************************************

            Cell       Library   References      Total Area

            VCC        flex10        1 x       1     1 VCC
            DFF        flex10       16 x       1    16 DFFs
                                               1    16 LCs


  Number of ports :                    33
  Number of nets :                     34
  Number of instances :                17
  Number of references to this view :   0

Total accumulated area :
  Number of VCC :                       1
  Number of LCs :                      16
  Number of DFFs :                     16
```

The preceding is the area report. The following is the timing report:

```
                Critical Path Report

Critical path #1, (unconstrained path)
NAME                                GATE     ARRIVAL    LOAD
-----------------------------------------------------------
clock information not specified
delay thru clock network                     0.00 (ideal)

q(0)/Q                              DFF      0.30 up    0.00
q(0)/                                        0.30 up    0.00
data arrival time                            0.30

data required time                           not specified
-----------------------------------------------------------
data required time                           not specified
data arrival time                            0.30
                                             ----------
                                             unconstrained path
-----------------------------------------------------------


Critical path #2, (unconstrained path)
NAME                                GATE     ARRIVAL    LOAD
-----------------------------------------------------------
```

```
clock information not specified
delay thru clock network                       0.00 (ideal)

q(1)/Q                            DFF          0.30 up    0.00
q(1)/ 0.30 up                                  0.00
data arrival time                              0.30

data required time                             not specified
-----------------------------------------------------------------
data required time                             not specified
data arrival time                              0.30
                                               ----------
                                               unconstrained path
-----------------------------------------------------------------
```

The **reg** entity is triggered by clock edges and produces only sequential logic because there is no other logic in the entity. This entity is very simply a 16-bit register and produces only a 16-bit register in the generated logic. Notice that the report file shows only flip-flops in the design. The design takes 16 LCs to implement, and the critical path through the design is .3 nanoseconds. Because the design is all registers with no other logic, the critical path is from the clock edge through to the **Q** output pin.

Regarray

The **regarray** entity implements an array or set of registers. To synthesize the design, the following commands are issued:

```
analyze cpu_lib.vhd
read regarray.vhd
set_attribute -port { .work.regarray.rtl.clock } -name
     CLOCK_CYCLE -value 25
optimize
```

Synthesizing the **regarray** entity produces the following reports:

```
**********************************************************
Cell: regarray     View: rtl    Library: work
**********************************************************
```

Cell	Library	References		Total Area	
TRI	flex10	16 x	1	16	TRIs
F1_LUT	flex10	16 x			
VCC	flex10	1 x	1	1	VCC
F4_LUT	flex10	48 x			

```
              F4_CAS         flex10         16 x
              F3_LUT         flex10          8 x
           CASCADE2          flex10         48 x
               DFFE          flex10        128 x         1      128 DFFs
                                                         1      128 LCs


  Number of ports :                         37
  Number of nets :                         302
  Number of instances :                    281
  Number of references to this view :        0

Total accumulated area :
  Number of TRIs :                          16
  Number of VCC :                            1
  Number of CASCADEs :                      64
  Number of LCs :                          200
  Number of DFFs :                         128
```

The preceding report is the area report. The following is the timing report:

```
                      Critical Path Report
Critical path #1, (unconstrained path)
NAME                          GATE            ARRIVAL        LOAD
------------------------------------------------------------------
sel(0)/                       0.00 up         0.00
ix1473_nx41/O                 F3_LUT          4.10 up        0.00
nx1628/O                      F4_CAS          9.30 up        0.00
nx1694/O                      CASCADE2       10.40 up        0.00
nx1695/O                      CASCADE2       11.50 up        0.00
nx1612/O                      CASCADE2       12.60 up        0.00
IN/O                          F1_LUT         12.60 up        0.00
q(15)/OUT                     TRI            15.60 up        0.00
q(15)/                                       15.60 up        0.00
data arrival time                            15.60

data required time                           not specified
------------------------------------------------------------------
data required time                           not specified
data arrival time                            15.60
                                             ----------
                                             unconstrained path
------------------------------------------------------------------


Critical path #2, (unconstrained path)
NAME                          GATE            ARRIVAL        LOAD
------------------------------------------------------------------
sel(1)/                       0.00 up         0.00
ix1473_nx41/O                 F3_LUT          4.10 up        0.00
nx1628/O                      F4_CAS          9.30 up        0.00
nx1694/O                      CASCADE2       10.40 up        0.00
nx1695/O                      CASCADE2       11.50 up        0.00
nx1612/O                      CASCADE2       12.60 up        0.00
```

```
IN/O                              F1_LUT          12.60 up      0.00
q(15)/OUT                         TRI             15.60 up      0.00
q(15)/                                            15.60 up      0.00
data arrival time                                 15.60

data required time                                not specified
-----------------------------------------------------------------
data required time                                not specified
data arrival time                                 15.60
                                                  ----------
                                                  unconstrained path
-----------------------------------------------------------------
```

The **regarray** entity is triggered by clock edges and produces sequential and combinational logic. The sequential logic is used to implement the register storage, and the combinational logic is used for the decode and muxing of input and output data. The report that follows shows a mix of LCs and flip-flops to implement the design. The design takes 200 LCs and 128 flip-flops to implement, and the critical path through the design is 15.6 nanoseconds.

Shift

The **shift** entity implements a number of shift operations. To synthesize the design, the following commands are issued:

```
analyze cpu_lib.vhd
read shift.vhd
optimize
```

Synthesizing the shift entity produces the following reports:

```
************************************************************

Cell: shift   View: rtl    Library: work

************************************************************
```

Cell	Library	References	Total Area
F1_LUT	flex10	16 x	
F4_CAS	flex10	16 x	
F3_LUT	flex10	16 x	
CASCADE2	flex10	16 x	
F2_LUT	flex10	1 x	

```
Number of ports :                       37
Number of nets :                        86
Number of instances :                   65
Number of references to this view :      0

Total accumulated area :
Number of CASCADEs :                    32
Number of LCs :                         33
```

The preceding report is the area report. The following is the delay report:

```
                      Critical Path Report
Critical path #1, (unconstrained path)
NAME                      GATE         ARRIVAL      LOAD
-----------------------------------------------------------
sel(1)/                                0.00 up      0.00
nx12/O                    F2_LUT       4.10 up      0.00
nx343/O                   F4_CAS       9.30 up      0.00
nx311/O                   CASCADE2    10.40 up      0.00
y(7)/O                    F1_LUT      10.40 up      0.00
y(7)/                                 10.40 up      0.00
data arrival time                     10.40

data required time                    not specified
-----------------------------------------------------------
data required time                    not specified
data arrival time                     10.40
                                      ----------
                                      unconstrained path
-----------------------------------------------------------


Critical path #2, (unconstrained path)
NAME                      GATE         ARRIVAL      LOAD
-----------------------------------------------------------
sel(3)/                                0.00 up      0.00
nx12/O                    F2_LUT       4.10 up      0.00
nx343/O                   F4_CAS       9.30 up      0.00
nx311/O                   CASCADE2    10.40 up      0.00
y(7)/O                    F1_LUT      10.40 up      0.00
y(7)/                                 10.40 up      0.00
data arrival time                     10.40

data required time                    not specified
-----------------------------------------------------------
data required time                    not specified
data arrival time                     10.40
                                      ----------
                                      unconstrained path
-----------------------------------------------------------
```

The **shift** entity is also not triggered by any clock edges and produces only combinational logic. Notice that the report file shows no flip-flops in the design, only gate level logic. The design takes 33 LCs to implement, and the critical path through the design is 10.4 nanoseconds.

Trireg

The **trireg** entity implements a tristate 16-bit register. The register can be loaded with data and drive a tristate bus when the output enable is a '1' value. When enable is a '0', the output from the register floats to a high impedance value. To synthesize the design, the following commands are issued:

```
analyze cpu_lib.vhd
read trireg.vhd
set_attribute -port { .work.trireg.rtl.clock } -name
      CLOCK_CYCLE -value 25
optimize
```

Synthesizing the **trireg** entity produces the following reports:

```
*********************************************************
Cell: trireg View: rtl    Library: work
*********************************************************
```

Cell	Library	References		Total	Area
TRI	flex10	16 x	1	16	TRIs
VCC	flex10	1 x	1	1	VCC
DFF	flex10	16 x	1	16	DFFs
			1	16	LCs

```
Number of ports :                          34
Number of nets :                           51
Number of instances :                      33
Number of references to this view :         0

Total accumulated area :
  Number of TRIs :                         16
  Number of VCC :                           1
  Number of LCs :                          16
  Number of DFFs :                         16
```

The preceding report shows the area for the design. The following report shows the timing of the design:

<pre><code>   Critical Path Report

Critical path #1, (unconstrained path)
NAME   GATE   ARRIVAL   LOAD
----
en/   0.00 up   0.00
q(5)/OUT   TRI   4.00 up   0.00
q(5)/   4.00 up   0.00
data arrival time   4.00

data required time   not specified
----
data required time   not specified
data arrival time   4.00
                                            ----
                                            unconstrained path
----

Critical path #2, (unconstrained path)
NAME   GATE   ARRIVAL   LOAD
----
en/   0.00 up   0.00
q(6)/OUT   TRI   4.00 up   0.00
q(6)/   4.00 up   0.00
data arrival time   4.00

data required time   not specified
----
data required time   not specified
data arrival time   4.00
                                            ----
                                            unconstrained path
----
</code></pre>

The **trireg** entity is loaded when a rising edge occurs on the clock input. Therefore, this entity produces sequential logic. The logic generated is 16 registers with tristate buffers on the outputs to control driving the data to the output bus. The design takes 16 LCs to implement, and the critical path through the design is 4 nanoseconds.

We have synthesized each of the pieces of the design separately to get an idea of how large and fast each of the blocks will be. Because this design is not very large, we could also synthesize the entire design and flatten the hierarchy to share logic across hierarchy boundaries. Most synthesis tools do not combine logic across hierarchy boundaries so that the user hierarchy is maintained. Flattening a small design (less than 20,000 gates) should produce a better output if the computer resources are sufficient to handle the large data structures that are generated.

To compile the design and flatten the design, the following script is used:

```
load_library flex10
analyze -work work cpulib.vhd
analyze -work work alu.vhd
analyze -work work comp.vhd
analyze -work work reg.vhd
analyze -work work shift.vhd
analyze -work work control.vhd
analyze -work work regarray.vhd
analyze -work work trireg.vhd
analyze -work work cpu.vhd
elaborate cpu -work work
ungroup -hierarchy -all
pre_optimize .work.cpu.rtl -common_logic -unused_logic -
    extract
load_modgen flex10
resolve_modgen .work.cpu.rtl -default_resolving
set tristate_map TRUE
optimize .work.cpu.rtl -target flex10 -effort Standard -
    chip -area -pass 1
report_area cpu.area -cell_usage
report_delay -num_paths 2
write -format EDIF cpu.edif
```

After running this script, the following report is generated:

```
Pass   LCs   Delay  DFFs   TRIs   PIs   POs   --CPU--
                                              min:sec
1      479   46     251    16     3     34    01:29


                      Critical Path Report
Critical path #1, (unconstrained path)
NAME                            GATE         ARRIVAL      LOAD
-------------------------------------------------------------
clock information not specified
delay thru clock network                     0.00 (ideal)

con1_current_state(15)/Q   DFF               0.30 up      0.00
nx3031/O                   F2_LUT            4.40 up      0.00
con1_nx3498/O              F4_LUT            8.50 up      0.00
nx2853/O                   F4_LUT           12.60 up      0.00
nx3026/O                   F2_LUT           16.70 up      0.00
nx3013/O                   F4_LUT           20.80 up      0.00
nx3021/O                   F2_LUT           24.90 up      0.00
ix1964_nx44/O              F2_LUT           29.00 up      0.00
nx2847/O                   F2_LUT           33.10 up      0.00
nx2972/O                   F4_CAS           38.30 up      0.00
nx2968/O                   CASCADE2         39.40 up      0.00
nx3093/O                   CASCADE2         40.50 up      0.00
nx3094/O                   CASCADE2         41.60 up      0.00
nx2831/O                   CASCADE2         42.70 up      0.00
nx3092/O                   F1_LUT           42.70 up      0.00
```

```
ix3152/OUT                    TRI              45.70 up    0.00
data(6)/                                       45.70 up    0.00
data arrival time                              45.70

data required time                             not specified
------------------------------------------------------------------
data required time                             not specified
data arrival time                              45.70
                                               ----------
                                               unconstrained path

------------------------------------------------------------------


Critical path #2, (unconstrained path)
NAME                          GATE             ARRIVAL     LOAD
------------------------------------------------------------------
clock information not specified
delay thru clock network                       0.00 (ideal)

con1_current_state(17)/Q      DFF              0.30 up     0.00
nx3031/O                      F2_LUT           4.40 up     0.00
con1_nx3498/O                 F4_LUT           8.50 up     0.00
nx2853/O                      F4_LUT           12.60 up    0.00
nx3026/O                      F2_LUT           16.70 up    0.00
nx3013/O                      F4_LUT           20.80 up    0.00
nx3021/O                      F2_LUT           24.90 up    0.00
ix1964_nx44/O                 F2_LUT           29.00 up    0.00
nx2847/O                      F2_LUT           33.10 up    0.00
nx2972/O                      F4_CAS           38.30 up    0.00
nx2968/O                      CASCADE2         39.40 up    0.00
nx3093/O                      CASCADE2         40.50 up    0.00
nx3094/O                      CASCADE2         41.60 up    0.00
nx2831/O                      CASCADE2         42.70 up    0.00
nx3092/O                      F1_LUT           42.70 up    0.00
ix3152/OUT                    TRI              45.70 up    0.00
data(6)/                                       45.70 up    0.00
data arrival time                              45.70

data required time                             not specified
------------------------------------------------------------------
data required time                             not specified
data arrival time                              45.70
                                               ----------
                                               unconstrained path

------------------------------------------------------------------


--Writing file cpu.edif
```

From the report, we can see that the design was reduced in area and
speed by flattening. However, as noted earlier, this is a small design, and
flattening does not work for large designs.

SUMMARY

In this chapter, we synthesized all of the VHDL RTL descriptions of the CPU and analyzed the results. In the next chapter, we read the synthesized netlist into the place and route tools and run the place and route to implement the design in the target technology.

16

Place and Route

This chapter discusses the process of implementing the synthesis netlist of the CPU design into a target FPGA device. The place and route tools read the netlist, extract the components and nets from the netlist, place the components on the target device, and interconnect the components using the specified interconnections. After the place and route process is complete, the designer has an implementation of the design in the target technology. The implementation still needs to be verified for logical and timing correctness.

Figure 16-1
Place and Route
Process.

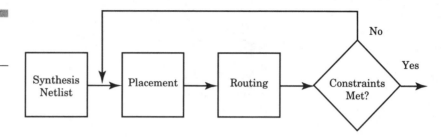

Place and Route Process

The place and route process places each macro from the synthesis netlist into an available location on the target silicon and connects the macros using routing resources available on the target silicon. The place and route process is shown in Figure 16-1.

The synthesis netlist is input to the placement process. The placement process analyzes all of the macros used in the design and their connectivity to try to determine an optimal placement for the macros. The placement algorithms take into account a number of technology specific factors of the target technology to determine whether a particular placement is good or not. After a trial placement and signal route is attempted, the design is analyzed with respect to timing constraints. If the timing constraints are not met, the place and route software continues to try different placements and signal routing to try to meet the constraints.

Typical target devices have areas of the chip where logical functions are placed, and areas where interconnect signals are routed to connect the logical functions. This is shown in Figure 16-2.

The device is split into a number of logic areas with routing channels that surround the logic areas. Logic areas contain the logical gates to implement the boolean function of the design. Routing channels contain the signals that are used to connect the logical gates together. For FPGA devices, the routing channels contain programmable interconnect wires. FPGA devices use an onboard RAM to store the value of programmable switches that are used to form the signal interconnections. By enabling the proper sets of pass transistor gates, signal interconnections between logic gates can be formed as shown in the example in Figure 16-3.

To make a connection from logic block 1 to logic block 3, all of the switches shown need to be enabled with a logic 1 value. The logic gates of the devices are connected to local routing signals that can be connected to more global routing signals by pass transistors that bridge the two signals. The control signals of the pass transistors are stored in a loadable

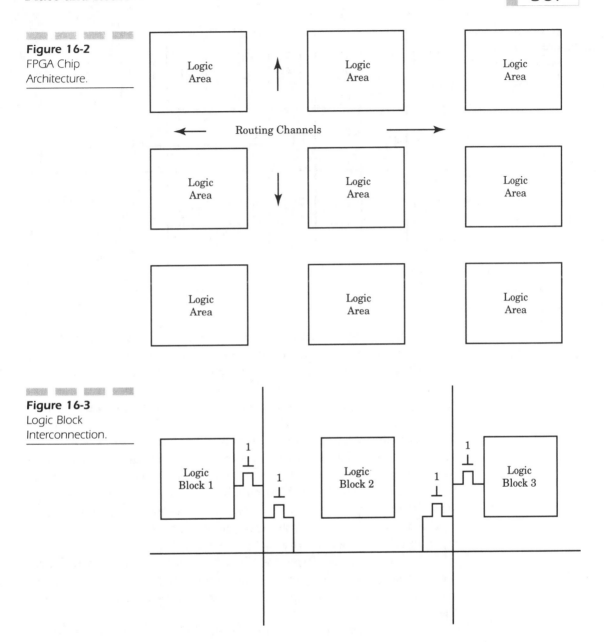

Figure 16-2
FPGA Chip
Architecture.

Figure 16-3
Logic Block
Interconnection.

RAM. The place and route tool generates the RAM image to be loaded into the RAM on the device.

The routing channels contain vertical and horizontal lines. The horizontal wires connect devices within a row, while the vertical lines allow connections across rows. Most routing channels contain wires of different

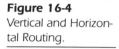

Figure 16-4
Vertical and Horizontal Routing.

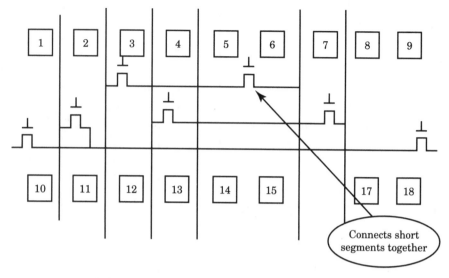

Connects short segments together

lengths that allow connections to adjacent logic areas. Sometimes, longer connections are needed, and either a longer line must be used or shorter lines must be connected together to form the connection. This is shown in Figure 16-4.

The job of the place and route tool is to create the programming files that will be used to specify the logic function of the logic macros in the logic areas and the switch programming of the wires used to connect the macros together. Too many switches on a routed signal can cause some negative performance effects. Each switch adds capacitance and resistance to the routed signal. After only a few connections, signals start to slow significantly because of the capacitance and resistance of the line.

The place and route tool, therefore, must try to minimize long connections and the number of switches for a particular signal to create designs with the highest speed. To get the highest utilization, the place and route tools need to pack as many of the logical functions into a logic area as possible and then use as much local routing resources as possible to connect these functions.

The place and route tools can make tradeoffs if the speed critical signals are known ahead of time and are implemented using the highest speed interconnect signals. The placement algorithm also tries to place logical gates on the critical path close to each other so that local interconnect can be used to connect the gates. Local interconnect is usually very fast because the wires are short. Short wires have less capacitance and resistance and, therefore, can operate at much higher speeds.

Placing and Routing the Device

The target device for the CPU design, as mentioned in earlier chapters, is an FPGA device. The device used is the Flex 10K architecture from Altera. The place and route tools used with the Flex 10K architecture are in the MaxPlus II toolset. MaxPlus II is a set of tools that includes not only place and route, but VHDL entry, VHDL simulation, gate level simulation, and timing analysis. The first step in the process is to compile the design into the place and route environment.

Setting up a project

Most tools that work on a design with multiple data descriptions have a project manager to keep all of the files for that design in one place. This facilitates file management of the design. The first step in the place and route process is to set up a project. In the case of the MaxPlus II environment, the project is usually named the same as the output EDIF file from synthesis. The project is named as shown in Figure 16-5.

Figure 16-5
Project Name
window.

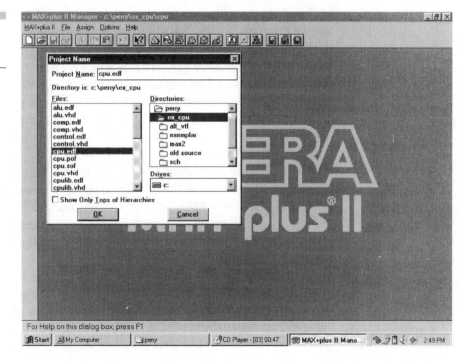

Figure 16-6

MaxPlus II database
with the Compiler
user interface.

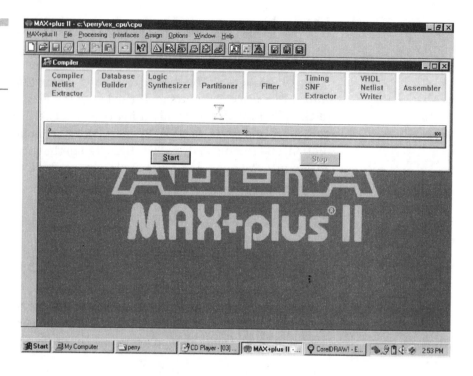

Figure 16-6

MaxPlus II database
with the Compiler
user interface.

Selecting the `File Project Name` menu item brings up a dialog box that
allows the user to pick the EDIF file for the project. In this example, the
design name is cpu, and the EDIF file to be read by the place and route tools
is named cpu.edf. After the EDIF file has been selected, the MaxPlus II
compiler is invoked. The compiler reads the EDIF file into the MaxPlus II
database with the compiler user interface, as shown in Figure 16-6.

Before the compiler can run, the MaxPlus II compiler needs to have a
number of options properly selected, as shown in Figure 16-7.

First, the target device family must be selected. In this example, the
FLEX10K family is selected. Below the device family, the actual target
device can be selected or the default used. This example design has very
few I/O pins so the smallest device can be selected.

Next, the options for the reader must be specified. First, the vendor
must be set to Exemplar so that the proper library is selected. This is
shown in Figure 16-8.

After the reader settings have been selected, the EDIF writer settings
must also be selected. The writer settings allow the synthesis tool to read
the post route timing back into the synthesis database where timing
analysis can be performed. The EDIF settings are set to EDIF 2 0 0 so
that the EDIF can be read back properly into the synthesis tool later. Also,

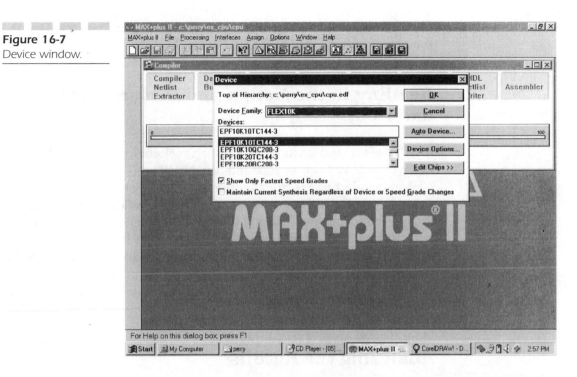

Figure 16-7
Device window.

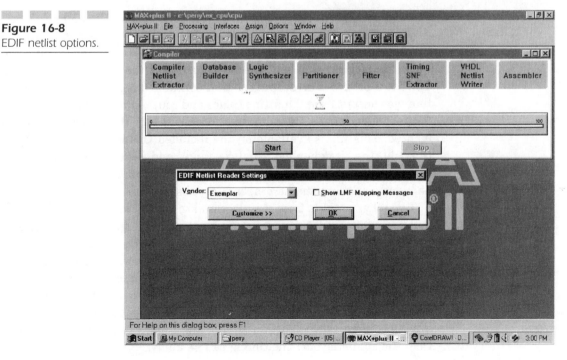

Figure 16-8
EDIF netlist options.

the SDF output file settings are set to SDF Version 2.1 again so that the
timing data can be read back into the synthesis tool.

Reading in the Netlist and Performing Place and Route

After all of the settings have been properly selected, the EDIF netlist from
the synthesis tool can be read into the MaxPlus II tools. This is accomplished by selecting the Start button from the Compiler menu.

This reads the EDIF description into the tool, builds the database, performs partitioning and fitting of the logic, runs timing analysis, and outputs the timing files, and creates the files necessary to program the device.
After this process has completed, all of the necessary files are created for
the designer to use the device in the hardware environment. However, the
designer must first analyze the results to determine if the desired results
were obtained.

Analyzing the Results

The compilation process produces a number of output files and messages.
Any compile errors or warnings from the compilation process appear in
the message output area. In this example, there are a couple of informational messages describing special behavior of the device during reset.
The fitter produces an output file that contains a report of all of the data
that was output from the fitting (place and route) process. This file is very
large and contains a lot of information about how the device was implemented. This file is available on the included CD as /cpu/cpu.rpt. From
this file, the designer can analyze input and output pin connections,
device utilization statistics, routing congestion, clock and clear signal connections, and whether the proper carry chains were generated.

Two other outputs from the place and route tool are the floorplan view
and timing report. The floorplan view gives the designer an idea of placement and routing congestion. From this view, the designer can trace critical paths and move macros around, as necessary, to try to improve timing, if needed. An example floorplan view is shown in Figure 16-9.

The timing report reveals to the designer the actual timing of the placed
and routed device. The timing numbers that we analyzed earlier during
synthesis were estimated timing, but this timing is the actual numbers
from the actual layout. The timing can be analyzed by the provided place

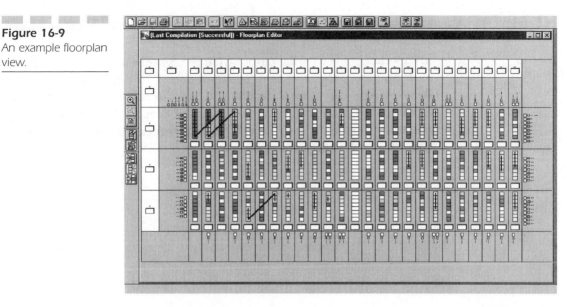

Figure 16-9
An example floorplan view.

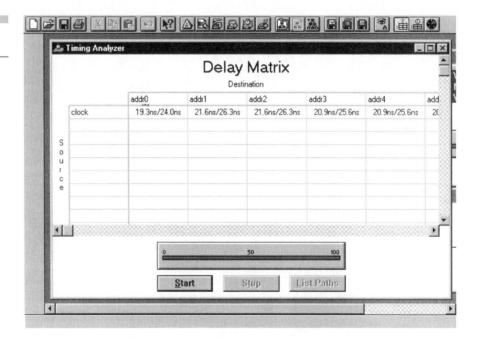

Figure 16-10
A timing report.

and route tools, or back annotated to the synthesis schematic. By using the synthesis schematic, critical paths can be analyzed using the schematic. A timing report from the CPU place and route is shown in Figure 16-10.

SUMMARY

In this chapter, the netlist output from the synthesis tool was read by the place and route tool, and an implementation of the netlist was generated. We examined the process required to run the place and route tool, the inputs to the place and route tool, and the outputs from it. In the next chapter, we examine how to verify that the design created from the place and route tool meets our requirements.

CPU: VITAL Simulation

The last step in the high density FPGA design process is to run gate level timing simulation of the design. Figure 17-1 shows the high density FPGA design flow. The place and route process produces a number of files that need to be verified before the design is implemented. The gate level timing simulation process verifies that the design from the place and route process is correct from a timing and functional point of view.

Within VHDL, this process is implemented using VITAL. VITAL is an IEEE standard that is used for modeling accurate timing at the gate level. VITAL is an acronym for the VHDL Initiative Toward ASIC Libraries. VITAL specifies a standard method of writing ASIC or FPGA libraries so that timing can be back annotated. VITAL libraries used in concert with a VITAL-compliant VHDL simulator can perform gate level timing simulation of the target design.

Figure 17-1
High Density Design
Flow.

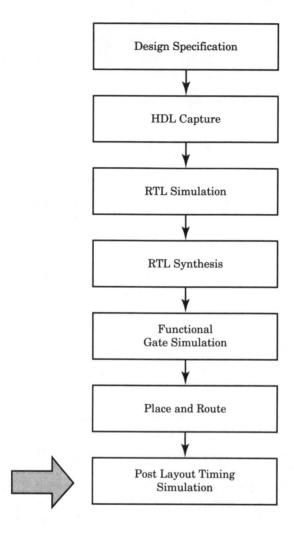

The VITAL process is shown in Figure 17-2.

The place and route tools generate two VITAL-compliant simulator input files. The first is a VHDL netlist that contains the interconnections of all of the entities used to model the design. The second is a timing accurate SDF back-annotation file used to input post-route timing into the VITAL simulation. There is a third input needed to the simulation process. The third input is the VITAL library that describes all of the behavior of the entities used to implement the design. In the next few sections, we examine each of these in more detail.

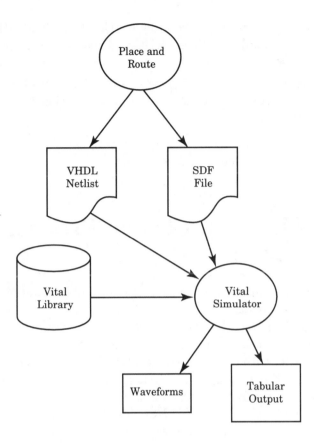

Figure 17-2
VITAL Data Flow.

VITAL Library

One of the reasons VITAL was developed was because there were no standard methods of describing timing behavior in VHDL. With no standard method of describing timing, there was also no standard method of describing timing back annotation. VHDL was also inefficient at modeling gate behavior when compared to gate level simulators optimized for gate level performance.

For all these reasons, VITAL was created to allow near gate level simulation performance with timing accurate models. Some of the features available with VITAL are as follows:

- *Accurate specification of delays*—Delays can be specified pin to pin, be dependent on state, or specified in relation to a particular occurrence of a condition.

■ *Accurate timing check support*—Checks include setup checks, hold checks, pulsewidth checks, period checks, and accurate glitch detection.

■ *Many ways to specify functionality*—Functionality can be specified with truth tables, state tables, boolean primitives, or a behavioral description.

All of these features give the designer the ability to create timing accurate FPGA or ASIC libraries.

VITAL Simulation Process Overview

The place and route tool generates a number of output files, as we saw in the last chapter. The VITAL simulation uses two of these files. The first is the VHDL netlist. This is a file containing component declarations, signals, and component instantiations that connect all of the components together with the declared signals to form the functionality of the design. This file is read by the VITAL simulator and used to create the component connectivity in the database.

The second file is an SDF (Standard Delay Format) file that describes the timing for the design. For each instance in the netlist, this file contains SDF statements that describe the delays and timing checks for the instance. This information is used during simulation to model the timing behavior.

To build the VITAL simulation database, the simulator needs to have a VITAL library that contains components for the target technology and the VHDL netlist and SDF timing file from the place and route tools. The simulator uses the netlist to instantiate the proper instances from the VITAL library in the internal database and then apply timing to the instances with the SDF file. Each of the instances contains a number of generics that receive the timing information. The timing data is used within the model to provide the correct behavior of the underlying device.

VITAL Implementation

VITAL descriptions follow a standard style and make use of standard functions and procedures from two VITAL packages. The VITAL Timing

Package contains procedures and functions for accurate delay modeling, timing checks, and timing error reporting. The VITAL Primitives Package contains built-in primitives that are optimized for simulator performance. Most VITAL-compliant simulators build the primitives package into the simulator for optimum performance.

VITAL contains two styles of modeling that can be back annotated with SDF timing data for timing accurate simulation. The first style, VITAL level 1, uses only VITAL primitives for modeling the behavior of the design. The second, VITAL level 0, has the capability to back annotate timing, but uses behavioral statements to describe the functionality of the design. VITAL level 1 descriptions can be accelerated by VITAL-compliant simulators because the constructs used are built in to the simulator. VITAL level 0 descriptions may not be accelerated because these descriptions use behavioral constructs which may not be built in.

Simple VITAL Model

To understand how the VITAL modeling process works, a simple VITAL model is examined. The model describes the behavior of a 2-input AND gate. The symbol for the AND gate is shown in Figure 17-3.

The AND gate has two inputs, **in1** and **in2**, and an output **y**. When modeled with VITAL, this device has an input delay on inputs **in1** and

Figure 17-3
VITAL AND Gate.

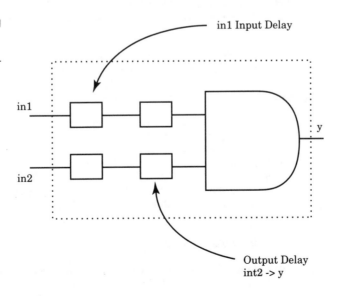

in2, and pin-to-pin delays from input **in1** to output **y** and from input **in2** to output **y**.

Following is the VITAL model that implements the functionality of the **AND2** device:

```
----- CELL AND2 -----
library IEEE;
use IEEE.STD_LOGIC_1164.all;
use IEEE.VITAL_Timing.all;
library alt_vtl;
use alt_vtl.SUPPORT.all;

-- entity declaration --
entity AND2 is
   generic(
       TimingChecksOn: Boolean := True;
       XGenerationOn: Boolean := False;
       InstancePath: STRING := "*";
       tpd_IN1_Y :  VitalDelayType01 := DefPropDelay01;
       tpd_IN2_Y :  VitalDelayType01 := DefPropDelay01;
       tipd_IN1  :  VitalDelayType01 := DefPropDelay01;
       tipd_IN2  :  VitalDelayType01 := DefPropDelay01);

   port(
      Y :   out   STD_LOGIC;
      IN1 :  in    STD_LOGIC;
      IN2 :  in    STD_LOGIC);
   attribute VITAL_LEVEL0 of AND2 : entity is TRUE;
end AND2;

-- architecture body --
architecture AltVITAL of AND2 is
   attribute VITAL_LEVEL1 of AltVITAL : architecture is
      TRUE;

   SIGNAL IN1_ipd : STD_ULOGIC := 'U';
   SIGNAL IN2_ipd : STD_ULOGIC := 'U';

begin

   --------------------
   --   INPUT PATH DELAYs
   --------------------
   WireDelay : block
   begin
   VitalWireDelay (IN1_ipd, IN1, tipd_IN1);
   VitalWireDelay (IN2_ipd, IN2, tipd_IN2);
   end block;
   --------------------
   --   BEHAVIOR SECTION
   --------------------
   VITALBehavior : process (IN1_ipd, IN2_ipd)
```

```
                         -- functionality results
                         VARIABLE Results : STD_LOGIC_VECTOR(1 to 1) :=
                                             (others => 'X');
                         ALIAS Y_zd : STD_ULOGIC is Results(1);

                         -- output glitch detection variables
                         VARIABLE Y_GlitchData : VitalGlitchDataType;

                         begin

                             --------------------------
                             --   Functionality Section
                             --------------------------
                             Y_zd := (IN2_ipd) AND (IN1_ipd);


                             -----------------------
                             --   Path Delay Section
                             -----------------------
                             VitalPathDelay01 (
                              OutSignal => Y,
                              OutSignalName => "Y",
                              OutTemp => Y_zd,
                              Paths => (0 => (IN1_ipd'last_event, tpd_IN1_Y,
                                              TRUE),
                                        1 => (IN2_ipd'last_event, tpd_IN2_Y,
                                              TRUE)),
                              GlitchData => Y_GlitchData,
                              Mode => DefGlitchMode,
                              XOn  => DefGlitchXOn);


                     end process;

                     end AltVITAL;

                     configuration CFG_AND2_VITAL of AND2 is
                         for AltVITAL
                         end for;
                     end CFG_AND2_VITAL;
```

The model looks like standard VHDL with some different packages included. In fact, the model is standard VHDL. The entity contains declarations for the **STD_1164** packages for the signal logic types, but also contains **USE** clauses for the VITAL timing package. The VITAL timing package is needed in the entity for **AND2** to provide the type declarations for the entity generics.

The **entity** statement contains four generics that are used to pass delay information to the model. Each of the generics has a prefix that represents the type of the delay. Generic **tipd_in1** is an input delay for input **in1**. Generic **tipd_in2** is an input delay for input **in2**. Generic

tpd_in1_y models the pin-to-pin delay from input **in1** to output **y**. Generic **tidp_in2_y** models the pin-to-pin delay from input **in2** to output **y**.

The timing information passed to these generics comes from the SDF file generated by the place and route tool. Each of the delays passed to the entity is instance specific.

Each of the generics has a type associated with it that represents how many delay values can be held. In this example, the generic contains two values. Delay **tr01** represents the delay value when the signal changes from a '0' to '1' value. Delay **tr10** represents the delay when the signal changes from a '1' to '0' value.

The entity also contains other generics that control functionality of the VITAL model. This example contains a generic called **TimingChecksOn** that controls whether or not the timing check functions in the VITAL model are executed or not. Finally, the entity contains the input and output ports for the model.

VITAL Architecture

The architecture for the VITAL model contains four distinct code areas. These are the wire delay section, the timing violation section, the function description section, and the path delay section. Not all models contain all of these sections. Some models are purely combinational and do not need timing check sections.

Wire Delay Section

The first section of the architecture is the wire delay section. The **AND2** architecture starts with a number of library declarations; but notice that the architecture also uses the VITAL primitives package. After the architecture statement, the architecture declares two local signals, **in1_ipd** and **in2_ipd**, and an attribute. The two signals are used to delay the input signals to the entity. The delay values applied to the two input signals represent the wiring delays to connect the physical gates together. For instance, in Figure 17-4, gate U1 drives gates U2 and U3. The wiring from gate U1 to gate U2 causes 8 nanoseconds of delay in the path, but the wiring from U1 to U3 causes 10 nanoseconds of delay in the path. With separate input delay values for each input, the wiring delays can be modeled correctly.

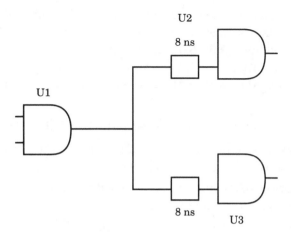

U2

8 ns

U1

8 ns

U3

Attribute **VitalLevel1** specifies that the VITAL model is Level 1 compliant. Level 1 models are modeled only with VITAL primitives and can be accelerated. Some simulators have compliance checkers that can validate level 1 compliance.

The architecture contains a block labeled **WireDelay** which contains the VHDL description that actually delays the input signals. The block contains a call to the **VitalWireDelay** procedure for each input port. The **VitalWireDelay** procedure delays the input ports by the value passed to the appropriate generic used in the procedure call. In this example, generic **tipd_in1** is used to delay input **in1**, and generic **tipd_in2** is used to delay input **in2**.

After the wire delay section is the timing check section. This example has no timing check section because it is a purely combinational gate model.

The next section is the functionality section. This section contains the statements that model the behavior of the device. This section starts with a process labeled **VitalBehavior**. Notice that the process is sensitive to the delayed versions of the two input signals, **in1_ipd** and **in2_ipd**. There are a number of local variables declared and a statement that performs an **AND** function of the two inputs. This **AND** function can be built in to the simulator so that execution can be accelerated.

The last section of the architecture starts with the **VitalPathDelay** procedure call. This section is the Path Delay section. This section schedules the new logic values calculated in the functionality section to occur after the appropriate delay. This section consists of a **VitalPathDelay01** procedure call for each output from the entity.

The `VitalPathDelay01` procedure has a number of parameters passed to it. These parameters are used to control what kind of glitch behavior is wanted, the delays to be used, and the temporary data used to store signal information.

In this example, the `VitalPathDelay` procedure is passed the following parameters:

- `OutSignal`—The signal to have the new value placed on it.

- `OutSignalName`—The name of the output signal to be used in glitch reporting or error reporting.

- `OutTemp`—A temporary signal used to store the current value of the signal for comparison.

- `Paths`—An array used to store delay information. There is a table entry for each delay arc through the device.

- `GlitchData`—A temporary storage area used to store signal state and transition information for use in calculating glitches.

- `Mode`—Specifies the type of glitch behavior wanted.

- `GlitchKind`—Specifies the kind of glitches generated, `OnEvent` or `OnDetect`.

Flip-Flop Example

In this next section, we examine another VITAL model with more complexity. This example shows the VITAL model for a DFF device. This device has sequential behavior and needs to have timing checks to check for illegal timing conditions:

```
----- CELL DFF -----
library IEEE;
use IEEE.STD_LOGIC_1164.all;
use IEEE.VITAL_Timing.all;
use IEEE.VITAL_Primitives.all;
library alt_vtl;
use alt_vtl.SUPPORT.all;

-- entity declaration --
entity DFF is
   generic(
       TimingChecksOn: Boolean := True;
       XGenerationOn: Boolean := False;
       InstancePath: STRING := "*";
       tpd_PRN_Q_negedge :  VitalDelayType01 :=
```

```
                        DefPropDelay01;
           tpd_CLRN_Q_negedge :  VitalDelayType01 :=
              DefPropDelay01;
           tpd_CLK_Q_posedge  :  VitalDelayType01 :=
              DefPropDelay01;
           tsetup_D_CLK_noedge_posedge :  VitalDelayType :=
              DefSetupHoldCnst;
           tsetup_D_CLK_noedge_negedge :  VitalDelayType :=
              DefSetupHoldCnst;
           thold_D_CLK_noedge_posedge  :   VitalDelayType :=
              DefSetupHoldCnst;
           thold_D_CLK_noedge_negedge  :   VitalDelayType :=
              DefSetupHoldCnst;
           tipd_D : VitalDelayType01 := DefPropDelay01;
           tipd_CLRN :  VitalDelayType01 := DefPropDelay01;
           tipd_PRN  :  VitalDelayType01 := DefPropDelay01;
           tipd_CLK  :  VitalDelayType01 := DefPropDelay01);

     port(
        Q :   out    STD_LOGIC;
        D :   in     STD_LOGIC;
        CLRN :  in     STD_LOGIC;
        PRN  :  in     STD_LOGIC;
        CLK  :  in     STD_LOGIC);
     attribute VITAL_LEVEL0 of DFF : entity is TRUE;
  end DFF;

  -- architecture body --

  architecture AltVITAL of DFF is
     attribute VITAL_LEVEL1 of AltVITAL : architecture is
        TRUE;

     SIGNAL D_ipd   : STD_ULOGIC := 'U';
     SIGNAL CLRN_ipd : STD_ULOGIC := 'U';
     SIGNAL PRN_ipd : STD_ULOGIC := 'U';
     SIGNAL CLK_ipd : STD_ULOGIC := 'U';

  begin

     ---------------------
     --   INPUT PATH DELAYs
     ---------------------
     WireDelay : block
     begin
     VitalWireDelay (D_ipd, D, tipd_D);
     VitalWireDelay (CLRN_ipd, CLRN, tipd_CLRN);
     VitalWireDelay (PRN_ipd, PRN, tipd_PRN);
     VitalWireDelay (CLK_ipd, CLK, tipd_CLK);
     end block;
     ---------------------
     --   BEHAVIOR SECTION
     ---------------------
```

```
VITALBehavior : process (D_ipd, CLRN_ipd, PRN_ipd,
                         CLK_ipd)

-- timing check results
VARIABLE Tviol_D_CLK : STD_ULOGIC := '0';
VARIABLE TimingData_D_CLK : VitalTimingDataType :=
        VitalTimingDataInit;

-- functionality results
VARIABLE Violation : STD_ULOGIC := '0';
VARIABLE PrevData_Q : STD_LOGIC_VECTOR(1 to 6) ;
VARIABLE D_delayed : STD_ULOGIC := 'U';
VARIABLE CLK_delayed : STD_ULOGIC := 'U';
VARIABLE Results : STD_LOGIC_VECTOR(1 to 1) :=
                   (others => 'X');

-- output glitch detection variables
VARIABLE Q_VitalGlitchData   : VitalGlitchDataType;

CONSTANT DFF_Q_tab : VitalStateTableType := (
-- Violation, CLRN_ipd, CLK_delayed, D_delayed, PRN_ipd,
   CLK_ipd
   ( L, L,  x,  x,  x,  x,  x,  L ),
   ( L, H,  L,  H,  x,  H,  x,  H ),
   ( L, H,  H,  x,  H,  x,  x,  S ),
   ( L, H,  x,  x,  L,  x,  x,  H ),
   ( L, H,  x,  x,  H,  L,  x,  S ),
   ( L, x,  L,  L,  H,  H,  x,  L )  );

begin

       ------------------------
       --   Timing Check Section
       ------------------------
       if (TimingChecksOn) then
          VitalSetupHoldCheck (
                  Violation        => Tviol_D_CLK,
                  TimingData       => TimingData_D_CLK,
                  TestSignal       => D_ipd,
                  TestSignalName   => "D",
                  RefSignal        => CLK_ipd,
                  RefSignalName    => "CLK",
                  SetupHigh        => tsetup_D_CLK_noedge_
                                      posedge,
                  SetupLow         => tsetup_D_CLK_noedge_
                                      posedge,
                  HoldHigh         => thold_D_CLK_noedge_
                                      posedge,
                  HoldLow          => thold_D_CLK_noedge_
                                      posedge,
                  CheckEnabled     => TO_X01(( (NOT PRN_ipd)
                                      ) OR ( (NOT
```

```
                                           CLRN_ipd) ) ) /= '1',
                RefTransition    => '/',
                HeaderMsg        => InstancePath & "/DFF",
                XOn              => DefTimingXon,
                MsgOn            => DefTimingMsgon );

        end if;

        -------------------------
        --  Functionality Section
        -------------------------
        Violation := Tviol_D_CLK;
        VitalStateTable(
          StateTable => DFF_Q_tab,
          DataIn => (
                    Violation, CLRN_ipd, CLK_delayed,
                    D_delayed, PRN_ipd, CLK_ipd),
          Result => Results,
          NumStates => 1,
          PreviousDataIn => PrevData_Q);
        D_delayed := D_ipd;
        CLK_delayed := CLK_ipd;

        ----------------------
        --  Path Delay Section
        ----------------------
        VitalPathDelay01 (
         OutSignal => Q,
         OutSignalName => "Q",
         OutTemp => Results(1),
         Paths => (0 => (PRN_ipd'last_event,
                    tpd_PRN_Q_negedge, TRUE),
                  1 => (CLRN_ipd'last_event,
                        tpd_CLRN_Q_negedge, TRUE),
                  2 => (CLK_ipd'last_event,
                        tpd_CLK_Q_posedge, TRUE),
                  3 => (D_ipd'last_event,
                        tpd_CLK_Q_posedge, TRUE)),
         GlitchData => Q_VitalGlitchData,
         Mode => DefGlitchMode,
         XOn  => DefGlitchXOn);

  end process;


  end AltVITAL;

  configuration CFG_DFF_VITAL of DFF is
      for AltVITAL
      end for;
  end CFG_DFF_VITAL;
```

The first thing to notice about this model is that there are quite a few more generics used to pass timing information to the model. This is because this model has more input ports; therefore, there are more input delay generics and the model contains timing checks that need timing information passed to them.

The wire delay section now delays four input ports instead of two. The **D**, **CLRN**, **PRN**, and **CLK** inputs are delayed in the wire delay section. The architecture for the DFF also contains a number of local signals and variables used to hold intermediate values for the timing check and functionality sections. The final declaration item in the architecture declaration section is a table that is used to model the behavior of the DFF. This DFF model uses a VITAL State Table procedure to model the behavior of the device. This table is used in the functionality section of the model by the **VitalStateTable** procedure call. The signal values of the signals passed to the **VitalStateTable** procedure call are compared to the values in the table, and the new values for the output signals and next state are predicted.

The timing check section for this example contains a **VitalSetupHold-Check** procedure call. This procedure checks the setup and hold of data changes versus the clock for the DFF device. The violation signal returned by the **VitalSetupHoldCheck** procedure is used to affect the behavior of the DFF device by the fact that its value is passed to the **VitalStateTable** that controls the behavior of the DFF device.

The functionality section of the DFF device contains the single call to the **VitalStateTable** procedure to calculate the value of the **Q** output based on the values of the input ports, the previous state, and the violation signal from the timing check procedures. Based on all of these inputs, a table row matches, and the new **Q** output is passed to the path delay section.

The path delay section looks very similar to the path delay section for the **AND2** device discussed previously. The path delay section contains a single call to the **VitalPathDelay01** procedure, which schedules output **Q** with the appropriate delay value.

To see how all of these VITAL functions and procedures are implemented, look at the VITAL packages included on the CD with the book or visit **www.vhdl.org/vital**

SDF File

The other piece of functionality needed to complete the VITAL simulation picture is the SDF back-annotation file. This file is generated by the place and route tools and contains accurate timing for the device. The SDF file contains timing information for all of the generics in the VITAL library that need data passed to them. Following is a sample SDF file:

```
(DELAYFILE
(SDFVERSION "2.1")
(DESIGN "cpu")
(DATE "10/25/97 10:59:58")
(VENDOR "Altera")
(PROGRAM "MAX+plus II")
(VERSION "Version 7.2 RC2 2/14/97")
(DIVIDER .)
(VOLTAGE  :5:) (PROCESS "typical") (TEMPERATURE  :25:)
(TIMESCALE 100ps)

(CELL
    (CELLTYPE "DFF")
    (INSTANCE DFF_457)
    (DELAY
      (ABSOLUTE
      (IOPATH (posedge CLK) Q (32:32:32) (32:32:32)))
      (ABSOLUTE
      (IOPATH (negedge PRN) Q (36:36:36) (36:36:36)))
      (ABSOLUTE
      (IOPATH (negedge CLRN) Q (37:37:37) (37:37:37)))
    )

    (TIMINGCHECK
      (SETUP D (posedge CLK) (2:2:2))
      (HOLD D (posedge CLK) (10:10:10))
    ))
)
```

The SDF file starts with a header section that describes the name of the design the file will back annotate, the vendor that generated the file, the environment used to generate the timing numbers, and so on. After the header, the file consists of a number of cells. Each cell in the SDF file represents an instance in the VHDL netlist produced by the place and route tools. Each cell contains the type of cell, the instance name in the netlist, and timing information to be back annotated to the design. The VITAL compliant simulator reads the SDF file and matches the generics in the VHDL source with the delay constructs in the SDF file. For instance, an **IOPATH** construct in the SDF file specifies the rising and falling delays from and input to an output signal. The **IOPATH** construct is converted into generic names and values to be applied to the VITAL simulation. The designer of the VITAL model must ensure that the names used in the SDF model and the names of the generics used in the VITAL model match so that the generics can be properly matched with proper timing values.

The last section is a timing check section that contains timing information for the timing checks of the cell, if they exist. The timing check section of the SDF file is read by the VITAL simulator and extracts timing information to plug in to generics of the VITAL model. The timing check generics control the timing values that are used in the timing checks of the VITAL model while simulation is progressing.

The cell description in the preceding example is for the DFF model that we looked at earlier. There are delay values for **CLK** to **Q**, **PRN** to **Q**, and **CLRN** to **Q**, and values for the setup and hold check.

VITAL Simulation

To run the VITAL simulation, the designer first compiles the VITAL library into a simulator library. The device manufacturers supply VITAL libraries for their devices. Next, the VITAL netlist is compiled to the working library, and, finally, the SDF file is read in to back annotate the timing data into the design. After these steps have been completed, the designer runs the VITAL simulation in the same manner as the RTL simulation that we ran earlier.

The first step is to compile the VITAL library into a simulator library so that it can be referenced. It is best if this library is compiled into the location specified by the netlist from the place and route tool so that no manual code modification is necessary. The following shows the first few lines of the VITAL netlist generated by the MaxPlusII place and route tool. The complete netlist is on the CD. Notice that the VITAL netlist expects the VITAL component declarations, package **VCOMPONENTS**, to be located in a library named **alt_vtl**:

```
-- MAX+plus II Version 7.2 RC2 2/14/97
-- Sat Oct 25 10:59:34 1997

--

LIBRARY IEEE;
USE IEEE.std_logic_1164.all;

LIBRARY alt_vtl;
USE alt_vtl.VCOMPONENTS.all;

--ENTITY cpu IS
--      PORT (
--          addr : OUT std_logic_vector(15 downto 0);
--          data : INOUT std_logic_vector(15 downto 0);
--          clock : IN std_logic;
```

```
--          ready : IN std_logic;
--          reset : IN std_logic;
--          rw : OUT std_logic;
--          vma : OUT std_logic);
--END cpu;

ARCHITECTURE EPF10K10TC144_a3 OF cpu IS

SIGNAL gnd : std_logic := '0';
SIGNAL vcc : std_logic := '1';
SIGNAL    ......
......... .
```

To compile the VCOMPONENTS package into library alt_vt1, the following commands are executed in Modelsim:

```
vlib alt_vt1
vcom -work alt_vt1 alt_vt1.cmp
```

Because there are no other library declarations for the actual vital library, the vital library entities need to be compiled into the working library to be visible. Following is the command to perform this step:

```
vcom alt_vt1.vhd
```

After these two files have been compiled, the VITAL netlist can be compiled into the working library. The following command compiles the netlist:

```
vcom cpuout.vhd
```

We still need to simulate design TOP to verify the gate level implementation of the CPU. However, this time, the CPU RTL description is replaced with a VITAL description of the CPU. This can be accomplished by two different methods. The first involves compilation order, and the second is by direct specification. Remember that the last architecture compiled is used by default for an entity. By compiling architecture EPF10K10TC144_a3 last, this architecture is used for entity cpu.

The other method is to write a configuration for architecture top that specifies exactly which architecture is to be used. The following example shows two configuration statements for the two different implementations of the CPU:

```
configuration topconrtl of top is
  for behave
    for  U1 : cpu use entity work.cpu(behave);
    end for;
```

```
        end for;
    end topconrtl;

    configuration topconstruct of top is
        for behave
            for  U1 : cpu use entity work.cpu(EPF10K10TC144_a3);
            end for;
        end for;
    end topconstruct;
```

Configuration **topconrtl** specifies the **rtl** implementation configuration for entity **top**, and configuration **topconstruct** specifies the structural implementation. Notice that the structural architecture was named the same as the device that was implemented by the place and route tools.

To complete the simulation setup process, the final compilations needed are shown here:

```
vcom top.vhd
vcom topconstruct.vhd
```

After these steps, the design is ready for simulation. To load the design into the simulator, the following command is executed:

```
vsim topconstruct
```

The simulator brings up its windows and begins the simulation. If the simulation is run ahead 500 nanoseconds, we can see the CPU start the reset sequence as instructions are fetched. This is shown in Figure 17-5.

Figure 17-5
The Simulator window.

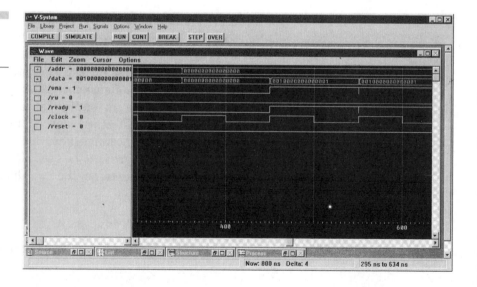

Running the simulation through the entire process verifies the functionality of the placed and routed design. To verify the timing and functionality, we need to back annotate the timing from place and route to the simulation.

Back-Annotated Simulation

To run timing back-annotated simulation, we don't need to recompile. We only need to specify to the simulator which SDF file to read. This is done by the following command:

```
vsim -sdfmax /u1=cpuout.sdf topconstruct
```

This command tells the simulator to back annotate the VITAL simulation of the CPU design with SDF file cpuout.sdf created by the place and route tools. After this command has executed, the simulation is invoked, and the SDF file is back annotated to component **U1** (cpu) and simulation started. Running the simulation produces the waveform shown in Figure 17-6.

The back-annotated delays are seen on the waveforms for **addr** and **data** around time 400 nanoseconds. Notice that, instead of one transition, the waveforms have a number of transitions that finally settle out. Using this timing information, the designer can now increase the clock speed un-

Figure 17-6
Simulation
Waveform.

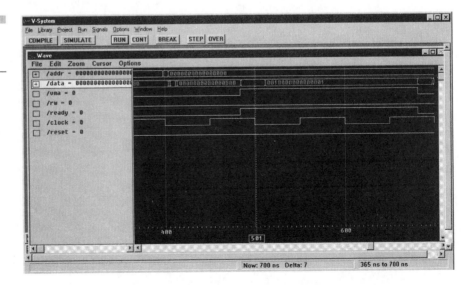

til the design stops working to determine the maximum speed that the design will run. By running the design through the entire simulation, the functionality and timing of the design can be verified for correctness. When the design meets the functionality and timing requirements, the design can be signed off and built.

SUMMARY

In this chapter, we examined VITAL simulation and how to perform VITAL simulation on the CPU design. The rest of the book contains useful appendices that describe some of the standard types, functions, and procedures used throughout the book.

Appendix A

Standard Logic Package

This is a copy of the IEEE 1164 standard logic package. It is used in all of the examples in the book and is listed here for reference.

```
-- -----------------------------------------------------------
--
--     Title        :   std_logic_1164 multi-value logic system
--     Library      :   This package shall be compiled into a
--                  :   library symbolically named IEEE.
--                  :
--     Developers   :   IEEE model standards group (par 1164)
--     Purpose      :   This packages defines a standard for
--                  :   designers to use in describing the
--                  :   interconnection data types used in vhdl
--                  :   modeling.
--                  :
--     Limitation   :   The logic system defined in this
--                  :   package may be insufficient for
--                  :   modeling switched transistors, since
--                  :   such a requirement is out of the scope
--                  :   of this effort. Furthermore,
--                  :   mathematics, primitives, timing
--                  :   standards, etc. are considered
--                  :   orthogonal issues as it relates to this
--                  :   package and are therefore beyond the
--                  :   scope of this effort.
--                  :
--     Note         :   No declarations or definitions shall be
--                  :   included in, or excluded from this
--                  :   package. The "package declaration"
--                  :   defines the types, subtypes and
--                  :   declarations of std_logic_1164. The
--                  :   std_logic_1164 package body shall be
--                  :   considered the formal definition of the
--                  :   semantics of this package. Tool
--                  :   developers may choose to implement the
--                  :   package body in the most efficient
--                  :   manner available to them.
--                  :
-- -----------------------------------------------------------
--    modification history :
-- -----------------------------------------------------------
--    version | mod. date:|
--     v4.200 | 01/02/92  |
-- -----------------------------------------------------------

PACKAGE std_logic_1164 IS
-- -----------------------------------------------------------
    -- logic state system  (unresolved)
-- -----------------------------------------------------------
    TYPE std_ulogic IS ( 'U',   -- Uninitialized
                         'X',   -- Forcing  Unknown
```

```
                    '0',   -- Forcing  0
                    '1',   -- Forcing  1
                    'Z',   -- High Impedance
                    'W',   -- Weak     Unknown
                    'L',   -- Weak     0
                    'H',   -- Weak     1
                    '-'    -- Don't care
               );
-------------------------------------------------------
-- unconstrained array of std_ulogic for use with the
-- resolution function
-------------------------------------------------------
TYPE std_ulogic_vector IS ARRAY ( NATURAL RANGE <> )
  OF std_ulogic;


-------------------------------------------------------
-- resolution function
-------------------------------------------------------
FUNCTION resolved ( s : std_ulogic_vector ) RETURN
  std_ulogic;
-------------------------------------------------------
-- *** industry standard logic type ***
-------------------------------------------------------
SUBTYPE std_logic IS resolved std_ulogic;


-------------------------------------------------------
-- unconstrained array of std_logic for use in
-- declaring signal arrays
-------------------------------------------------------
TYPE std_logic_vector IS ARRAY ( NATURAL RANGE <>) OF
  std_logic;


-------------------------------------------------------
-- common subtypes
-------------------------------------------------------
SUBTYPE X01      IS resolved std_ulogic RANGE 'X' TO
                 '1'; -- ('X','0','1')
SUBTYPE X01Z     IS resolved std_ulogic RANGE 'X' TO
                 'Z'; -- ('X','0','1','Z')
SUBTYPE UX01     IS resolved std_ulogic RANGE 'U' TO
                 '1'; -- ('U','X','0','1')
SUBTYPE UX01Z    IS resolved std_ulogic RANGE 'U' TO
                 'Z'; -- ('U','X','0','1','Z')


-------------------------------------------------------
-- overloaded logical operators
-------------------------------------------------------

FUNCTION "and"       ( l : std_ulogic; r : std_ulogic )
  RETURN UX01;
FUNCTION "nand"      ( l : std_ulogic; r : std_ulogic )
  RETURN UX01;
FUNCTION "or"        ( l : std_ulogic; r : std_ulogic )
  RETURN UX01;
FUNCTION "nor"       ( l : std_ulogic; r : std_ulogic )
  RETURN UX01;
FUNCTION "xor"       ( l : std_ulogic; r : std_ulogic )
  RETURN UX01;
--   function "xnor" ( l : std_ulogic; r : std_ulogic )
--   return ux01;
```

```
FUNCTION "not"        ( l : std_ulogic                    )
  RETURN UX01;


-------------------------------------------------------------
-- vectorized overloaded logical operators
-------------------------------------------------------------
FUNCTION "and"  ( l, r : std_logic_vector  ) RETURN
  std_logic_vector;
FUNCTION "and"  ( l, r : std_ulogic_vector ) RETURN
  std_ulogic_vector;

FUNCTION "nand" ( l, r : std_logic_vector  ) RETURN
  std_logic_vector;
FUNCTION "nand" ( l, r : std_ulogic_vector ) RETURN
  std_ulogic_vector;

FUNCTION "or"   ( l, r : std_logic_vector  ) RETURN
  std_logic_vector;
FUNCTION "or"   ( l, r : std_ulogic_vector ) RETURN
  std_ulogic_vector;

FUNCTION "nor"  ( l, r : std_logic_vector  ) RETURN
  std_logic_vector;
FUNCTION "nor"  ( l, r : std_ulogic_vector ) RETURN
  std_ulogic_vector;

FUNCTION "xor"  ( l, r : std_logic_vector  ) RETURN
  std_logic_vector;
FUNCTION "xor"  ( l, r : std_ulogic_vector ) RETURN
  std_ulogic_vector;

-- -------------------------------------------------------
-- Note : The declaration and implementation of the "xnor"
-- function is specifically commented until at which time
-- the VHDL language has been officially adopted as
-- containing such a function. At such a point, the
-- following comments may be removed along with this
-- notice without further "official" ballotting of this
-- std_logic_1164 package. It is the intent of this effort
-- to provide such a function once it becomes available
-- in the VHDL standard.
-- -------------------------------------------------------
-- function "xnor" ( l, r : std_logic_vector  ) return
-- std_logic_vector;
-- function "xnor" ( l, r : std_ulogic_vector ) return
-- std_ulogic_vector;

FUNCTION "not"  ( l : std_logic_vector  ) RETURN
  std_logic_vector;
FUNCTION "not"  ( l : std_ulogic_vector ) RETURN
  std_ulogic_vector;


-------------------------------------------------------------
-- conversion functions
-------------------------------------------------------------
FUNCTION To_bit        ( s : std_ulogic;         xmap :
                         BIT := '0') RETURN BIT;
FUNCTION To_bitvector ( s : std_logic_vector ; xmap :
                         BIT := '0') RETURN BIT_VECTOR;
```

```
FUNCTION To_bitvector  ( s : std_ulogic_vector; xmap :
                        BIT := '0') RETURN BIT_VECTOR;

FUNCTION To_StdULogic           ( b : BIT             )
  RETURN std_ulogic;
FUNCTION To_StdLogicVector      ( b : BIT_VECTOR      )
  RETURN std_logic_vector;
FUNCTION To_StdLogicVector      ( s : std_ulogic_vector )
  RETURN std_logic_vector;
FUNCTION To_StdULogicVector     ( b : BIT_VECTOR      )
  RETURN std_ulogic_vector;
FUNCTION To_StdULogicVector     ( s : std_logic_vector  )
  RETURN std_ulogic_vector;


------------------------------------------------------------
-- strength strippers and type convertors
------------------------------------------------------------

FUNCTION To_X01  ( s : std_logic_vector            )
  RETURN  std_logic_vector;
FUNCTION To_X01  ( s : std_ulogic_vector           )
  RETURN  std_ulogic_vector;
FUNCTION To_X01  ( s : std_ulogic                  )
  RETURN  X01;
FUNCTION To_X01  ( b : BIT_VECTOR                  )
  RETURN  std_logic_vector;
FUNCTION To_X01  ( b : BIT_VECTOR                  )
  RETURN  std_ulogic_vector;
FUNCTION To_X01  ( b : BIT                         )
  RETURN  X01;

FUNCTION To_X01Z ( s : std_logic_vector            )
  RETURN  std_logic_vector;
FUNCTION To_X01Z ( s : std_ulogic_vector           )
  RETURN  std_ulogic_vector;
FUNCTION To_X01Z ( s : std_ulogic                  )
  RETURN  X01Z;
FUNCTION To_X01Z ( b : BIT_VECTOR                  )
  RETURN  std_logic_vector;
FUNCTION To_X01Z ( b : BIT_VECTOR                  )
  RETURN  std_ulogic_vector;
FUNCTION To_X01Z ( b : BIT                         )
  RETURN  X01Z;

FUNCTION To_UX01                ( s : std_logic_vector  )
  RETURN  std_logic_vector;
FUNCTION To_UX01                ( s : std_ulogic_vector )
  RETURN  std_ulogic_vector;
FUNCTION To_UX01                ( s : std_ulogic      )
  RETURN  UX01;
FUNCTION To_UX01                ( b : BIT_VECTOR      )
  RETURN  std_logic_vector;
FUNCTION To_UX01                ( b : BIT_VECTOR      )
  RETURN  std_ulogic_vector;
FUNCTION To_UX01                ( b : BIT             )
  RETURN  UX01;


------------------------------------------------------------
-- edge detection
------------------------------------------------------------
```

```
                      FUNCTION rising_edge  (SIGNAL s : std_ulogic) RETURN
                        BOOLEAN;
                      FUNCTION falling_edge (SIGNAL s : std_ulogic) RETURN
                        BOOLEAN;


                      -----------------------------------------------------
                      -- object contains an unknown
                      -----------------------------------------------------
                      FUNCTION Is_X ( s : std_ulogic_vector ) RETURN
                        BOOLEAN;
                      FUNCTION Is_X ( s : std_logic_vector  ) RETURN
                        BOOLEAN;
                      FUNCTION Is_X ( s : std_ulogic         ) RETURN
                        BOOLEAN;

                  END std_logic_1164;


         -- -----------------------------------------------------------
         --
         --    Title      :  std_logic_1164 multi-value logic system
         --    Library    :  This package shall be compiled into a
         --               :  library symbolically named IEEE.
         --               :
         --    Developers :  IEEE model standards group (par 1164)
         --    Purpose    :  This packages defines a standard for
         --               :  designers to use in describing the
         --               :  interconnection data types used in vhdl
         --               :  modeling.
         --               :
         --    Limitation :  The logic system defined in this
         --               :  package may be insufficient for modeling
         --               :  switched transistors, since such a
         --               :  requirement is out of the scope of this
         --               :  effort. Furthermore, mathematics,
         --               :  primitives, timing standards, etc. are
         --               :  considered orthogonal issues as it
         --               :  relates to this package and are
         --               :  therefore beyond the scope of this
         --               :  effort.
         --               :
         --    Note       :  No declarations or definitions shall be
         --               :  included in, or excluded from this
         --               :  package. The "package declaration"
         --               :  defines the types, subtypes and
         --               :  declarations of std_logic_1164. The
         --               :  std_logic_1164 package body shall be
         --               :  considered the formal definition of the
         --               :  semantics of this package. Tool
         --               :  developers may choose to implement the
         --               :  package body in the most efficient
         --               :  manner available to them.
         --               :
         -- -----------------------------------------------------------
         --    modification history :
         -- -----------------------------------------------------------
         --    version | mod. date:|
         --    v4.200  | 01/02/91  |
         -- -----------------------------------------------------------

         PACKAGE BODY std_logic_1164 IS
```

```
-------------------------------------------------------
-- local types
-------------------------------------------------------
TYPE stdlogic_1d IS ARRAY (std_ulogic) OF std_ulogic;
TYPE stdlogic_table IS ARRAY(std_ulogic, std_ulogic) OF
  std_ulogic;

-------------------------------------------------------
-- resolution function
-------------------------------------------------------
CONSTANT resolution_table : stdlogic_table := (
--  -------------------------------------------------
--| U    X    0    1    Z    W    L    H    -        |  |
--  -------------------------------------------------
  ( 'U', 'U', 'U', 'U', 'U', 'U', 'U', 'U', 'U' ), --  | U |
  ( 'U', 'X', 'X', 'X', 'X', 'X', 'X', 'X', 'X' ), --  | X |
  ( 'U', 'X', '0', 'X', '0', '0', '0', '0', 'X' ), --  | 0 |
  ( 'U', 'X', 'X', '1', '1', '1', '1', '1', 'X' ), --  | 1 |
  ( 'U', 'X', '0', '1', 'Z', 'W', 'L', 'H', 'X' ), --  | Z |
  ( 'U', 'X', '0', '1', 'W', 'W', 'W', 'W', 'X' ), --  | W |
  ( 'U', 'X', '0', '1', 'L', 'W', 'L', 'W', 'X' ), --  | L |
  ( 'U', 'X', '0', '1', 'H', 'W', 'W', 'H', 'X' ), --  | H |
  ( 'U', 'X', 'X', 'X', 'X', 'X', 'X', 'X', 'X' )  --  | - |
);
    FUNCTION resolved ( s : std_ulogic_vector ) RETURN
      std_ulogic IS
    VARIABLE result : std_ulogic := 'Z';   -- weakest state
      default
    BEGIN
        -- the test for a single driver is essential
        -- otherwise the loop would return 'X' for a
        -- single driver of '-' and that would conflict
        -- with the value of a single driver unresolved
        -- signal.
        IF    (s'LENGTH = 1) THEN    RETURN s(s'LOW);
        ELSE
            FOR i IN s'RANGE LOOP
                result := resolution_table(result, s(i));
            END LOOP;
        END IF;
        RETURN result;
    END resolved;


-------------------------------------------------------
-- tables for logical operations
-------------------------------------------------------
-- truth table for "and" function
CONSTANT and_table : stdlogic_table := (
--  -------------------------------------------------
--|U    X    0    1    Z    W    L    H    -          |  |
--  -------------------------------------------------
  ( 'U', 'U', '0', 'U', 'U', 'U', '0', 'U', 'U' ), --  | U |
  ( 'U', 'X', '0', 'X', 'X', 'X', '0', 'X', 'X' ), --  | X |
  ( '0', '0', '0', '0', '0', '0', '0', '0', '0' ), --  | 0 |
  ( 'U', 'X', '0', '1', 'X', 'X', '0', '1', 'X' ), --  | 1 |
  ( 'U', 'X', '0', 'X', 'X', 'X', '0', 'X', 'X' ), --  | Z |
  ( 'U', 'X', '0', 'X', 'X', 'X', '0', 'X', 'X' ), --  | W |
  ( '0', '0', '0', '0', '0', '0', '0', '0', '0' ), --  | L |
```

```
     ( 'U', 'X', '0', '1', 'X', 'X', '0', '1', 'X' ),   -- | H |
     ( 'U', 'X', '0', 'X', 'X', 'X', '0', 'X', 'X' )    -- | - |
);

-- truth table for "or" function
CONSTANT or_table : stdlogic_table := (
--  -------------------------------------------------------------
--  |U    X    0    1    Z    W    L    H    -             |    |
--  -------------------------------------------------------------
     ( 'U', 'U', 'U', '1', 'U', 'U', 'U', '1', 'U' ),   -- | U |
     ( 'U', 'X', 'X', '1', 'X', 'X', 'X', '1', 'X' ),   -- | X |
     ( 'U', 'X', '0', '1', 'X', 'X', '0', '1', 'X' ),   -- | 0 |
     ( '1', '1', '1', '1', '1', '1', '1', '1', '1' ),   -- | 1 |
     ( 'U', 'X', 'X', '1', 'X', 'X', 'X', '1', 'X' ),   -- | Z |
     ( 'U', 'X', 'X', '1', 'X', 'X', 'X', '1', 'X' ),   -- | W |
     ( 'U', 'X', '0', '1', 'X', 'X', '0', '1', 'X' ),   -- | L |
     ( '1', '1', '1', '1', '1', '1', '1', '1', '1' ),   -- | H |
     ( 'U', 'X', 'X', '1', 'X', 'X', 'X', '1', 'X' )    -- | - |
);

-- truth table for "xor" function
CONSTANT xor_table : stdlogic_table := (
--  -------------------------------------------------------------
--  |U    X    0    1    Z    W    L    H    -             |    |
--  -------------------------------------------------------------
     ( 'U', 'U', 'U', 'U', 'U', 'U', 'U', 'U', 'U' ),   -- | U |
     ( 'U', 'X', 'X', 'X', 'X', 'X', 'X', 'X', 'X' ),   -- | X |
     ( 'U', 'X', '0', '1', 'X', 'X', '0', '1', 'X' ),   -- | 0 |
     ( 'U', 'X', '1', '0', 'X', 'X', '1', '0', 'X' ),   -- | 1 |
     ( 'U', 'X', 'X', 'X', 'X', 'X', 'X', 'X', 'X' ),   -- | Z |
     ( 'U', 'X', 'X', 'X', 'X', 'X', 'X', 'X', 'X' ),   -- | W |
     ( 'U', 'X', '0', '1', 'X', 'X', '0', '1', 'X' ),   -- | L |
     ( 'U', 'X', '1', '0', 'X', 'X', '1', '0', 'X' ),   -- | H |
     ( 'U', 'X', 'X', 'X', 'X', 'X', 'X', 'X', 'X' )    -- | - |
);

-- truth table for "not" function
CONSTANT not_table: stdlogic_1d :=
--  -------------------------------------------------------------
--  |U    X    0    1    Z    W    L    H    -      |
--  -------------------------------------------------------------
     ( 'U', 'X', '1', '0', 'X', 'X', '1', '0', 'X' );


-- -----------------------------------------------------------------
-- overloaded logical operators ( with optimizing hints )
-- -----------------------------------------------------------------

    FUNCTION "and"  ( l : std_ulogic; r : std_ulogic )
      RETURN UX01 IS
    BEGIN
        RETURN (and_table(l, r));
    END "and";

    FUNCTION "nand" ( l : std_ulogic; r : std_ulogic )
      RETURN UX01 IS
    BEGIN
        RETURN  (not_table ( and_table(l, r)));
    END "nand";
```

```
    FUNCTION "or"     ( l : std_ulogic; r : std_ulogic )
      RETURN UX01 IS
    BEGIN
        RETURN (or_table(l, r));
    END "or";

    FUNCTION "nor"    ( l : std_ulogic; r : std_ulogic )
      RETURN UX01 IS
    BEGIN
        RETURN  (not_table ( or_table( l, r )));
    END "nor";

    FUNCTION "xor"    ( l : std_ulogic; r : std_ulogic )
      RETURN UX01 IS
    BEGIN
        RETURN (xor_table(l, r));
    END "xor";

--  function "xnor"   ( l : std_ulogic; r : std_ulogic )
--  return ux01 is
--  begin
--      return not_table(xor_table(l, r));
--  end "xnor";

    FUNCTION "not"   ( l : std_ulogic ) RETURN UX01 IS
    BEGIN
        RETURN (not_table(l));
    END "not";

    -----------------------------------------------------------
    -- and
    -----------------------------------------------------------
    FUNCTION "and"   ( l,r : std_logic_vector ) RETURN
      std_logic_vector IS
        ALIAS lv : std_logic_vector ( 1 TO l'LENGTH ) IS l;
        ALIAS rv : std_logic_vector ( 1 TO r'LENGTH ) IS r;
        VARIABLE result : std_logic_vector ( 1 TO
                          l'LENGTH );
    BEGIN
        IF ( l'LENGTH /= r'LENGTH ) THEN
            ASSERT FALSE
            REPORT "arguments of overloaded 'and' operator
              are not of the same length"
            SEVERITY FAILURE;
        ELSE
            FOR i IN result'RANGE LOOP
                result(i) := and_table (lv(i), rv(i));
            END LOOP;
        END IF;
        RETURN result;
    END "and";
    -----------------------------------------------------------
    FUNCTION "and"   ( l,r : std_ulogic_vector ) RETURN
      std_ulogic_vector IS
        ALIAS lv : std_ulogic_vector ( 1 TO l'LENGTH )
                   IS l;
        ALIAS rv : std_ulogic_vector ( 1 TO r'LENGTH )
                   IS r;
        VARIABLE result : std_ulogic_vector ( 1 TO
                          l'LENGTH );
```

```
   BEGIN
       IF ( l'LENGTH /= r'LENGTH ) THEN
           ASSERT FALSE
           REPORT "arguments of overloaded 'and' operator
               are not of the same length"
           SEVERITY FAILURE;
       ELSE
           FOR i IN result'RANGE LOOP
               result(i) := and_table (lv(i), rv(i));
           END LOOP;
       END IF;
       RETURN result;
   END "and";
   --------------------------------------------------------------
   -- nand
   --------------------------------------------------------------
   FUNCTION "nand"   ( l,r : std_logic_vector ) RETURN
     std_logic_vector IS
       ALIAS lv : std_logic_vector ( 1 TO l'LENGTH )
                   IS l;
       ALIAS rv : std_logic_vector ( 1 TO r'LENGTH )
                   IS r;
       VARIABLE result : std_logic_vector ( 1 TO
                           l'LENGTH );
   BEGIN
       IF ( l'LENGTH /= r'LENGTH ) THEN
           ASSERT FALSE
           REPORT "arguments of overloaded 'nand'
               operator are not of the same length"
           SEVERITY FAILURE;
       ELSE
           FOR i IN result'RANGE LOOP
               result(i) := not_table(and_table (lv(i),
                   rv(i)));
           END LOOP;
       END IF;
       RETURN result;
   END "nand";
   --------------------------------------------------------------
   FUNCTION "nand"   ( l,r : std_ulogic_vector ) RETURN
     std_ulogic_vector IS
       ALIAS lv : std_ulogic_vector ( 1 TO l'LENGTH )
                   IS l;
       ALIAS rv : std_ulogic_vector ( 1 TO r'LENGTH )
                   IS r;
       VARIABLE result : std_ulogic_vector ( 1 TO
                           l'LENGTH );
   BEGIN
       IF ( l'LENGTH /= r'LENGTH ) THEN
           ASSERT FALSE
           REPORT "arguments of overloaded 'nand'
               operator are not of the same length"
           SEVERITY FAILURE;
       ELSE
           FOR i IN result'RANGE LOOP
               result(i) := not_table(and_table (lv(i),
                   rv(i)));
           END LOOP;
       END IF;
       RETURN result;
```

```
END "nand";
----------------------------------------------------------
-- or
----------------------------------------------------------
FUNCTION "or"    ( l,r : std_logic_vector ) RETURN
  std_logic_vector IS
     ALIAS lv : std_logic_vector ( 1 TO l'LENGTH )
                 IS l;
     ALIAS rv : std_logic_vector ( 1 TO r'LENGTH )
                 IS r;
     VARIABLE result : std_logic_vector ( 1 TO
                         l'LENGTH );
BEGIN
     IF ( l'LENGTH /= r'LENGTH ) THEN
         ASSERT FALSE
         REPORT "arguments of overloaded 'or' operator
           are not of the same length"
         SEVERITY FAILURE;
     ELSE
         FOR i IN result'RANGE LOOP
             result(i) := or_table (lv(i), rv(i));
         END LOOP;
     END IF;
     RETURN result;
END "or";
----------------------------------------------------------
FUNCTION "or"    ( l,r : std_ulogic_vector ) RETURN
  std_ulogic_vector IS
     ALIAS lv : std_ulogic_vector ( 1 TO l'LENGTH )
                 IS l;
     ALIAS rv : std_ulogic_vector ( 1 TO r'LENGTH )
                 IS r;
     VARIABLE result : std_ulogic_vector ( 1 TO
                         l'LENGTH );
BEGIN
     IF ( l'LENGTH /= r'LENGTH ) THEN
         ASSERT FALSE
         REPORT "arguments of overloaded 'or' operator
           are not of the same length"
         SEVERITY FAILURE;
     ELSE
         FOR i IN result'RANGE LOOP
             result(i) := or_table (lv(i), rv(i));
         END LOOP;
     END IF;
     RETURN result;
END "or";
----------------------------------------------------------
-- nor
----------------------------------------------------------
FUNCTION "nor"    ( l,r : std_logic_vector ) RETURN
  std_logic_vector IS
     ALIAS lv : std_logic_vector ( 1 TO l'LENGTH )
                 IS l;
     ALIAS rv : std_logic_vector ( 1 TO r'LENGTH )
                 IS r;
     VARIABLE result : std_logic_vector ( 1 TO
                         l'LENGTH );
BEGIN
```

```
            IF ( l'LENGTH /= r'LENGTH ) THEN
                ASSERT FALSE
                REPORT "arguments of overloaded `nor' operator
                   are not of the same length"
                SEVERITY FAILURE;
            ELSE
                FOR i IN result'RANGE LOOP
                    result(i) := not_table(or_table (lv(i),
                    rv(i)));
                END LOOP;
            END IF;
            RETURN result;
    END "nor";
-------------------------------------------------------------
    FUNCTION "nor"  ( l,r : std_ulogic_vector ) RETURN
      std_ulogic_vector IS
        ALIAS lv : std_ulogic_vector ( 1 TO l'LENGTH )
                    IS l;
        ALIAS rv : std_ulogic_vector ( 1 TO r'LENGTH )
                    IS r;
        VARIABLE result : std_ulogic_vector ( 1 TO
                          l'LENGTH );
    BEGIN
        IF ( l'LENGTH /= r'LENGTH ) THEN
            ASSERT FALSE
            REPORT "arguments of overloaded `nor' operator
               are not of the same length"
            SEVERITY FAILURE;
        ELSE
            FOR i IN result'RANGE LOOP
                result(i) := not_table(or_table (lv(i),
                rv(i)));
            END LOOP;
        END IF;
        RETURN result;
    END "nor";
-------------------------------------------------------------
    -- xor
-------------------------------------------------------------
    FUNCTION "xor"  ( l,r : std_logic_vector ) RETURN
      std_logic_vector IS
        ALIAS lv : std_logic_vector ( 1 TO l'LENGTH )
                    IS l;
        ALIAS rv : std_logic_vector ( 1 TO r'LENGTH )
                    IS r;
        VARIABLE result : std_logic_vector ( 1 TO
                          l'LENGTH );
    BEGIN
        IF ( l'LENGTH /= r'LENGTH ) THEN
            ASSERT FALSE
            REPORT "arguments of overloaded `xor' operator
               are not of the same length"
            SEVERITY FAILURE;
        ELSE
            FOR i IN result'RANGE LOOP
                result(i) := xor_table (lv(i), rv(i));
            END LOOP;
        END IF;
        RETURN result;
```

```
                      END "xor";
                      ----------------------------------------------------
                      FUNCTION "xor"   ( l,r : std_ulogic_vector ) RETURN
                        std_ulogic_vector IS
                          ALIAS lv : std_ulogic_vector ( 1 TO l'LENGTH )
                                     IS l;
                          ALIAS rv : std_ulogic_vector ( 1 TO r'LENGTH )
                                     IS r;
                          VARIABLE result : std_ulogic_vector ( 1 TO
                                               l'LENGTH );
                      BEGIN
                          IF ( l'LENGTH /= r'LENGTH ) THEN
                              ASSERT FALSE
                              REPORT "arguments of overloaded 'xor' operator
                                are not of the same length"
                              SEVERITY FAILURE;
                          ELSE
                              FOR i IN result'RANGE LOOP
                                  result(i) := xor_table (lv(i), rv(i));
                              END LOOP;
                          END IF;
                          RETURN result;
                      END "xor";
--    ----------------------------------------------------------
--    -- xnor
--    ----------------------------------------------------------
--    ----------------------------------------------------------
--    Note : The declaration and implementation of the "xnor"
--    function is specifically commented until at which time
--    the VHDL language has been officially adopted as
--    containing such a function. At such a point, the
--    following comments may be removed along with this
--    notice without further "official" balloting of this
--    std_logic_1164 package. It is the intent of this effort
--    to provide such a function once it becomes available
--    in the VHDL standard.
--    ----------------------------------------------------------
--    function "xnor"   ( l,r : std_logic_vector ) return
--      std_logic_vector is
--          alias lv : std_logic_vector ( 1 to l'length )
--                     is l;
--          alias rv : std_logic_vector ( 1 to r'length )
--                     is r;
--          variable result : std_logic_vector ( 1 to
--                               l'length );
--    begin
--        if ( l'length /= r'length ) then
--            assert false
--            report "arguments of overloaded 'xnor'
--            operator are not of the same length"
--            severity failure;
--        else
--            for i in result'range loop
--                result(i) := not_table(xor_table (lv(i),
--                rv(i)));
--            end loop;
--        end if;
--        return result;
--    end "xnor";
```

```
--     ------------------------------------------------------
--     function "xnor"  ( l,r : std_ulogic_vector ) return
--         std_ulogic_vector is
--         alias lv : std_ulogic_vector ( 1 to l'length )
--                     is l;
--         alias rv : std_ulogic_vector ( 1 to r'length )
--                     is r;
--         variable result : std_ulogic_vector ( 1 to
--                             l'length );
--     begin
--         if ( l'length /= r'length ) then
--             assert false
--             report "arguments of overloaded `xnor'
--             operator are not of the same length"
--             severity failure;
--         else
--             for i in result'range loop
--                 result(i) := not_table(xor_table (lv(i),
--                 rv(i)));
--             end loop;
--         end if;
--         return result;
--     end "xnor";


    ------------------------------------------------------
    -- not
    ------------------------------------------------------
    FUNCTION "not"  ( l : std_logic_vector ) RETURN
        std_logic_vector IS
        ALIAS lv : std_logic_vector ( 1 TO l'LENGTH )
                    IS l;
        VARIABLE result : std_logic_vector ( 1 TO
                            l'LENGTH ) := (OTHERS => 'X');
    BEGIN
        FOR i IN result'RANGE LOOP
            result(i) := not_table( lv(i) );
        END LOOP;
        RETURN result;
    END;
    ------------------------------------------------------
    FUNCTION "not"  ( l : std_ulogic_vector ) RETURN
        std_ulogic_vector IS
        ALIAS lv : std_ulogic_vector ( 1 TO l'LENGTH )
                    IS l;
        VARIABLE result : std_ulogic_vector ( 1 TO
                            l'LENGTH ) := (OTHERS => 'X');
    BEGIN
        FOR i IN result'RANGE LOOP
            result(i) := not_table( lv(i) );
        END LOOP;
        RETURN result;
    END;
    ------------------------------------------------------
    -- conversion tables
    ------------------------------------------------------
    TYPE logic_x01_table IS ARRAY (std_ulogic'LOW TO
        std_ulogic'HIGH) OF X01;
    TYPE logic_x01z_table IS ARRAY (std_ulogic'LOW TO
```

```
    std_ulogic'HIGH) OF X01Z;
TYPE logic_ux01_table IS ARRAY (std_ulogic'LOW TO
  std_ulogic'HIGH) OF UX01;
-------------------------------------------------------
-- table name : cvt_to_x01
--
-- parameters :
--          in : std_ulogic   -- some logic value
-- returns    : x01           -- state value of logic
                                    value
-- purpose    : to convert state-strength to state
                 only
--
-- example    : if (cvt_to_x01 (input_signal) = `1' )
                 then ...
--
-------------------------------------------------------
CONSTANT cvt_to_x01 : logic_x01_table := (
                       `X',   -- `U'
                       `X',   -- `X'
                       `0',   -- `0'
                       `1',   -- `1'
                       `X',   -- `Z'
                       `X',   -- `W'
                       `0',   -- `L'
                       `1',   -- `H'
                       `X'    -- `-'
                       );
-------------------------------------------------------
-- table name : cvt_to_x01z
--
-- parameters :
--          in : std_ulogic   -- some logic value
-- returns    : x01z          -- state value of logic
                                    value
-- purpose    : to convert state-strength to state
                 only
--
-- example    : if (cvt_to_x01z (input_signal) = `1' )
                 then ...
--
-------------------------------------------------------
CONSTANT cvt_to_x01z : logic_x01z_table := (
                       `X',   -- `U'
                       `X',   -- `X'
                       `0',   -- `0'
                       `1',   -- `1'
                       `Z',   -- `Z'
                       `X',   -- `W'
                       `0',   -- `L'
                       `1',   -- `H'
                       `X'    -- `-'
                       );

-------------------------------------------------------
-- table name : cvt_to_ux01
--
-- parameters :
--          in : std_ulogic   -- some logic value
```

```
-- returns       : ux01           -- state value of logic
                                           value
-- purpose       : to convert state-strength to state
                   only
--
-- example       : if (cvt_to_ux01 (input_signal) = '1' )
                   then ...
--
------------------------------------------------------------
CONSTANT cvt_to_ux01 : logic_ux01_table := (
                         'U',   -- 'U'
                         'X',   -- 'X'
                         '0',   -- '0'
                         '1',   -- '1'
                         'X',   -- 'Z'
                         'X',   -- 'W'
                         '0',   -- 'L'
                         '1',   -- 'H'
                         'X'    -- '-'
                         );

------------------------------------------------------------
-- conversion functions
------------------------------------------------------------
FUNCTION To_bit          ( s : std_ulogic;        xmap :
 BIT := '0') RETURN BIT IS
BEGIN
        CASE s IS
            WHEN '0' | 'L' => RETURN ('0');
            WHEN '1' | 'H' => RETURN ('1');
            WHEN OTHERS => RETURN xmap;
        END CASE;
END;
------------------------------------------------------------
FUNCTION To_bitvector ( s : std_logic_vector ; xmap :
                        BIT := '0') RETURN BIT_VECTOR
                        IS
    ALIAS sv : std_logic_vector ( s'LENGTH-1 DOWNTO
            0 ) IS s;
    VARIABLE result : BIT_VECTOR ( s'LENGTH-1 DOWNTO
                    0 );
BEGIN
    FOR i IN result'RANGE LOOP
        CASE sv(i) IS
            WHEN '0' | 'L' => result(i) := '0';
            WHEN '1' | 'H' => result(i) := '1';
            WHEN OTHERS => result(i) := xmap;
        END CASE;
    END LOOP;
    RETURN result;
END;
------------------------------------------------------------
FUNCTION To_bitvector ( s : std_ulogic_vector; xmap :
                        BIT := '0') RETURN BIT_VECTOR
                        IS
    ALIAS sv : std_ulogic_vector ( s'LENGTH-1 DOWNTO
            0 ) IS s;
    VARIABLE result : BIT_VECTOR ( s'LENGTH-1 DOWNTO
                    0 );
BEGIN
```

```
        FOR i IN result'RANGE LOOP
            CASE sv(i) IS
                WHEN '0' | 'L' => result(i) := '0';
                WHEN '1' | 'H' => result(i) := '1';
                WHEN OTHERS => result(i) := xmap;
            END CASE;
        END LOOP;
        RETURN result;
    END;
---------------------------------------------------------------
    FUNCTION To_StdULogic        ( b : BIT                 )
                                    RETURN std_ulogic IS
    BEGIN
        CASE b IS
            WHEN '0' => RETURN '0';
            WHEN '1' => RETURN '1';
        END CASE;
    END;
---------------------------------------------------------------
    FUNCTION To_StdLogicVector  ( b : BIT_VECTOR          )
                                    RETURN std_logic_vector
                                    IS
        ALIAS bv : BIT_VECTOR ( b'LENGTH-1 DOWNTO 0 )
                    IS b;
        VARIABLE result : std_logic_vector ( b'LENGTH-1
                        DOWNTO 0 );
    BEGIN
        FOR i IN result'RANGE LOOP
            CASE bv(i) IS
                WHEN '0' => result(i) := '0';
                WHEN '1' => result(i) := '1';
            END CASE;
        END LOOP;
        RETURN result;
    END;
---------------------------------------------------------------
    FUNCTION To_StdLogicVector  ( s : std_ulogic_vector )
                                    RETURN std_logic_vector
                                    IS
        ALIAS sv : std_ulogic_vector ( s'LENGTH-1 DOWNTO
                    0 ) IS s;
        VARIABLE result : std_logic_vector ( s'LENGTH-1
                        DOWNTO 0 );
    BEGIN
        FOR i IN result'RANGE LOOP
            result(i) := sv(i);
        END LOOP;
        RETURN result;
    END;
---------------------------------------------------------------
    FUNCTION To_StdULogicVector ( b : BIT_VECTOR          )
                                    RETURN std_ulogic_vector
                                    IS
        ALIAS bv : BIT_VECTOR ( b'LENGTH-1 DOWNTO 0 )
                    IS b;
        VARIABLE result : std_ulogic_vector ( b'LENGTH-1
                        DOWNTO 0 );
    BEGIN
        FOR i IN result'RANGE LOOP
            CASE bv(i) IS
```

```
                        WHEN '0' => result(i) := '0';
                        WHEN '1' => result(i) := '1';
                END CASE;
            END LOOP;
            RETURN result;
        END;
        ----------------------------------------------------------
        FUNCTION To_StdULogicVector ( s : std_logic_vector )
          RETURN std_ulogic_vector IS
            ALIAS sv : std_logic_vector ( s'LENGTH-1 DOWNTO 0
              ) IS s;
            VARIABLE result : std_ulogic_vector ( s'LENGTH-1
              DOWNTO 0 );
        BEGIN
            FOR i IN result'RANGE LOOP
                result(i) := sv(i);
            END LOOP;
            RETURN result;
        END;


        ----------------------------------------------------------
        -- strength strippers and type convertors
        ----------------------------------------------------------
        -- to_x01
        ----------------------------------------------------------
        FUNCTION To_X01  ( s : std_logic_vector ) RETURN
          std_logic_vector IS
            ALIAS sv : std_logic_vector ( 1 TO s'LENGTH ) IS s;
            VARIABLE result : std_logic_vector ( 1 TO s'LENGTH
          );
        BEGIN
            FOR i IN result'RANGE LOOP
                result(i) := cvt_to_x01 (sv(i));
            END LOOP;
            RETURN result;
        END;
        ----------------------------------------------------------
        FUNCTION To_X01  ( s : std_ulogic_vector ) RETURN
          std_ulogic_vector IS
            ALIAS sv : std_ulogic_vector ( 1 TO s'LENGTH ) IS
              s;
            VARIABLE result : std_ulogic_vector ( 1 TO
              s'LENGTH );
        BEGIN
            FOR i IN result'RANGE LOOP
                result(i) := cvt_to_x01 (sv(i));
            END LOOP;
            RETURN result;
        END;
        ----------------------------------------------------------
        FUNCTION To_X01  ( s : std_ulogic ) RETURN  X01 IS
        BEGIN
            RETURN (cvt_to_x01(s));
        END;
        ----------------------------------------------------------
        FUNCTION To_X01  ( b : BIT_VECTOR ) RETURN
          std_logic_vector IS
            ALIAS bv : BIT_VECTOR ( 1 TO b'LENGTH ) IS b;
            VARIABLE result : std_logic_vector ( 1 TO b'LENGTH
              );
```

```vhdl
BEGIN
    FOR i IN result'RANGE LOOP
        CASE bv(i) IS
            WHEN '0' => result(i) := '0';
            WHEN '1' => result(i) := '1';
        END CASE;
    END LOOP;
    RETURN result;
END;
----------------------------------------------------------
FUNCTION To_X01  ( b : BIT_VECTOR ) RETURN
  std_ulogic_vector IS
    ALIAS bv : BIT_VECTOR ( 1 TO b'LENGTH ) IS b;
    VARIABLE result : std_ulogic_vector ( 1 TO b'LENGTH
  );
BEGIN
    FOR i IN result'RANGE LOOP
        CASE bv(i) IS
            WHEN '0' => result(i) := '0';
            WHEN '1' => result(i) := '1';
        END CASE;
    END LOOP;
    RETURN result;
END;
----------------------------------------------------------
FUNCTION To_X01  ( b : BIT ) RETURN  X01 IS
BEGIN
        CASE b IS
            WHEN '0' => RETURN('0');
            WHEN '1' => RETURN('1');
        END CASE;
END;
----------------------------------------------------------
-- to_x01z
----------------------------------------------------------
FUNCTION To_X01Z  ( s : std_logic_vector ) RETURN
  std_logic_vector IS
    ALIAS sv : std_logic_vector ( 1 TO s'LENGTH ) IS s;
    VARIABLE result : std_logic_vector ( 1 TO s'LENGTH
  );
BEGIN
    FOR i IN result'RANGE LOOP
        result(i) := cvt_to_x01z (sv(i));
    END LOOP;
    RETURN result;
END;
----------------------------------------------------------
FUNCTION To_X01Z  ( s : std_ulogic_vector ) RETURN
  std_ulogic_vector IS
    ALIAS sv : std_ulogic_vector ( 1 TO s'LENGTH ) IS
      s;
    VARIABLE result : std_ulogic_vector ( 1 TO s'LENGTH
      );
BEGIN
    FOR i IN result'RANGE LOOP
        result(i) := cvt_to_x01z (sv(i));
    END LOOP;
    RETURN result;
END;
----------------------------------------------------------
```

```
FUNCTION To_X01Z   ( s : std_ulogic ) RETURN   X01Z IS
BEGIN
    RETURN (cvt_to_x01z(s));
END;
-----------------------------------------------------------
FUNCTION To_X01Z   ( b : BIT_VECTOR ) RETURN
   std_logic_vector IS
    ALIAS bv : BIT_VECTOR ( 1 TO b'LENGTH ) IS b;
    VARIABLE result : std_logic_vector ( 1 TO b'LENGTH
   );
BEGIN
    FOR i IN result'RANGE LOOP
        CASE bv(i) IS
            WHEN '0' => result(i) := '0';
            WHEN '1' => result(i) := '1';
        END CASE;
    END LOOP;
    RETURN result;
END;
-----------------------------------------------------------
FUNCTION To_X01Z   ( b : BIT_VECTOR ) RETURN
   std_ulogic_vector IS
    ALIAS bv : BIT_VECTOR ( 1 TO b'LENGTH ) IS b;
    VARIABLE result : std_ulogic_vector ( 1 TO b'LENGTH
   );
BEGIN
    FOR i IN result'RANGE LOOP
        CASE bv(i) IS
            WHEN '0' => result(i) := '0';
            WHEN '1' => result(i) := '1';
        END CASE;
    END LOOP;
    RETURN result;
END;
-----------------------------------------------------------
FUNCTION To_X01Z   ( b : BIT ) RETURN   X01Z IS
BEGIN
        CASE b IS
            WHEN '0' => RETURN('0');
            WHEN '1' => RETURN('1');
        END CASE;
END;
-----------------------------------------------------------
-- to_ux01
-----------------------------------------------------------
FUNCTION To_UX01   ( s : std_logic_vector ) RETURN
   std_logic_vector IS
    ALIAS sv : std_logic_vector ( 1 TO s'LENGTH ) IS s;
    VARIABLE result : std_logic_vector ( 1 TO s'LENGTH
     );
BEGIN
    FOR i IN result'RANGE LOOP
        result(i) := cvt_to_ux01 (sv(i));
    END LOOP;
    RETURN result;
END;
-----------------------------------------------------------
FUNCTION To_UX01   ( s : std_ulogic_vector ) RETURN
   std_ulogic_vector IS
```

```vhdl
    ALIAS sv : std_ulogic_vector ( 1 TO s'LENGTH ) IS
        s;
    VARIABLE result : std_ulogic_vector ( 1 TO s'LENGTH
        );
BEGIN
    FOR i IN result'RANGE LOOP
        result(i) := cvt_to_ux01 (sv(i));
    END LOOP;
    RETURN result;
END;
```

```vhdl
FUNCTION To_UX01   ( s : std_ulogic ) RETURN   UX01 IS
BEGIN
    RETURN (cvt_to_ux01(s));
END;
```

```vhdl
FUNCTION To_UX01   ( b : BIT_VECTOR ) RETURN
   std_logic_vector IS
    ALIAS bv : BIT_VECTOR ( 1 TO b'LENGTH ) IS b;
    VARIABLE result : std_logic_vector ( 1 TO b'LENGTH
        );
BEGIN
    FOR i IN result'RANGE LOOP
        CASE bv(i) IS
            WHEN '0' => result(i) := '0';
            WHEN '1' => result(i) := '1';
        END CASE;
    END LOOP;
    RETURN result;
END;
```

```vhdl
FUNCTION To_UX01   ( b : BIT_VECTOR ) RETURN
   std_ulogic_vector IS
    ALIAS bv : BIT_VECTOR ( 1 TO b'LENGTH ) IS b;
    VARIABLE result : std_ulogic_vector ( 1 TO b'LENGTH
        );
BEGIN
    FOR i IN result'RANGE LOOP
        CASE bv(i) IS
            WHEN '0' => result(i) := '0';
            WHEN '1' => result(i) := '1';
        END CASE;
    END LOOP;
    RETURN result;
END;
```

```vhdl
FUNCTION To_UX01   ( b : BIT ) RETURN   UX01 IS
BEGIN
        CASE b IS
            WHEN '0' => RETURN('0');
            WHEN '1' => RETURN('1');
        END CASE;
END;
```

```vhdl
-- edge detection
```

```
FUNCTION rising_edge   (SIGNAL s : std_ulogic) RETURN
   BOOLEAN IS
BEGIN
    RETURN (s'EVENT AND (To_X01(s) = '1') AND
                        (To_X01(s'LAST_VALUE) = '0'));
END;


FUNCTION falling_edge (SIGNAL s : std_ulogic) RETURN
   BOOLEAN IS
BEGIN
    RETURN (s'EVENT AND (To_X01(s) = '0') AND
                        (To_X01(s'LAST_VALUE) = '1'));
END;


-------------------------------------------------------
-- object contains an unknown
-------------------------------------------------------
FUNCTION Is_X ( s : std_ulogic_vector ) RETURN  BOOLEAN
   IS
BEGIN
    FOR i IN s'RANGE LOOP
        CASE s(i) IS
            WHEN 'U' | 'X' | 'Z' | 'W' | '-' => RETURN
               TRUE;
            WHEN OTHERS => NULL;
        END CASE;
    END LOOP;
    RETURN FALSE;
END;
-------------------------------------------------------
FUNCTION Is_X ( s : std_logic_vector  ) RETURN  BOOLEAN
   IS
BEGIN
    FOR i IN s'RANGE LOOP
        CASE s(i) IS
            WHEN 'U' | 'X' | 'Z' | 'W' | '-' => RETURN
               TRUE;
            WHEN OTHERS => NULL;
        END CASE;
    END LOOP;
    RETURN FALSE;
END;
-------------------------------------------------------
FUNCTION Is_X ( s : std_ulogic        ) RETURN  BOOLEAN
   IS
BEGIN
    CASE s IS
        WHEN 'U' | 'X' | 'Z' | 'W' | '-' => RETURN
           TRUE;
        WHEN OTHERS => NULL;
    END CASE;
    RETURN FALSE;
END;

END std_logic_1164;
```

Appendix B

VHDL Reference Tables

This appendix focuses on tables of information that are useful when writing VHDL descriptions. Most of the information in the tables is available in the text of the book, however, these tables consolidate the information into one area for easy reference.

Table B-1 lists all of the different kinds of statements alphabetically and includes an example usage.

Table B-1

Statement or Clause	Example(s)
Access Type	TYPE access_type IS ACCESS type_to_be_accessed;
Aggregate	record_type := (first, second, third);
Alias	ALIAS opcode : BIT_VECTOR(0 TO 3) IS INSTRUCTION(10TO 13);

Architecture	ARCHITECTURE architecture_name OF entity
	    name IS
	    -- declare some signals here
	BEGIN
	    -- put some concurrent statements here
	END architecture_name;

Array Type	TYPE array_type IS ARRAY (0 TO 7) OF BIT;
Assert	ASSERT x > 10 REPORT "x is too small" SEVERITY ERROR;
Attribute Declaration	ATTRIBUTE attribute_name : attribute_type;
Attribute Specification	ATTRIBUTE attribute_name OF entity_name : entity_class IS value;

Block Statement	block_name : BLOCK
	    -- declare some stuff here
	BEGIN
	    - put some concurrent statements here
	END BLOCK block_name;

Case Statement	CASE some_expression IS
	    WHEN some_value =>
	        -- do_some_stuff
	    WHEN some_other_value =>
	        -- do_some_other_stuff
	    WHEN OTHERS =>
	        -- do_some_default_stuff
	END CASE;

Table B-1

Continued.

Statement or Clause	Example(s)

Component Declaration	COMPONENT component_name
	  PORT(port1_name : port1_type;
	            port2_name : port2_type;
	            port3_name : port3_type);
	END COMPONENT;

Component Instantiation	instance_name : component_name PORT MAP
	  (first_port, second_port, third_port);
	instance_name : component_name PORT MAP
	  (formal1 => actual1, formal2 => actual2);

Conditional Signal Assignment	target <= first_value WHEN (x = y) ELSE
	            second_value WHEN a >= b ELSE
	            third_value;

Configuration Declaration	CONFIGURATION configuration_name OF
	  entity_name IS
	  FOR architecture_name
	    FOR instance_name : entity_name USE
	    ENTITY
	      library_name.entity_name
	      (architecture_name);
	    END FOR;
	    FOR instance_name : entity_name USE
	    CONFIGURATION
	      library_name.configuration_name;
	    END FOR;
	  END FOR;
	END configuration_name;

Constant Declaration	CONSTANT constant_name : constant_type := value;

Entity Declaration	ENTITY entity_name IS
	  PORT(port1 : port1_type;
	      port2 : port2_type);
	END entity_name;

Exit Statement	EXIT; EXIT WHEN a <= b; EXIT loop_label WHEN x = z;
File Type Declaration	TYPE file_type_name IS FILE OF data_type;
File Object Declaration	FILE file_object_name : file_type_name IS IN "/absolute/path/name";
For Loop	FOR loop_variable IN start TO end LOOP -- do_some_stuff END LOOP;

Table B-1

Statement or Clause	Example(s)
Function Declaration	FUNCTION function_name (parameter1 : parameter1_type; parameter2 : parameter2_type) RETURN return_type;
Function Body	FUNCTION function_name(parameter1 : parameter1_type; parameter2 : parameter2_type) RETURN return_type IS BEGIN -- do some stuff END function_name;
Generate Statement	generate_label : FOR gen_var IN start TO end GENERATE label : component_name PORT MAP(.........); END GENERATE;
Generic Declaration	GENERIC (generic1_name : generic1_type; generic2_name : generic2_type);
Generic Map	GENERIC MAP(generic1_name => value1, value2);

Guarded Signal Assignment	g1 : BLOCK(clk = '1' AND clk'EVENT)
	BEGIN
	  q <= GUARDED d AFTER 5 NS;
	END BLOCK;

IF Statement	IF x <= y THEN
	  -- some statements
	END IF;
	
	IF z > w THEN
	  -- some statements
	ELSIF q < r THEN
	  -- some more statements
	END IF;
	
	IF a = b THEN
	  -- some statements
	ELSIF c = d THEN
	  -- some more statements
	ELSE
	  -- even more statements
	END IF;

Incomplete Type	TYPE type_name;

Table B-1

Continued.

Statement or Clause	Example(s)
Library Declaration	LIBRARY library_name;

Loop Statement	FOR loop_variable IN start TO end LOOP
	  -- do lots of stuff
	END LOOP;
	
	WHILE x < y LOOP
	  -- modify x and y and do other stuff
	END LOOP;

Next Statement	IF i < 0 THEN
	  NEXT;
	END IF;

Others Clause	WHEN OTHERS => -- do some stuff
Package Declaration	PACKAGE package_name IS -- declare some stuff END PACKAGE;
Package Body	PACKAGE BODY package_name IS --put subprogram bodies here END package_name;

Physical Type	TYPE physical_type_name IS RANGE start TO end
	  UNITS
	    unit1 ;
	    unit2 = 10 unit1;
	    unit3 = 10 unit2;
	  END UNITS;

Port Clause	PORT(port1_name : port1_type; port2_name : port2_type);
Port Map Clause	PORT MAP (port1_name => signal1, signal2);

Procedure Declaration	PROCEDURE procedure_name(parm1 : in
	                  parm1_type; parm2 : out
	                  parm2_type; parm3 :
	                  inout parm3_type);

Procedure Body	PROCEDURE procedure_name(parm1 : in
	                  parm1_type; parm2 : out
	                  parm1_type; parm3 :
	                  inout parm3_type) IS
	BEGIN
	  -- do some stuff
	END procedure_name;

Table B-1

Statement or Clause	Example(s)
Process Statement	PROCESS(signal1, signal2, signal3) -- declare some stuff BEGIN -- do some stuff END PROCESS;

Record Type	TYPE record_type IS
	    RECORD
	        field1 : field1_type;
	        field2 : field2_type;
	    END RECORD;

Report Clause	ASSERT x = 10 REPORT "some string";
Return Statement	RETURN; RETURN (x + 10);

Selected Signal Assignment	WITH z SELECT
	    x <= 1 AFTER 5 NS WHEN 0,
	            2 AFTER 5 NS WHEN 1,
	            3 AFTER 5 NS WHEN OTHERS;

Severity Clause	ASSERT x > 5 REPORT "some string" SEVERITY ERROR;
Signal Assignment	a <= b AFTER 20 NS;
Signal Declaration	SIGNAL x : xtype;
Subtype Declaration	SUBTYPE bit8 IS INTEGER RANGE 0 TO 255;
Transport Signal Assignment	x <= TRANSPORT y AFTER 50 NS;
Type Declaration	TYPE color is (red, yellow, blue, green, orange); TYPE small_int is 0 to 65535;
Use Clause	USE WORK.my_package.all;
Variable Declaration	VARIABLE variable_name : variable_type;Wait Statement WAIT ON a, b, c; WAIT UNTIL clock'EVENT AND clock = '1'; WAIT FOR 100 NS; WAIT ON a, b UNTIL b > 10 FOR 50 NS;
While Loop	WHILE x > 15 LOOP -- do some stuff END LOOP;

Table B-2 lists all of the predefined attributes that retrieve information about VHDL type data. The descriptions are necessarily terse to fit into the table cells; see Chapter 6, "Predefined Attributes" for more detailed information.

Table B-2

Attribute	Explanation	Examples
`T'BASE`	Returns the base type of datatype it is attached to	`NATURAL'BASE returns INTEGER`
`T'LEFT`	Returns left value specified in type declaration	`INTEGER'LEFT is -2147483647` `BIT'LEFT is '0'`
`T'RIGHT`	Returns right value specified in type declaration	`INTEGER'RIGHT is 2147483647` `BIT'RIGHT is '1'`
`T'HIGH`	Returns largest value specified in declaration	`TYPE bit8 is 255 downto 0` `bit8'HIGH is 255`
`T'LOW`	Returns smallest value specified in declaration	`TYPE bit8 is 255 downto 0` `bit8'LOW is 0`
`T'POS(X)`	Returns position number of argument in type(first position is 0)	`TYPE color IS (red, green, blue, orange);` `color'POS(green) is 1`
`T'VAL(X)`	Returns value in type at specified position number	`TYPE color IS (red, green, blue, orange);` `color'VAL(2) is blue`
`T'SUCC(X)`	Returns the successor to the value passed in	`TYPE color IS (red, green, blue, orange);` `color'SUCC(green) is blue`
`T'PRED(X)`	Returns the predecessor to the value passed in	`TYPE color IS (red, green, blue, orange);` `color'PRED(blue) is green`
`T'LEFTOF(X)`	Returns the value to the left of the value passed in	`TYPE color IS (red, green, blue, orange);` `color'LEFTOF(green) is red`
`T'RIGHTOF(X)`	Returns the value to the right of the value passed in	`TYPE color IS (red, green, blue, orange);` `color'RIGHTOF(blue) is orange`

Table B-3 lists all predefined attributes that return information about array datatypes. The **N** parameter for all attributes specifies to which particular range the attribute is being applied. This only makes sense for multidimensional arrays. For single dimensional arrays, the parameter can be ignored. For more detailed information, see Chapter 6, "Predefined Attributes."

Table B-3

Attribute	Explanation	Example
A'LEFT(N)	Returns left array bound of selected index range	a_type'LEFT(1) is 0 a_type'LEFT(2) is 7
A'RIGHT(N)	Returns right array bound of selected index range	a_type'RIGHT(1) is 3 a_type'RIGHT(2) is 0
A'HIGH(N)	Returns largest array bound value of selected index range	a_type'HIGH(1) is 3 a_type'HIGH(2) is 7
A'LOW(N)	Returns smallest array bound value of selected index range	a_type'LOW(1) is 0 a_type'LOW(2) is 0
A'RANGE(N)	Returns selected index range	a_type'RANGE(1) is 0 TO 3 a_type'RANGE(2) is 7 DOWNTO 0
A'REVERSE_RANGE(N)	Returns selected index range reversed	a_type'REVERSE_RANGE(1) is 3 DOWNTO 0 a_type'REVERSE_RANGE(2) is 0 TO 7
A'LENGTH(N)	Returns size of selected index range	a_type'LENGTH(1) is 4 a_type'LENGTH(2) is 8

All of the next examples apply to the following declaration:

```
TYPE a_type IS ARRAY(0 TO 3, 7 DOWNTO 0) OF BIT;
```

Table B-4 lists all predefined attributes that return information about signals or create new signals. For more detailed information, see Chapter 6, "Predefined Attributes."

Table B-5 lists all of the operators and their relative precedence.

Table B-6 lists all of the different types of literals and a sample usage. In all cases, the _ character is ignored when interpreting the value of a literal. The base that the exponent in the based integer and based real examples is applied to is the base specified for interpreting the number. Bit String literals are used to specify values for types that resemble the **BIT_VECTOR** type.

Table B-4

Attribute	Explanation	Example
S'DELAYED(T)	Creates a new signal delayed by T	clock'DELAYED(10 ns)
S'QUIET(T)	Creates a new signal that is true when signal **s** has had no transactions for time **T**; otherwise, false	reset'QUIET(5 ns)
S'STABLE(T)	Creates a new signal that is true when signal **s** has had no events for time **T**; otherwise, false	clock'STABLE(1 ns)
S'TRANSACTION	Creates a signal of type **BIT** that toggles for every transaction on signal **s**	load'TRANSACTION
S'EVENT	Returns true when an event has occurred for signal **s** this delta	clock'EVENT
S'ACTIVE	Returns true when a transaction has occurred for signal **s** this delta	load'ACTIVE
S'LAST_EVENT	Returns the elapsed time since the last event on signal **s**	data'LAST_EVENT
S'LAST_ACTIVE	Returns the elapsed time since the last transaction on signal **s**	clock'LAST_ACTIVE
S'LAST_VALUE	Returns the previously assigned value of signal **s**	data'LAST_VALUE

Table B-5

Precedence	Operator Class	Operator
Highest	Miscellaneous	**, ABS, NOT
	Multiplying	*, /, MOD, REM
	Sign	+, -
	Adding	+, -, &
	Relational	=, /=, <, <=, >, >=
Lowest	Logical	AND, OR, NAND, NOR, XOR

Table B-6

Literal Type	Example
Decimal Integer	52
	0
	3E3 -- equals 3000
	1_000_000 -- equals 1 million
Decimal Real	52.0
	0.0
	.178
	1.222_333
Decimal Real with Exponent	1.2E+10
	4.6E-9
Based Integer	16#FF# -- equals 255
	8#777# -- equals 511
	2#1101_0101# -- equals 213
	16#FF#E1 -- equals 4080
Based Real	2#11.11#
	16#AB.CD#E+2
	8#77.66#E-10
Character	'a'
	'*'
	' ' -- the space character
String	"this is a string"
	" " -- empty string
	"ABC" & "CDE" -- concatenation
Bit String	X"FFEF"
	O"770770"
	B"1111_0000_1111"

Appendix C

Reading VHDL BNF

After the basic concepts of VHDL are understood, the designer might want to try to write VHDL in a more elegant manner. To fully understand how to apply all of the syntactic constructs available in VHDL, it is helpful to know how to read the VHDL Bachus-Naur format (BNF) of the language. This format is in Appendix A of the IEEE Std 1076-1987 VHDL Language Reference Manual (LRM), pages A−1 to A−17. This format specifies which constructs are necessary versus optional, or repeatable versus singular, and how constructs can be associated.

BNF is basically a hierarchical description method, where complex constructs are made of successive specifications of lower-level constructs. Our purpose for examining BNF is not to understand every nuance of the BNF but to put the basics to use to help build complex VHDL constructs. To this end, let us examine some BNF and discuss what it means.

Following is the BNF for the **IF** statement:

```
if_statement ::=
  IF condition THEN
    sequence_of_statements
  {ELSIF condition THEN
    sequence_of_statements}
  [ELSE
    sequence_of_statements]
  END IF;
```

The first line of the BNF description specifies the name of the construct being described. This line is read as follows: "The **IF** statement consists of," or "The **IF** statement is constructed from." The rest of the description represents the rules for constructing an **IF** statement.

The second line of the description specifies that the **IF** statement starts with the keyword **IF**, is followed by a condition construct, and ends the clause with the keyword **THEN**. The next line contains the construct **SEQUENCE_OF_STATEMENTS** (which is discussed later in this appendix). All of the constructs discussed so far are required for the **IF** statement because the constructs are not enclosed in any kind of punctuation.

Statements enclosed in brackets [ ], as in lines 6 and 7, are optional constructs. An optional construct can be specified or left out depending on the functionality required. The **ELSE** clause of the **IF** statement is an example of an optional construct. A legal **IF** statement may or may not have an **ELSE** clause.

Statements enclosed in curly braces { }, as in lines 4 and 5, are optional and repeatable constructs. An optional and repeatable construct can either be left out or have one or more of the construct exist. The **ELSIF** clause is an example of an optional and repeatable construct. The **IF** statement can be constructed without an **ELSIF** clause, or have one or more **ELSIF** clauses, depending on the desired behavior.

The last line of the **IF_STATEMENT** description contains the **END IF** clause. This is a required clause because it is not optional [ ] and is not optional and repeatable { }.

The **IF** statement contains two other constructs that need more description: the **SEQUENCE_OF_STATEMENTS** and the **CONDITION**. The **SEQUENCE_OF_STATEMENTS** construct is described by the BNF shown here:

```
sequence_of_statements ::=
  {sequential_statement}
```

The **SEQUENCE_OF_STATEMENTS** construct is described by one or more sequential statements, where a sequential statement is described in the following:

```
sequential_statement ::=
  wait_statement
  | assertion_statement
  | signal_assignment_statement
  | variable_assignment_statement
  | procedure_call_statement
  | if_statement
  | case_statement
  | loop_statement
  | next_statement
  | exit_statement
  | return_statement
  | null_statement
```

The | character means **OR**, such that a sequential statement can be a **WAIT** statement, or an **ASSERT** statement, or a **SIGNAL ASSIGNMENT** statement, and so on. From this description, we can see that the statement part of the **IF** statement can contain one or more sequential statements, such as **WAIT** statements, **ASSERT** statements, and so on.

The **CONDITION** construct is specified with the BNF description shown here:

```
condition ::= boolean_expression
```

Notice that the keyword **boolean** is italic. The italic indicates the type of the expression required for the **CONDITION**. If a designer looks for a boolean expression construct to describe the syntax required, none will be found. The reason is that all expressions share the same syntax description. For our purposes, the boolean type of the expression is ignored, and the construct description can be found under the following description:

```
expression ::=
  relation {and relation}
  |relation {or relation}
  |relation {xor relation}
  | relation [nand relation]
  |relation [nor relation]
```

To summarize, curly braces { } are optional and repeatable constructs, square brackets [ ] are optional constructs, and italic pieces of a construct can be ignored for purposes of finding descriptions.

Appendix D

VHDL93 Updates

Early in 1993 the VHDL Language Standard was updated to reflect a number of shortcomings with the VHDL 1076-1987 standard and to add some new features to the language. This new standard is called VHDL 1076-1993. In this appendix the 1987 standard will be referred to as VHDL87 and the 1993 standard as VHDL93.

The goal of this appendix is not to give the user a complete description of every new or changed feature, but to give the reader an idea of the scope of these changes and what effect they will have on future VHDL modeling efforts.

The goal of the update was to remain compatible with VHDL87 so that VHDL87 models would work in a VHDL93 environment. This goal was not entirely achieved as some of the new features were no longer compatible. The main reason for the incompatibility was the use of new keywords in VHDL93, that may have been used as identifiers in VHDL87, and a major update of TEXTIO.

The rest of this appendix includes discussions of the VHDL87 features that have either been added or changed. They are listed in alphabetical order for easier access.

Alias

The **alias** clause has been generalized for VHDL93. In VHDL87 an **alias** was used to give an alternate name to an object (See Chapter 8, "Advanced Topics"). In VHDL93 the **alias** construct has been generalized to allow aliasing not only types but functions and procedures as well.

A typical **alias** in VHDL87 would look as follows:

```
ALIAS opcode : BIT_VECTOR( 3 DOWNTO 0) IS instruction(31
    DOWNTO 28);
```

Notice that the type of the opcode needed to be specified **(BIT_VECTOR(3 DOWNTO 0)**. In VHDL93 the type is now optional. This same **alias** can be written in VHDL93 as follows:

```
ALIAS opcode IS instruction(27 DOWNTO 22);
```

Not only can objects be aliased in VHDL93 but functions as well. To specify a function **alias** requires a subprogram signature specification. The signature specifies the types of the input parameters as well as the type of the return parameter. An example is shown here:

```
ALIAS sub IS "-" [STD_LOGIC_VECTOR, STD_LOGIC_VECTOR,
     RETURN STD_LOGIC_VECTOR];
```

This statement creates an **alias** called **sub** for an overloaded operator function call that has two **std_logic_vector** input arguments and returns a **std_logic_vector**.

Attribute Changes

There have been a number of new attributes added to VHDL93. They reflect added functionality that was either difficult in VHDL87 or not possible. The following attributes have been added to VHDL93:

```
`ASCENDING
`DRIVING_VALUE
`IMAGE
`VALUE
`PATHNAME
`INSTANCE_NAME
`SIMPLE_NAME
```

`ASCENDING**   In VHDL87 it was tedious to find if a particular range was ascending or descending. The `high** and `**low** attributes of the type had to be compared to determine if the range was truly ascending, a null range, or a single value. Attribute `**ascending** will return true if the range is ascending or false if not. An example is shown here:

```
SUBTYPE descend IS STD_LOGIC_VECTOR( 7 DOWNTO 0);
SUBTYPE ascend IS STD_LOGIC_VECTOR(0 TO 7);

descend`ASCENDING --> false
ascend`ASCENDING  --> true
```

`DRIVING_VALUE   In VHDL87 the value of an output port could not be read. To do this required the port mode of the port to be inout, or the use of an internal signal. These workarounds caused an increase in complexity that typically was not warranted and therefore to get around this inconvenience VHDL93 adds attribute `**driving_value**. Attribute `**driving_value** allows the ability to read the value component of the resolved value that a particular driver is driving so that it can be further used in the model. In the example shown here the second component instantiation statement would cause an error because input port a of U2 is trying to read the current value of **dout**. In VHDL93 the `` `driving_value `` attribute gets around this problem by reading the driving value of **dout**.

```
ENTITY invert IS
  PORT( w: IN STD_LOGIC;
                dout, doutb : OUT STD_LOGIC);
END invert;
ARCHITECTURE struct OF invert IS
  COMPONENT inv
    PORT( a : IN STD_LOGIC;
                q : OUT STD_LOGIC);
  END COMPONENT;
BEGIN
  u1 : inv PORT MAP(a => w, q => dout);
--u2 : inv PORT MAP(a => dout, q => doutb);
-- won't work because port
-- dout cannot be read

  u2 : inv PORT MAP(a => dout`DRIVING_VALUE, q => doutb);
     -- In VHDL93 this
                                        -- will work
END struct;
```

`` `IMAGE AND `VALUE `` In VHDL87 it was difficult for an error message to display the actual error value of a signal or a variable in a string. In VHDL93 the attributes `` `IMAGE `` and `` `VALUE `` allow the modeler to convert to and from type values into string values. Attribute `` `IMAGE `` converts a type value into a string, and attribute `` `VALUE `` converts a string to a type value.

`` `PATHNAME, `INSTANCE_NAME, AND `SIMPLE_NAME `` The other difficulty in VHDL87 of model error reporting was to uniquely determine exactly which instance of a model was generating a message. Most VHDL simulators had some mechanism of reporting the instance information to the modeler, but this information was simulator specific and not standard. In VHDL93 three new attributes allow the modeler access to all parts of the pathname that describes which instance a particular message is generated from.

▨ `` `SIMPLE_NAME ``—returns a string which is the local name of the calling entity.

▨ `` `PATH_NAME ``—returns a string that describes the path to the entity starting at the root of the design. The `` `PATH_NAME `` attribute does not include the names of instantiated entities (`` `INSTANCE_NAME `` does)

■ `` `INSTANCE_NAME ``—returns a string that describes the path to the entity starting at the root of the design. The `` `INSTANCE_NAME `` attribute also includes the names of instantiated entities. These entities are specified using a **label@entity(architecture)** syntax.

Bit String Literal

Bit string literals are a handy way in VHDL87 to assign **bit_vector** values. For instance instead of having to explicitly enumerate each bit value when assigning to a **bit_vector** an octal or hexadecimal notation can be used as shown here:

```
SUBTYPE bit16 IS STD_LOGIC_VECTOR(15 DOWNTO 0);
..
VARIABLE bus_value : bit16;

-- these won't work with VHDL87
bus_value := "0101010101010101"; --- or
bus_value := O"052525"; -- or
bus_value := X"5555";
```

In VHDL93 this concept is extended to types **std_logic_vector**.

DELAY_LENGTH Subtype

In VHDL87 most time delays were specified with a type **TIME**. Type **TIME** included negative and positive time values. Most uses of **TIME** required only positive values of **TIME**. Therefore in VHDL93 a new type in package **STANDARD** has been created, and called **DELAY_LENGTH**. It's definition is shown here:

```
SUBTYPE DELAY_LENGTH IS TIME RANGE 0 FS TO TIME`HIGH;
```

As can be seen this type only includes the positive values of **TIME**. Compiler writers can optimize the compilation and simulation processes more with this knowledge.

Direct Instantiation

In VHDL87 an entity from a particular library could not be directly instantiated in an architecture. A component was declared, instantiated, and bound to an entity with a configuration. The component could have been directly or implicitly configured.

In VHDL93 entities can be directly instantiated if they are visible. In the example here **entity adder** from library work is directly instantiated and configured in architecture struct.

```
ENTITY direct IS
  PORT( i1 : IN STD_LOGIC;
                o1 : OUT STD_LOGIC);
END ENTITY direct;

ARCHITECTURE struct OF direct IS
  SIGNAL s1, s2 : STD_LOGIC;
BEGIN
  U1 : ENTITY work.adder(behave)
                        GENERIC MAP(out_delay : delay_type)
                        PORT MAP(s1, s2, i1, o1);
END ARCHITECTURE struct;
```

A separate configuration is not necessary as the entity is uniquely specified. This makes it very easy to describe designs structurally and with a lot less lines of VHDL code. However it can make design reuse more difficult.

Extended Identifiers

In VHDL87 identifiers were limited to only characters a-z, A-Z, and 0-9. This limited the number of identifiers that could be created. For manually created VHDL this was not a major problem, but for VHDL that was translated from some other format this caused some major problems. Certain netlist formats contain identifiers that consist of operator symbols, or start with a number. With VHDL93 the extended identifier allows the user to specify identifiers in a much less restricted manner. Extended identifiers can start with numbers or contain operator symbols.

Extended identifiers are specified by backslashes(\..\) around an identifier. Extended identifiers can be used anywhere a normal identifier can be used. An example using extended identifiers is shown here:

```
entity \741s163\ is
  port (clk : in std_logic;
             \1n1\ : in std_logic;
          reset : in std_logic;
          q1 : out std_logic;
          q2 : out std_logic;
          q3 : out std_logic);
end \741s163\;
```

In this example the entity name(`\741s163\`), and one of the input ports(`\1n1\`) are extended identifiers.

File Operations

One of the most welcome additions to VHDL93 is the ability to open and close files. In VHDL87 files were declared in declarations and opened implicitly by the elaboration process. VHDL93 adds the ability to specifically open and close files. This allows one subprogram or entity to create a file which another subprogram or entity can read. In VHDL87 the modeler would declare a file type to define the type, and later a file declaration that would ultimately open the file. This is shown here:

```
TYPE int_file IS FILE OF INTEGER; -- VHDL87
--
FILE infile: int_file IS IN "/doug/test/example3";
     -- VHDL87 declares
     -- and opens file
```

In VHDL93 the file type declarations remains the same, but the modeler has a couple of ways to actually open the file. Probably the most common will be to call the explicit **FILE_OPEN** procedure as shown here:

```
PROCEDURE FILE_OPEN(FILE infile: int_file;
               EXTERNAL_NAME : IN "/doug/test/example3";
               OPEN_KIND : IN READ_MODE);
```

This will open the file for reading. If the file cannot be opened for some reason a runtime error will be generated. An alternate way to open the file is to call a different version of the **FILE_OPEN** procedure as shown here:

```
PROCEDURE FILE_OPEN(FILE_STATUS: FILE_OPEN_STATUS;
               FILE : int_file;
               EXTERNAL_NAME : IN "/doug/test/example3";
               OPEN_KIND : IN READ_MODE);
```

This procedure returns an output parameter called **FILE_STATUS** that contains the status of the **FILE_OPEN** procedure call. A status value of **OPEN_OK** means that the file is open and ready to be read. A value of **STATUS_ERROR** means that the file object already has an external file associated with it. A value of **NAME_ERROR** means that the external file does not exist. A value of **MODE_ERROR** means that the external file cannot be opened using the mode passed.

An alternate way of opening the file without calling the explicit **FILE_OPEN** procedure is similar to the method used in VHDL87. This method uses a file declaration similar to the one in VHDL87, that specifies the name of the file object, the mode of the file object, and the external filename to be associated with the file object as shown here:

```
FILE infile : int_file OPEN READ_MODE IS "/doug/test/
    example3";
```

This effectively calls the **FILE_OPEN** procedure as follows:

```
FILE_OPEN(infile, "/doug/test/example3", READ_MODE);
```

When a file type declaration of a particular **type_mark** is declared the following declarations are implicitly declared.

```
TYPE FT IS FILE OF type_mark;

PROCEDURE FILE_OPEN( FILE  F : FT;
                     EXTERNAL_NAME : IN STRING;
                     OPEN_KIND : IN FILE_OPEN_KIND :=
                       READ_MODE);

PROCEDURE FILE_OPEN( STATUS : OUT FILE_OPEN_STATUS;
                     FILE  F : FT;
                     EXTERNAL_NAME : IN STRING;
                     OPEN_KIND : IN FILE_OPEN_KIND :=
                       READ_MODE);

PROCEDURE FILE_CLOSE( FILE F : FT);

PROCEDURE READ( FILE F : FT; VALUE : OUT type_mark);

PROCEDURE WRITE( FILE F : FT; VALUE : OUT type_mark);

PROCEDURE ENDFILE( FILE F : FT) RETURN BOOLEAN;
```

The file type declaration declares a file of type **type_mark**. With the file type declaration all of the above procedures are implicitly declared. Once these procedures are declared they can be used to read and write files of the **type_mark**.

Foreign Interface

In VHDL87 it was possible to call functions and procedures that were not described using VHDL. It was possible but limited in scope and not very well defined. The VHDL93 package standard now contains an attribute called **FOREIGN** whose value is a string. This string value describes the interface to the external function, procedure, or entity. The value of this string is not standardized and depends on the type of the external code being called. An example might look as follows:

```
FUNCTION beep( length : INTEGER) IS
   ATTRIBUTE FOREIGN OF beep : FUNCTION IS
      "sysbeep(length)";
BEGIN
END FUNCTION beep;
```

In this example a function called **beep** is declared that contains a **FOREIGN** attribute. The **FOREIGN** attribute specifies that the body of this function will be implemented by code other than VHDL. The string value of the attribute declares the interface expected between function **beep** and the foreign code to implement the function. However the string value is not defined in VHDL93 to be anything more than just a string.

Generate Statement Changes

In a minor addition, VHDL93 adds a declaration section to the generate statement. Any declarations before the begin clause are local only to the generate statement.

```
g1: FOR k IN 0 TO 3 GENERATE
   SIGNAL reset : STD_LOGIC;
BEGIN
   dffx : dff PORT MAP( z(i), reset, clk, z(i + 1));
END GENERATE;
```

The generate statement above declares local signal reset. This signal is local only to the generate statement.

Globally Static Assignment

VHDL93 adds a new feature that allows globally static values to be assigned to port maps. In VHDL87 port maps could only bind formal parameters to signals. In VHDL93 this has been generalized to include expressions as well. These expressions have to be globally static, or known at elaboration time.

```
u1: mux4 PORT MAP( k0 => s0, k1 => s1, k2 => s2, en =>
      '1', q => outp);
```

In the example above the value *1* is mapped to port en. In VHDL87 a separate signal would have to be created, assigned to the value *1*, and then mapped to port en.

The globally static value does not have to be just a simple value, it can be any expression known at compile time that matches the type of the port.

Groups

It is sometimes useful while modeling to declare an attribute that is to apply to more than one object. Especially in writing synthesizable models some attributes are useful to describe behavior for an entire section of a model. In VHDL87 there was no way to describe this type of attribute structure. VHDL93 has the concept of groups which allows an attribute to pertain to all objects in the group.

A group starts with a group template declaration such as shown here:

```
GROUP timing_arc IS (SIGNAL, SIGNAL);
```

This describes a group template called **timing_arc** that is a group of two signal objects. After the group template is declared a group declaration can be declared as shown here:

```
GROUP clk_to_q : timing_arc(clk, q);
GROUP rst_to_q : timing_arc(rst, q);
GROUP set_to_q : timing_arc(set, q);
```

These declarations show three separate group declarations named **clk_to_q**, **rst_to_q**, and **set_to_q**. Each of these groups describe a group object with two signals in the group. Once declared these groups can be operated on as a single object. For instance, if the following attribute is declared:

```
ATTRIBUTE prop_delay IS DELAY_LENGTH;
```

then the following attributes can be applied to the group.

```
ATTRIBUTE prop_delay OF clk_to_q : GROUP IS 2.3 NS;
ATTRIBUTE prop_delay OF rst_to_q : GROUP IS 3.1 NS;
ATTRIBUTE prop_delay OF set_to_q : GROUP IS 2.7 NS;
```

These attributes act on both signals in the group.

Another way to describe a group, especially a group that varies in size, is shown here:

```
GROUP timing_arc IS (SIGNAL <>);
```

This syntax is similar to an unconstrained array and describes a group consisting of one or more signal objects.

Incremental Binding

In VHDL87 the rules about binding were very restrictive. If a component was bound in a configuration specification, it could not be bound in a configuration declaration. This made back annotation of timing delays rather difficult because the back annotation program had to generate not only the generic parameter values, but also the proper entity use clauses. What the modeler would like to do is pick the proper entity to use with a configuration specification in the architecture of the containing entity, and use a configuration declaration to specify the values for the back annotated timing.

In VHDL87 this was not possible because the component could be configured in either place, but not both. In VHDL93 the incremental binding feature allows the modeler to create models that behave as wanted.

An example is shown here:

```
ENTITY dff IS
  GENERIC( delay : TIME;
  PORT( din, clock : IN STD_LOGIC;
              dout : OUT STD_LOGIC);
END ENTITY dff;

ENTITY top IS
  PORT( z, clock : IN STD_LOGIC; qout : OUT STD_LOGIC);
END ENTITY top;

ARCHITECTURE struct OF top IS
  COMPONENT dff IS
    PORT( d, clk : IN STD_LOGIC;
                q : OUT STD_LOGIC);
  END COMPONENT dff;
  FOR d1: dff USE ENTITY WORK.dff(behave)
                GENERIC MAP (clk_to_q => 5.2 NS)
                PORT MAP( d => din, clk => clock, q =>
    open );

  SIGNAL
```

```
BEGIN
    -- ..
    -- ..
    d1 : dff PORT MAP( z, clock, qout);

END ARCHITECTURE struct;

CONFIGURATION topcon OF top IS
  FOR struct
    FOR d1 : dff GENERIC MAP( clk_to_q => 8.1 NS) PORT
      MAP( q => dout );
    END FOR;
  END FOR;
END CONFIGURATION topcon;
```

In this example a **dff** component is instantiated in **entity top**. A configuration specification in the architecture declaration section specifies a value for the **clk_to_q** generic of 5.2 NS, and maps ports *d* and *clk*. Port *q* is not mapped but is left open. After the end of the architecture a configuration declaration specifies a new value for the **clk_to_q** generic of 8.1 NS, and maps port *q* to dout. The new **clk_to_q** generic value will override the previous value specified in the configuration specification. The mapping of port *q* mapped to open in the configuration specification is also overriden with the new value **dout** as specified in the configuration declaration.

Postponed Process

In VHDL93 a new type of process has been added, the **postponed process**. A **postponed process** is executed after all of the delta cycles have been processed so that each signal receives the final value of a simulation time. A typical use for such a process is to perform timing checks. There are cases in performing timing checks where the input signals need to stabilize before the timing checks are performed. The **postoned process** will allow all of the input signals to stabilize, and finally the **postponed process** will be executed.

A **postponed process** is specified.

```
p1: POSTPONED PROCESS( clk, reset) IS
    -- postponed process declaration section
BEGIN
    IF reset = '1' THEN
    --
    END IF;
END PROCESS p1;
```

The keyword **POSTPONED** is specified before the **PROCESS** keyword to specify a **postponed process**.

Pure and Impure Functions

Functions in VHDL87 were very restrictively defined. The input mode of all input arguments were constant, and only input arguments were allowed. The function could have no side effects such as modifying a value outside the function. The only information returned from the function was through the return value. In VHDL93 this type of function is known as a **pure function**. VHDL93 also contains **impure functions** which can modify data outside their own scope. These functions must be explicitly declared as being impure as shown here:

```
FILE bit_file : TEXT OPEN READ_OPEN IS "ram_data";

IMPURE FUNCTION get_val RETURN BIT IS
  VARIABLE myline : LINE;
  VARIABLE result : BIT;
BEGIN
  READLINE( bit_file, myline );
  READ ( myline, result );
  RETURN result;
END get_val;
```

This function is used to read a set of bits from a file. Function **get_val** is declared impure so that it has access to data outside the function. The file **bit_file** is opened externally to function **get_val** but since the function is impure, access to file **bit_file** is possible.

Functions in VHDL87 do not have access to data outside of the function so this function would not work. In VHDL87 the file would have to be declared within the function declaration section, and implicitly opened and closed from within the function. In VHDL93 the file can be opened external to the function and an **impure function** can access the file.

Pulse Reject

In VHDL87 there were two types of delay categories, inertial and transport. Chapter 2, "Behavioral Modeling" talks about the differences between them. The VHDL87 inertial delay will reject pulses smaller than the **inertial delay** specified. In some cases this is too pesimistic. In VHDL93 the modeler has the ability to specify a **pulse reject** limit which can be less than or equal to the **inertial delay** through the device.

```
s1 <= REJECT 5 NS INERTIAL newval AFTER 15 NS;
```

In this example the **inertial delay** through the device is 15 ns, but the reject limit is 5 ns. Any pulses of 5 ns or less will be rejected but pulses greater than 5 ns will be passed through with a 15 ns delay.

Report Statement

In VHDL87 the **report** clause could only be used within the **assert** statement, in VHDL93 a **report** clause can exist separately. In VHDL87 if a modeler wanted to issue a message to inform the designer that a particular piece of a model was executing, the following statement would have been required:

```
ASSERT FALSE REPORT "entered file procedure read";
```

The **report** statement would have to be called from an **assert** statement, and the **assert** condition would need to return false to trigger the **assert** statement. In VHDL93 the **report** statement can exist separately so the following would also work:

```
REPORT "entered file procedure read";
```

A **report** statement still has the ability to specify the severity level of the message. In the above cases the severity level defaulted to error. If some other severity was wanted, it could be specified as follows:

```
REPORT "entered timing check code" SEVERITY NOTE;
```

The severity clause at the end of the report statement allows the modeler to specify any legal level of severity.

Shared Variables

A shared variable is one that is accessible by any design unit that includes the package where the variable is declared. In VHDL87 variables could only be declared in processes and were therefore local to the process. In VHDL93 variables are now able to be declared in packages, and therefore become global. Any design unit that includes the package can access the variable. In the example here package share has shared variable `timing_checks_on` declared in it.

```
PACKAGE share IS
  VARIABLE timing_checks_on : BOOLEAN := TRUE;
END PACKAGE share;

USE WORK.share.ALL;
ENTITY dff IS
  PORT( din : IN STD_LOGIC;
              clk : IN STD_LOGIC;
              q : OUT STD_LOGIC);
END ENTITY dff;

ARCHITECTURE behave OF dff IS
BEGIN
  PROCESS(clk) IS
  BEGIN
    IF timing_checks_on THEN
      -- timing check statements
    END IF;
    -- other statements
  END PROCESS;
END ARCHITECTURE behave;

USE WORK.share.ALL;
ENTITY jkff IS
  PORT( j, k, clk, se, clr : IN STD_LOGIC;
              q, qb : OUT STD_LOGIC);
END jkff;

ARCHITECTURE behave OF jkff IS
BEGIN
  PROCESS(clk, set, clr) IS
    IF timing_checks_on THEN
      -- timing check statements
    END IF;
  END PROCESS;
END ARCHITECTURE behave;
```

`Package share` is included by entities `dff` and `jkff`, making the variable `timing_checks_on` globally accessible by both entities. Global variables are very useful for passing information which is not really part of the design functionality, but affect the simulation or synthesis operation. In this example global variable `timing_checks_on` allows the ability to turn off and on timing check operation. This does not affect the actual functionality of the behavior of the models except to disable timing check reporting. Another use for global variables is to use them to pass input and output file handles.

Shift Operators

VHDL87 did not contain operators to allow shifting or rotating. Most of these functions were built by VHDL standard package creators. Without the built in operators however, overloaded shift and rotate operators were not possible. VHDL93 includes built in shift and rotate operators: **sll**(shift left logical), **srl**(shift right logical), **sla**(shift left arithmetic), **sra**(shift right arithmetic), **rol**(rotate left), and **ror**(rotate right). These operators allow shifting and rotating operations for any one dimensional array type. These operators work as follows:

SLL—shift left logical

```
q <= a SLL b;
```

q equals A shifted left by B bits and filled on the right with the value *A'left*. If B is negative then A is shifted right.

SRL—shift right logical

```
q <= a SRL b;
```

q equals A shifted right by B bits and filled on the left with the value *A'left*. If B is negative then A is shifted left.

SLA—shift left arithmetic

```
q <= a SLA b;
```

q equals A shifted left by B bits and filled on the right with *A(A'RIGHT)*. If B is negative then A is shifted right.

SRA—shift right arithmetic

```
q <= a SRA b;
```

q equals A shifted right by B bits and filled on the left with *A(A'LEFT)*. If B is negative then A is shifted left.

ROL—rotate left

```
q <= a ROL b;
```

q equals *A* rotated left by *B* bits. Instead of filling the right *B* bits with a value, the *B* bits that were shifted off the left end are copied to the right of the shifted *A* bits. An array that originally contained **ABCD** and is rotated left one bit will become **BCDA**.

ROR—rotate right

```
q <= a ROR b;
```

q equals *A* rotated right by *B* bits. Instead of filling the left *B* bits with a value, the *B* bits that were shifted off the right end are copied to the left of the shifted *A* bits. An array that originally contained **ABCD** and is rotated right one bit will become **DABC**.

These operators can now be overloaded to work with user defined types.

Syntax Consistency

As part of the VHDL93 syntax update a number of the language end clauses were modified to become more consistent. All of the clauses now include the beginning clause identifier. For instance in VHDL87 the **entity** clause was as shown here:

```
ENTITY test IS
---
END test;
```

In VHDL93 the same construct can optionally look as follows:

```
ENTITY test IS
---
END ENTITY test;
```

Notice that the end clause contains the starting **ENTITY** clause. Including the keyword **ENTITY** in the **END** clause is optional but allowed in VHDL93. The same holds true for the architecture, **package**, **package body**, **configuration**, **component**, **block**, **process**, **record**, **case**, **if**, **procedure**, and **generate** statement. Examples are shown here:

```
ARCHITECTURE behave OF test IS
BEGIN
---
END ARCHITECTURE behave;

PACKAGE mypack IS
---
END PACKAGE mypack;

PACKAGE BODY mypack IS
---
END PACKAGE BODY mypack;

CONFIGURATION chip OF processor IS
---
END CONFIGURATION chip;

COMPONENT memory IS -- notice addition of IS at end of
       component clause
--
END COMPONENT memory;

block1 : BLOCK IS
BEGIN
--
END BLOCK block1;

proc1: PROCESS(clk, din) IS -- notice addition of IS at end
       of process clause
BEGIN
--
END PROCESS proc1;

RECORD myrec IS
--
END RECORD myrec;

CASE selector IS
--
END CASE selector;

lab: IF expr THEN
--
END IF lab;

PROCEDURE convertval(...) IS
BEGIN
--
END PROCEDURE convertval;

g1: FOR k IN 0 TO 7 GENERATE
BEGIN
--
END GENERATE g1;

loop1: FOR k IN 0 TO 7 LOOP
--
END LOOP loop1;
```

Unaffected

In VHDL87 it was sometimes difficult to describe exactly the behavior required with a concurrent signal assignment statement. For instance there are cases where the modeler wants the value to remain unchanged if certain conditions are met. In VHDL87 this can be accomplished with the following statement:

```
new_state <= state5 WHEN current_state =  state1 AND input
                = 7 ELSE
              state6 WHEN current_state = state5 AND input
                = 8 ELSE
              state7 WHEN current_state = state2 AND input
                = 9 ELSE
              new_state;
```

If none of the above conditions are met the designer has to assign the current value, `new_state` to `new_state` to ensure no value change. This produces the correct value but has a side effect that a transaction is generated on signal `new_state`. Any behaviors sensitive to transactions on `new_state` will be evaluated and may update their values causing further activity when none was wanted.

VHDL93 has the new keyword **UNAFFECTED** that allows no change on a signal. Unaffected causes no value changes or transactions on the signal. The same statement above rewritten to include this new feature looks as follows:

```
new_state <= state5 WHEN current_state =  state1 AND input
                = 7 ELSE
              state6 WHEN current_state = state5 AND input
                = 8 ELSE
              state7 WHEN current_state = state2 AND input
                = 9 ELSE
              UNAFFECTED;
```

Now if none of the conditions are met nothing will be assigned and no transactions will be generated.

XNOR Operator

VHDL87 contained a number of operators such as **OR**, **NOR**, **AND**, etc. but did not contain the **XNOR** operator. Therefore the **XNOR** operator could not be overloaded. VHDL93 adds the **XNOR** operator to the list of operators built into the language and therefore in VHDL93 it can be used and over-loaded. The example here shows a use of the **XNOR** operator.

```
ENTITY xnor2 IS
   GENERIC( delay : TIME);
   PORT( a, b : IN STD_LOGIC;
                  q : OUT STD_LOGIC);
END ENTITY xnor2;

ARCHITECTURE behave OF xnor2 IS
BEGIN
   a <= a XNOR b AFTER delay;
END ARCHITECTURE behave;
```

This example shows a two input **xnor** gate using the built in **XNOR** operator. The **XNOR** operator can be overloaded to work with any types. In this example two **STD_LOGIC** type values are **xnor**'ed together to form the final result.

Index

About the Author

Douglas L. Perry is a senior applications engineer with Exemplar Logic Inc., and has been active in the CAE field for more than 15 years. He is the author of the first two editions of *VHDL* and has co-chaired the VHDL User's Group. Perry earned a B.S. degree in Electrical Engineering at South Dakota State University and undertook graduate studies at the University of Santa Clara. He lives in San Ramon, California.